The Genius of Euler

Reflections on his Life and Work

Leonhard Euler (1707–1783)
Portrait by Zh. Darbesa, 1770

The Genius of Euler

Reflections on his Life and Work

Edited by

William Dunham
Truman Koehler Professor of Mathematics
Muhlenberg College

Published and Distributed by
The Mathematical Association of America

For three of the best:

Doris Schattschneider,
Don Albers,
Jerry Alexanderson

SPECTRUM SERIES

The Spectrum Series of the Mathematical Association of America was so named to reflect its purpose: to publish a broad range of books including biographies, accessible expositions of old or new mathematical ideas, reprints and revisions of excellent out-of-print books, popular works, and other monographs of high interest that will appeal to a broad range of readers, including students and teachers of mathematics, mathematical amateurs, and researchers.

777 Mathematical Conversation Starters, by John de Pillis

99 Points of Intersection: Examples—Pictures—Proofs, by Hans Walser. Translated from the original German by Peter Hilton and Jean Pedersen

All the Math That's Fit to Print, by Keith Devlin

Carl Friedrich Gauss: Titan of Science, by G. Waldo Dunnington, with additional material by Jeremy Gray and Fritz-Egbert Dohse

The Changing Space of Geometry, edited by Chris Pritchard

Circles: A Mathematical View, by Dan Pedoe

Complex Numbers and Geometry, by Liang-shin Hahn

Cryptology, by Albrecht Beutelspacher

The Early Mathematics of Leonhard Euler, by C. Edward Sandifer

The Edge of the Universe: Celebrating 10 Years of Math Horizons, edited by Deanna Haunsperger and Stephen Kennedy

Five Hundred Mathematical Challenges, Edward J. Barbeau, Murray S. Klamkin, and William O. J. Moser

The Genius of Euler: Reflections on his Life and Work, edited by William Dunham

The Golden Section, by Hans Walser. Translated from the original German by Peter Hilton, with the assistance of Jean Pedersen.

I Want to Be a Mathematician, by Paul R. Halmos

Journey into Geometries, by Marta Sved

JULIA: a life in mathematics, by Constance Reid

The Lighter Side of Mathematics: Proceedings of the Eugène Strens Memorial Conference on Recreational Mathematics & Its History, edited by Richard K. Guy and Robert E. Woodrow

Lure of the Integers, by Joe Roberts

Magic Numbers of the Professor, by Owen O'Shea and Underwood Dudley

Magic Tricks, Card Shuffling, and Dynamic Computer Memories: The Mathematics of the Perfect Shuffle, by S. Brent Morris

Martin Gardner's Mathematical Games: The entire collection of his Scientific American columns

The Math Chat Book, by Frank Morgan

Mathematical Adventures for Students and Amateurs, edited by David Hayes and Tatiana Shubin. With the assistance of Gerald L. Alexanderson and Peter Ross

Mathematical Apocrypha, by Steven G. Krantz

Mathematical Apocrypha Redux, by Steven G. Krantz

Mathematical Carnival, by Martin Gardner

Mathematical Circles Vol I: In Mathematical Circles Quadrants I, II, III, IV, by Howard W. Eves

Mathematical Circles Vol II: Mathematical Circles Revisited and Mathematical Circles Squared, by Howard W. Eves

Mathematical Circles Vol III: Mathematical Circles Adieu and Return to Mathematical Circles, by Howard W. Eves

Mathematical Circus, by Martin Gardner

Mathematical Cranks, by Underwood Dudley

MAA Service Center
P.O. Box 91112
Washington, DC 20090-1112
800-331-1622   FAX 301-206-9789

Photos on pages 17, 45, 65, 91, 134 are courtesy of Hans Walser

ISBN 10: 0-88385-558-5
ISBN 13: 978-0-88385-558-4

Printed in the United States of America

Current Printing (last digit):
10 9 8 7 6 5 4 3 2 1

Acknowledgments

As the dedication suggests, this book owes its existence to Doris Schattschneider, Don Albers, and Jerry Alexanderson, three mathematicians who have given vast amounts of time and talent to the MAA's publication enterprise.

But if Doris, Don, and Jerry were the visionaries behind this Euler tribute volume, it was Beverly Ruedi who made the book a reality. Working under tight deadlines, she was indefatigable in pushing the project to completion. She left the rest of us breathless just trying to keep up.

Others deserve words of appreciation for their generous assistance: my wife and colleague Penny Dunham, Elaine Pedreira at the MAA's Washington offices, Leonard Klosinski in sunny California, and a trio of friends who provided illustrations—Hans Walser of the University of Basel, Lee Stemkoski of Adelphi University, and Mike Huber of Muhlenberg College.

Most of all, thanks are due to the 34 authors whose papers fill this volume. Although from different places and different times, these individuals are united in their admiration for Leonhard Euler, an admiration evident to all who read the pages that follow.

All celebrated mathematicians now alive are his disciples: there is no one who ... is not guided and sustained by the genius of Euler.

From the Marquis de Condorcet's
Eulogy to Euler, 1783

Preface

In the mathematical firmament, the star of Leonhard Euler (1707–1783) shines with a special brilliance. His contributions, astonishing in both quantity and quality, have left him with few peers in the long history of mathematics.

It was thus appropriate when, on the bicentennial of his death, the MAA journal *Mathematics Magazine* devoted an entire issue (November, 1983) to Euler and his work. Under the editorship of Doris Schattschneider, the magazine included articles on a host of topics from a who's who of mathematicians. Many grateful readers put the issue aside as a "keeper."

One such reader was Don Albers, Director of Publications for the Mathematical Association of America. A fan of Euler and an admirer of first-rate mathematical writing, Don went so far as to have his copy *bound*, making it both a prominent and a permanent item on his bookshelves.

This year, the tercentenary of Euler's birth, we have the perfect occasion to pay homage once again. In mulling over how this could be done, Don's gaze fell upon his bound copy, and—bingo!—he realized that the MAA could produce a book-length tribute by resurrecting those papers from *Mathematics Magazine* and augmenting them with articles from other sources. He pitched his idea to me, with the request that I be the volume's editor, although "assembler" might be a more apt description. Call it what you will, this was an opportunity too good to pass up.

And so it was that I found myself re-reading the 1983 issue and grazing through the mathematical literature for appropriate companion pieces. My search was guided by a few basic principles. First, additions had to be primarily about Euler or his work. I eliminated papers giving, for instance, a new proof of an old Eulerian result. Or, if an article promised to survey a topic "from Euler to X and Y," I insisted that Euler be featured in a prominent role so as to maintain a distinction between the book's star (Euler) and its supporting players (X and Y).

Second, because these articles had been published elsewhere, there was less concern about their individual quality than about their interrelationship. I scouted for papers that fit together in a reasonably coherent way—not always an easy outcome to achieve. I concede that any book of this sort lacks the unity of vision that would result from a single author's treatment of the material, but I hope that the dangers of publishing a hodge-podge are offset by the charm of variety.

And variety we have. More than thirty authors are represented across these pages. Their papers span a significant time period, with the oldest dating from 1872 and the most recent being a 2006 article from *The College Mathematics Journal*. As a consequence, the reader will find not just a range of topics but a range of styles and eras, a diversity meant to enhance rather than diminish the experience.

A third and more mundane requirement was that these pieces be readily available. Under our tight deadline, there simply was not time to obtain publication rights in a leisurely manner. We thus decided to limit our choices, insofar as possible, to papers from MAA journals. This I have done with two exceptions: Clifford Truesdell's biographical work "Leonhard Euler, Supreme Geometer" and J. W. L. Glaisher's article on Euler's constant from *The Messenger of Mathematics*.

Even with these constraints, there was no shortage of material (mathematicians, it seems, have been writing about Euler in very large numbers for a very long time). Ultimately, the articles divided themselves into two groups: those with a more qualitative flavor that provided biography and background; and those with a more technical bent that examined in detail some of Euler's mathematical achievements. The book has been organized according to this natural dichotomy. If I have chosen wisely and assembled well, the reader should be in for a treat.

So, Happy 300th, Uncle Leonhard!

William Dunham
Allentown, Pennsylvania

About the Authors

The authors whose papers appear in this volume are a distinguished, if eclectic, group. Among them we find research mathematicians, historians of mathematics, and college professors, not to mention a high school teacher (Charlie Marion) and a pair of graduate students (Dominic Klyve and Lee Stemkoski). This diverse cast is united by two characteristics: an enthusiasm for Euler's work and the ability to write engagingly about it.

Many of our contributors are currently active with (one hopes!) long careers ahead of them. For instance, we have articles from Jerry Alexanderson, the MAA's president from 1997 to 1998 and a wise counsel for its many publishing ventures; Victor Katz, whose textbook *A History of Mathematics* is a widely used introduction to the subject; Robin Wilson, popular lecturer and author of *Introduction to Graph Theory* and *Four Colors Suffice*; Underwood Dudley, winner of the MAA's Trevor Evans Award for expository excellence; and Philip J. Davis, co-author with Reuben Hersh of *The Mathematical Experience*, which received an American Book Award in 1983. Clearly, these contemporary writers are an accomplished lot.

But some of our authors, equally accomplished, flourished long ago. For the benefit of younger readers, we provide a few words of introduction.

- **J. W. L. Glaisher** (1848–1928): a Fellow of the Royal Society who taught at Cambridge University, where he served as president of the Antiquarian Society and the Bicycle Club and still found time to become one of the world's foremost collectors of pottery.

- **W. W. Rouse Ball** (1850–1925): another son of Cambridge and author of the 1892 classic *Mathematical Recreations and Essays* (still in print!).

- **Florian Cajori** (1859–1930): MAA president from 1917 to 1918 and one of the great math historians of his day, whose two-volume *A History of Mathematical Notations* remains the seminal work on the subject.

- **Benjamin Finkel** (1865–1947): founder in 1894 of *The American Mathematical Monthly*, which—as the MAA's flagship periodical—has become the world's most widely read mathematics journal.

- **George Pólya** (1887–1985): not only a research mathematician with interests ranging from complex analysis to combinatorics but also an inspirational teacher and writer whose book *How to Solve It* stands in a class by itself.

- **Carl Boyer** (1906–1976): author of the 1949 work *The Concepts of the Calculus* (now available as *The History of the Calculus and its Conceptual Development*) that has been mandatory reading for all those interested in the origins of calculus.

- **André Weil** (1906–1998): a renowned number theorist, group theorist, and algebraic geometer who wrote *Number Theory: An Approach through History from Hammurapi to Legendre* with its extensive discussion of Euler's contributions to the subject.

- **Morris Kline** (1908–1992): mathematician and author of more than a dozen books, from the provocative *Why the Professor Can't Teach* to the two-volume *Mathematical Thought from Ancient to Modern Times*, the standard reference for historians of mathematics everywhere.

- **Paul Erdős** (1913–1996): the world-famous, itinerant mathematician whose productivity rivaled that of Euler himself and whose multitude of co-authors generated the delightful concept of "Erdős number."

- **Clifford Truesdell** (1919–2000): an applied mathematician, author of numerous books and papers in rational mechanics, founder of the journal *Archive for History of Exact Sciences*, and a life-long admirer of Euler.

Contents

Part I
Biography and Background

Introduction to Part I

This section of the book opens with three biographies. The first is B. F. Finkel's overview of Euler's life and work, published in *The American Mathematical Monthly* of 1897. Given that Finkel founded the *Monthly*, it is fitting that he appear in this volume.

The second piece is Clifford Truesdell's classic, "Leonhard Euler, Supreme Geometer," which mixes mathematical detail, biographical information, and biting wit (not necessarily in that order). A pleasure to read, Truesdell's article is a colorful departure from stereotypical academic writing.

Then comes a short biography from the 20th century mathematician André Weil. It begins by placing Euler in a long line of predecessors and ends with a letter from Euler to Lagrange about the former's deteriorating vision.

A significant portion of Euler's career was spent at the Berlin Academy during the reign of Frederick the Great, so readers may enjoy a 1927 article by Florian Cajori about the monarch's relationship with his most illustrious academician. This is followed by B. H. Brown's debunking of an oft-recounted exchange between Euler and Diderot.

Next are two articles about Euler's expository skills. G. L. Alexanderson's paper, "*Ars Expositionis*: Euler as Writer and Teacher," is taken from the 1983 issue of *Mathematics Magazine* that gave rise to the present volume, after which historian Carl Boyer presents a compelling argument that Euler's *Introductio in Analysin Infinitorum* is, hands down, the greatest math textbook of modern times.

We survey Euler's legacy in J. J. Burckhardt's article, also drawn from the 1983 tribute issue, touching on Euler's contributions to number theory, analysis, and applied mathematics. Then W. W. Rouse Ball addresses the sheer scope of Euler's achievement and concludes about Euler's *œuvre*: "There is nothing like this, except this, in the history of science."

Three poems bring this section to an end. The first, by Marta Sved and Dave Logothetti, puts into verse the despair of mathematicians who learn too late that they have been scooped by Euler. The other two are from the *Mathematics Magazine* of 1997. In the June issue of that year, J. D. Memory published "Bell's Conjecture," a poem about E. T. Bell's ranking of Archimedes, Newton, and Gauss as the three greatest mathematicians of all time. In reply, the December issue carried "A Response to 'Bell's Conjecture,'" hastily composed during the doggerel days of summer by Charlie Marion and William Dunham as a plea for Euler's inclusion on any short list of this kind. After all, there are four Horsemen of the Apocalypse, four presidents on Mount Rushmore, and four Beatles. Surely mathematicians can fall in line behind the quartet of Archimedes, Newton, Gauss ... and Euler.

Leonhard Euler[1]

B. F. Finkel

Leonhard Euler (oi′ler), one of the greatest and most prolific mathematicians that the world has produced, was born at Basel, Switzerland, on the 15th day of April, 1707, and died at St. Petersburg, Russia, November the 18th (N. S.), 1783. Euler received his preliminary instruction in mathematics from his father who had considerable attainments as a mathematician, and who was a Calvinistic[2] pastor of the village of Riechen, which is not far from Basel. He was then sent to the University of Basel where he studied mathematics under the direction of John Bernoulli, with whose two sons, Daniel and Nicholas, he formed a life-long friendship. Geometry soon became his favorite study. His genius for analytical science soon gained for him a high place in the esteem of his instructor, John Bernoulli, who was at the time one of the first mathematicians of Europe. Having taken his degree as Master of Arts in 1723, Euler afterwards applied himself, at his father's desire, to the study of theology and the Oriental languages, with the view of entering the ministry, but, with his father's consent, he returned to his favorite pursuit, the study of mathematics. At the same time, by the advice of the younger Bernoullis, who had removed to St. Petersburg in 1725, he applied himself to the study of physiology, to which he made useful applications of his mathematical knowledge; he also attended the lectures of the most eminent professors of Basel. While he was eagerly engaged in physiological researches, he composed a dissertation on the nature and propagation of sound. In his nineteenth year he also composed a dissertation in answer to a prize-question concerning the masting of ships, for which he received the second prize from the French Academy of Sciences.

---

[1] Reprinted from the *American Mathematical Monthly*, Vol. 4, December, 1897, pp. 297–302.

[2] The *Encyclopædia Brittanica* says Euler's father was a Calvinistic minister, while W. W. R. Ball, in his *History of Mathematics*, says he was a Lutheran minister. Euler himself was a Calvinist in doctrine, as the following, which is his apology for prayer, indicates: "I remark, first, that when God established the course of the universe, and arranged all the events which must come to pass in it, he paid attention to all the circumstances which should accompany each event; and particularly to the dispositions, to the desires, and prayers of every intelligent being; and that the arrangement of all events was disposed in perfect harmony with all these circumstances. When, therefore, a man addresses God a prayer worthy of being heard it must not be imagined that such a prayer came not to the knowledge of God till the moment it was formed. That prayer was already heard from all eternity; and if the Father of Mercies deemed it worthy of being answered, he arranged the world expressly in favor of that prayer, so that the accomplishment should be a consequence of the natural course of events. It is thus that God answers the prayers of men without working a miracle."

Leonhard Euler in a portrait by Emanuel Handmann (1756)

When his two close friends, Daniel and Nicholas Bernoulli, went to Russia, they induced Catherine I, in 1727, to invite Euler to St. Petersburg, where Daniel, in 1733, was assigned to the chair of mathematics. Euler took up his residence in St. Petersburg, and was made an associate of the Academy of Sciences. In 1730 he became professor of physics, and in 1733 he succeeded his friend Daniel Bernoulli, who resigned on a plea of ill health.

At the commencement of his astonishing career, he enriched the Academical collection with many memoirs, which excited a noble emulation between him and the Bernoullis, though this did not in any way affect their friendship. It was at this time that he carried the integral calculus to a higher degree of perfection, invented the calculation of sines, reduced analytical operations to greater simplicity, and threw new light on nearly all parts of pure or abstract mathematics. In 1735, an astronomical problem proposed by the Academy, for the solution of which several eminent mathematicians had demanded several months' time, was solved by Euler in three days with the aid of improved methods of his own, but the effort threw him into a fever which endangered his life and deprived him of his right eye, his eyesight having been impaired by the severity of the climate. With still superior methods, this same problem was solved later by the illustrious German mathematician, Gauss.

In 1741, at the request, or rather command, of Frederick the Great, he moved to Berlin, where he was made a member of the Academy of Sciences, and Professor of Mathematics. He enriched the last volume of the *Mélanges* or Miscellanies of Berlin, with five memoirs, and these were followed, with astonishing rapidity, by a great number of important researches, which were scattered throughout the annual memoirs of the Prussian Academy. At the same time, he continued his philosophical contributions to the Academy of St. Petersburg, which granted him a pension in 1742.

The respect in which he was held by the Russians was strikingly shown in 1760, when a farm he occupied near Charlottenburg happened to be pillaged by the invading Russian army. On its being ascertained that the farm belonged to Euler, the general immediately ordered compensation to be paid, and the Empress Elizabeth sent an additional sum of four thousand crowns. The despotism of Anne I caused Euler, who was a very timid man, to shrink from public affairs, and to devote all his time to science. After his call to Berlin, the Queen of Prussia who received him kindly, wondered how so distinguished a scholar should be so timid and reticent. Euler replied, "Madam, it is because I come from a country where, when one speaks, one is hanged."

In 1766, Euler, with difficulty, obtained permission from the King of Prussia to return to St. Petersburg, to which he had been originally called by Catherine II. Soon after returning to St. Petersburg a cataract formed in his left eye, which ultimately deprived him of sight, but this did not stop his wonderful literary productiveness, which continued for seventeen years—until the day of his death. It was under these circumstances that he dictated to his amanuensis, a tailor's apprentice who was absolutely devoid of mathematical knowledge, his *Anleitung zur Algebra*, or *Elements of Algebra*, 1770, a work which, though purely elementary, displays the mathematical genius of its author, and is still considered one of the best works of its class. Euler was one of the very few great mathematicians who did not deem it beneath the dignity of genius to give some attention to the recasting of elementary processes and the perfecting of elementary textbooks, and it is not improbable that modern mathematics is as greatly indebted to him for his work along this line as for his original creative work.

Another task to which he set himself soon after returning to St. Petersburg was the preparation of his *Lettres à une Princesse d'Allemagne sur divers sujects de Physique et de Philosophie*, (3 vols. 1768–72). These letters were written at the request of the princess of Anhalt-Dessau, and contain an admirably clear exposition of the principal facts of mechanics, optics, acoustics, and physical astronomy. Theory, however, is frequently unsoundly applied in it, and it is to be observed generally that Euler's strength lay rather in pure than in applied mathematics. In 1755, Euler had been elected a foreign member of the Academy of Sciences at Paris, and sometime afterwards the academical prize was adjudged to three of his memoirs *Concerning the Inequalities in the Motions of the Planets*. The two prize-problems proposed by the same Academy in 1770 and 1772 were designed to obtain a more perfect theory of the moon's motion. Euler, assisted by his eldest son, Johann Albert was a competitor for these prizes and obtained both. In his second memoir, he reserved for further consideration the several inequalities of the moon's motion, which he could not determine in his first theory on account of the complicated calculations in which the method he then employed had engaged him. He afterward reviewed his whole theory with the assistance of his son and Krafft and Lexell, and pursued his researches until he had constructed the new tables, which appeared with the great work in 1772. Instead of confining himself, as before, to the fruitless integration of three differential equations of the second degree, which are furnished by mathematical principles, he reduced them to three ordinates which determine the place of the moon; and he divides into classes all the inequalities of that planet, as far as they depend either on the elongation of the sun and moon, or upon the eccentricity, or the parallax, or the inclination of the linear orbit. The inherent difficulties of this task were immensely enhanced by the fact that Euler was virtually blind, and had to carry all the elaborate computations involved in his memory. A further difficulty arose from the burning of his house and the destruction of a greater part of his property in 1771. His manuscripts were fortunately preserved. His own life only was saved by the courage of a native of Basel, Peter Grimmon, who carried him out of the burning house.

Some time after this, the celebrated Wenzell, by couching the cataract, restored his sight; but a too harsh use of the recovered faculty, together with some carelessness on the part of the surgeons, brought about a relapse. With the assistance of his sons, and of Krafft and Lexell, however, he continued his labors, neither the loss of his sight nor the infirmities of an advanced age being sufficient to check his activity. Having engaged to furnish the Academy of St. Petersburg with as many memoirs as would be sufficient to complete its acts for twenty years after his death, he in seven years transmitted to the Academy above seventy memoirs, and left above two hundred more, which were revised and completed by another hand.

Euler's knowledge was more general than might have been expected in one who had pursued with such unremitting ardor, mathematics and astronomy, as his favorite studies. He had made considerable progress in medicine, botany, and chemistry, and he was an excellent classical scholar and extensively read in general literature. He could repeat the Æneid of Virgil from the beginning to the end without hesitation, and indicate the first and last line of every page of the edition which he used. But such lines from Virgil as, "The anchor drops, the rushing keel is staid," always suggested to him a problem and he could not help enquiring what would be the ship's motion in such a case.

Euler's constitution was uncommonly vigorous and his general health was always good. He was enabled to continue his labors to the very close of his life so that it was said of him, that he ceased to calculate and to breath at nearly the same moment. His last subject of investigation was the motions of balloons, and the last subject on which he conversed was the newly discovered planet Herschel [Ed. Note: Uranus].

On the 18th of September, 1783, while he was amusing himself at tea with one of his grandchildren, he was struck with apoplexy, which terminated the illustrious career of this wonderful genius, at the age of seventy-six. His works, if printed in their completeness, would occupy from 60 to 80 quarto volumes. However, no complete edition of Euler's writings has been published, though the work has been begun twice. [Ed. note: Since 1911, over 70 volumes of Euler's collected works, the Opera Omnia, have appeared in print. See the Appendix to Dunham's Euler: The Master of Us All (MAA, 1999) for details about the individual volumes.]

He was simple, upright, affectionate, and had a strong religious faith. His single and unselfish devotion to the truth and his joy at the discoveries of science whether made by himself or others, were striking attributes of his character. He was twice married. His second wife being a half-sister of his first, and he had a numerous family, several of whom attained to distinction. His *éloge* was written for the French Academy by Condorcet, and an account of his life, with a list of his works, was written by Von Fuss, the secretary of the Imperial Academy of St. Petersburg.

As has been said, Euler wrote an immense number of works, chief of which are the following: *Introductio in Analysin infinitorum*, 1748, which was intended to serve as an introduction to pure analytical mathematics. This work produced a revolution in analytical mathematics, as the subject of which it treated had hitherto never been presented in so general and systematic a manner. The first part of the *Analysis Infinitorum* contains the bulk of the matter which is to be found in modern textbooks on algebra, theory of equations, and trigonometry. In the algebra, he paid particular attention to the expansion of various functions in series, and to the summation of given series; and pointed out explicitly that an infinite series can not be safely employed in mathematical investigations unless it is convergent. In trigonometry he introduced (simultaneously with Thomas Simpson in England) the now current abbreviations for trigonometric functions, and simplified formulae by the simple expedient of designating the angles of a triangle by A, B, C, and the opposite sides by a, b, c. He also showed that the trigonometrical and exponential functions are connected by the relation $\cos\theta + i\sin\theta = e^{i\theta}$. Here too we meet the symbol e used to denote the base of the Naperian logarithms, namely the incommensurable number 2.7182818…and the symbol π used to denote the incommensurable number 3.14159265…. The use of a single symbol to denote the number 2.7182818…seems to be due to Cotes, who denoted it by M. Newton was probably the first to employ the literal exponential notation, and Euler, using the form a^z, had taken a as the base of any system of logarithms. It is probable that the choice of e for a particular base was determined by its being the vowel consecutive to a, or, still more probable because e is the initial of the word *exponent*.

The use of a single symbol to denote 3.14159265…appears to have been introduced by John Bernoulli, who represented it by c. Euler in 1734 denoted it by p, and in a letter of 1736 in which he enunciated the theorem that the sum of the square of the reciprocals of the natural numbers is $\frac{1}{6}\pi^2$, he uses the letter c. Chr. Goldbach in 1742 used π, and after the publication of Euler's *Analysis* the symbol π was generally employed, the choice of π being determined by the initial of the word, $\pi\varepsilon\rho\iota\phi\varepsilon'\rho\varepsilon\iota\alpha$—*periphereia*.

The second part of the *Analysis Infinitorum* is on analytical geometry. Euler begins this part by dividing curves into algebraic and transcendental, and establishes a number of propositions which are true for all algebraic curves. He then applied these to the general equation of the second degree in two dimensions, showed that it represents the various conic sections, and deduced most of their properties from the general equation. He also considered the classification of cubic, quartic, and other algebraic curves. He next discussed the question as to what surfaces are represented by the general equation of the second degree in three dimensions, and how they may be discriminated one from the other. Some of these surfaces had not been previously investigated. In this work he also laid down the rules for the transformation of coordinates in space. Here also we find the first attempt to bring the curvature of surfaces within the domain of mathematics, and the first complete discussion of tortuous curves.

In 1755 appeared *Institutiones Calculi Differentialis*, to which the *Analysis Infinitorum* was intended as an introduction. This is the first textbook on the differential calculus which

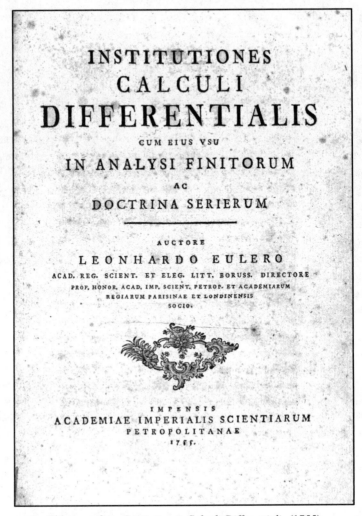

INSTITUTIONES
CALCULI
DIFFERENTIALIS
CUM EIUS VSU
IN ANALYSI FINITORUM
AC
DOCTRINA SERIERUM

AUCTORE
LEONHARDO EULERO
ACAD. REG. SCIENT. ET ELEG. LITT. BORUSS. DIRECTORE
PROF. HONOR. ACAD. IMP. SCIENT. PETROP. ET ACADEMIARUM
REGIARUM PARISINAE ET LONDINENSIS
SOCIO.

IMPENSIS
ACADEMIAE IMPERIALIS SCIENTIARUM
PETROPOLITANAE
1755.

Title page from *Institutiones Calculi Differentialis* (1755)

has any claim to be regarded as complete, and it may be said that most modern treatises on the subject are based upon it. At the same time, the exposition of the principles of the subject is often prolix and obscure, and sometimes not quite accurate.

This series of works was completed by the publication in three volumes in 1768 to 1770 of the *Institutionum Calculi Integralis*, in which the results of several of Euler's earlier memoirs on the same subjects and on differential equations are included. In this treatise as in the one on the differential calculus was summed up all that was at that time known on the subject. The beta and gamma functions were invented by Euler, and are discussed here, but only as methods of reduction and integration. His treatment of elliptic integrals is superficial. The classic problems on isoperimetrical curves, the brachistochrone in a resisting medium, and theory of geodesics had engaged Euler's attention at an early date, and the solving of which led him to the calculus of variations. The general idea of this was laid down in his *Methodus Inveniendi Lineas Curvas*, published in 1744, but the complete development of the new calculus was first effected by Lagrange in 1759. The

INSTITVTIONVM
CALCVLI INTEGRALIS
VOLVMEN PRIMVM
IN QVO METHODVS INTEGRANDI A PRIMIS PRIN-
CIPIIS VSQVE AD INTEGRATIONEM AEQVATIONVM DIFFE-
RENTIALIVM PRIMI GRADVS PERTRACTATVR.

AVCTORE
LEONHARDO EVLERO
ACAD. SCIENT. BORVSSIAE DIRECTORE VICENNALI ET SOCIO
ACAD. PETROP. PARISIN. ET LONDIN.

PETROPOLI
Impenfis Academiae Imperialis Scientiarum
1768.

Title page from *Institutionum Calculi Integralis* (1768)

method used by Lagrange is described in Euler's integral calculus, and is the same as that given in most modern textbooks on the subject.

In 1770, Euler published the *Anleitung zur Algebra* in two volumes. The first volume treats of determinate algebra. This work includes the proof of the binomial theorem for any index, which is still known by Euler's name. The proof, which is not accurate according to the modern views of infinite series, depends upon the principle of the permanence of equivalent forms, and may be seen in C. Smith's *Treatise on Algebra*, pages 336–337. Euler's proof with important additions due to Cauchy, may be seen in G. Chrystal's *Algebra*, Part II.

It is a fact worthy of note that Euler made no attempt to investigate the convergency of the series, though he clearly recognized the necessity of considering the convergency of infinite series. While Euler recognized the convergency of series, his conclusions in reference to infinite series are not always sound. In his time no clear notion as to what constitutes a convergent series existed, and the rigid treatment to which infinite series are now subjected was undreamed of. Euler concluded that the sum of the oscillating series $1-1+1-1+1-1+\cdots = \frac{1}{2}$, for the reason, that by stopping with an even number of terms the sum is 0, and by stopping with an odd number of terms the sum is 1. Hence, the sum of the series is $\frac{1}{2}(0+1) = \frac{1}{2}$. Guido Grandi went so far as to conclude that $\frac{1}{2} = 0+0+0+0+\cdots$. The paper in which Euler cautions against divergent series contains the proof that ...

$$\cdots + \frac{1}{n^2} + \frac{1}{n} + 1 + n + n^2 + n^3 + \cdots = 0.$$

His proof is as follows,

$$n + n^2 + n^3 + \cdots = \frac{n}{1-n},$$

$$1 + \frac{1}{n} + \frac{1}{n^2} + \cdots = \frac{n}{n-1},$$

and $$\frac{n}{n-1} + \frac{n}{1-n} = 0.$$

Euler had no hesitation in writing $1-3+5-7+9-\cdots = 0$, and he confidently believed that $\sin\phi - 2\sin 2\phi + 3\sin 3\phi - \cdots = 0$.

A remarkable development, due to Euler, is what he named the hypergeometrical series, the summation of which he observed to be dependent upon the integration of linear differential equations of the second order, but it remained for Gauss to point out that for special values of the letters, this series represented nearly all the functions then known. By giving the factors 641×6700417 of the number $2^{2^n} + 1 = 4294967297$ when $n = 5$, he pointed out the fact that this expression did not always represent primes, as was supposed by Fermat.

The sources from which this biography has been obtained are Cajori's and Ball's *History of Mathematics*, and the *Encyclopædia Britannica*.

Leonard Euler, Supreme Geometer[1]

C. Truesdell

On 23 August 1774, within a month of his appointment as Ministre de la Marine and the day before he was made Comptrolleur Général of France, Turgot wrote as follows to Louis XVI:

> The famous Leonard Euler, one of the greatest mathematicians of Europe, has written two works which could be very useful to the schools of the Navy and the Artillery. One is a Treatise on the Construction and Manoeuver of Vessels; the other is a commentary on the principles of artillery of Robins ...I propose that Your Majesty order these to be printed; ....

> It is to be noted that an edition made thus without the consent of the author injures somewhat the kind of ownership he has of his work. But it is easy to recompense him in a manner very flattering for him and glorious to Your Majesty. The means would be that Your Majesty would vouchsafe to authorize me to write on Your Majesty's part to the lord Euler and to cause him to receive a gratification equivalent to what he could gain from the edition of his book, which would be about 5,000 francs. This sum will be paid from the secret accounts of the Navy.

"The famous Leonard Euler," then sixty-nine years old and blind, was the principal light of Catherine II's Academy of Sciences in Petersburg. His name had figured before in the correspondence between Turgot, the economist and politician, and Condorcet, the prolific if rather superficial mathematician and littérateur soon to become Perpetual Secretary of the Paris Academy of Sciences, and later first an architect and then a victim of the Revolution. Just twenty years afterward Condorcet was to die because his hands had been found to be uncalloused and his pocket to contain a volume of Horace, but in 1774 equality, while already advocated and projected by Turgot, had not progressed so far. In a France threatened by bankruptcy a minister of state could still find time to write in letters to a friend his opinions and doubts and conjectures about everything from literature to manufacture, and by the way the solution of algebraic equations. It was such a minister

[1]Reprinted with permission from *Studies in Eighteenth Century Culture, Vol 2: Irrationalism in the Eighteenth Century* Case Western University Press, 1972.

Leonhard Euler (1737)

who asked whether "this Euler, who lets nothing slip by unnoticed, might have treated in his mechanics or elsewhere" the most advantageous height for wagon wheels.[2]

In a time when intelligence was the highest virtue, when even men and women then thought to be lazy and stupid (and today proved by their words and deeds to have been lazy and stupid) were portrayed with little wrinkles of alertness around their sparkling, comprehending eyes, the name of Leonard Euler, the greatest mathematician of the century in which mathematics was almost unexceptionally regarded as the summit of knowledge, was better known than those of the literary and musical geniuses, for example Swift and Bach. In the firmament of letters only Voltaire outshone Euler. True, in all the world there were but seven or eight men who could enter into discourse with him, Voltaire certainly not being one of them, and most of what he wrote could be understood in detail by only two or three hundred, Voltaire not being one of these either, but pinnacles could then still be admired from below. In the volume for 1754 of *The Gentleman's Magazine*, a British periodical of general interest the contents of which ranged from heraldry to midwifery, we find an article entitled "Of the general and fundamental principles of all mechanics, wherein all other principles relative to the motion of solids or fluids should be established, by M. Euler, extracted from the last Berlin Memoirs." The anonymous extractor concludes that Euler's principle "comprises in itself all the principles which can contribute to the knowledge of the motion of all bodies, of what nature soever they be." This principle we call today the *principle of linear momentum*. There are in fact two further general principles of motion, the *principle of rotational momentum* and the *principle of energy*. The former of these Euler himself evolved and enounced twenty-five years later; it was the culmination of his researches on special cases of rotation that had extended over half of the eighteenth century. The latter principle was left for physicists of the next century to discover.

An entire volume is required to contain the list of Euler's publications. Approximately one third of the entire corpus of research on mathematics and mathematical physics and engineering mechanics published in the last three quarters of the eighteenth century is by him. From 1729 onward he filled about half of the pages of the publications of the Petersburg Academy, not only until his death in 1783 but on and on over fifty years afterward. (Surely a record for slow publication was won by the memoir presented by him to that academy in 1777 and published by it in 1830.) From 1746 to 1771 Euler filled approximately half of the scientific pages of the proceedings of the Berlin Academy also. He wrote for other periodicals as well, but in addition he gave some of his papers to booksellers for issue in volumes consisting wholly of his work. By 1910 the number of his publications had reached 866, and five volumes of his manuscript remains, a mere beginning, have been printed in the last ten years. There is almost no duplication of material from one paper to another in any one decade, and even most of his expository books, some twenty-five volumes ranging from algebra and analysis and geometry through

---

[2]This remark is enlightening. The book to which Turgot refers is Euler's famous *Mechanica,* published in 1738. One of the most abstract works of the century, it never comes near anything concerning a wheel, let alone a wagon. Respect unsupported by even vague familiarity with the contents of this book is not limited to statesmen but is shown even by modern general histories of science or mathematics, which regularly and in positive terms provide it with a purely imaginary description as the "analytical translation" of Newton's *Principia*. In fact, it is a treatise on the motion of a single point whose acceleration is induced by a rule of one of several simple kinds. Were it not for the headings, only an initiate would be able to recognize the contents as being mechanics.

mechanics and optics to philosophy and music, include matter he had not published elsewhere. The modern edition of Euler's collected works was begun in 1911 and is not yet quite complete; although mainly limited to republication of works which were published at least once before 1910, it will require seventy-four large quarto parts, each containing 300 to 600 pages. Euler left behind him also 3000 pages of clearly and consecutively written mathematical notebooks and early draughts of several books.[3] A whole volume is filled by the catalogue of the manuscripts preserved in Russia. Euler corresponded with savants and administrators all over Europe; the topics of his letters range more widely than his papers, going into geography, chemistry, machines and processes, exploration, physiology, and economics. About 3000 letters from or to Euler are presently known; the catalogue of these, too, occupies a large volume; nearly one-third of them have been printed, usually in volumes consisting of particular correspondences. The first such volume, published in 1843, was of great importance for its impetus to developments in the theory of numbers in the nineteenth century, more than fifty years after all the principals in the correspondence had died. This kind of permanence, difficult for literary men and historians and physicists to comprehend, is typical of sound mathematics.

In modern usage Euler's name is attached as a designation to dozens of theorems scattered over every part of mathematical science cultivated in his time. Even more astonishing than this broad though vague and incomplete tradition is the influence Euler's own writings continue to exert upon current research. The *Science Citation Index* for 1975 through 1979 lists roughly 200 citations of some 100 of Euler's publications; most of the works in which these citations occur are contributions to modern science, not historical studies.

It was Euler who first in the western world wrote mathematics openly, so as to make it easy to read. He taught his era that the infinitesimal calculus was something any intelligent person could learn, with application, and use. He was justly famous for his clear style and for his honesty to the reader about such difficulties as there were. While most of his writings are dense with calculations, four of his books are elementary. One of these is a textbook for the Russian schools; one is the naval manual which Turgot caused to be reprinted in France; one is a treatise on algebra which begins with counting and ends with subtle problems in the theory of numbers; and the fourth, called *Letters to a Princess of Germany on Different Subjects in Natural Philosophy*, is a survey of general physics and metaphysics. This last is the most widely circulated book on physics written before the recent explosion of science and schooling. It was translated into eight languages; the English text was published ten times, each time revised so as to bring the contents somewhat up to date; six of the editions were American, the last one in 1872, a date only a little further from the present day than from 1768, when the original first appeared.

While Euler is known today primarily as a mathematician, he was also the greatest

---

[3]There are also four classes of manuscripts of memoirs and books:

1. Manuscripts from which, perhaps with some correction the works were set in type in Euler's lifetime.

2. Manuscripts intended for Publication and published in the regular volumes of the Petersburg Academy after Euler's death.

3. Manuscripts which Euler withheld from publication but which were published in the collections entitled *Commentationes arithmeticae collectae* (St. Petersburg, 1849) and *Opera Postuma*, 2 vols. (St. Petersburg, 1862).

4. Manuscripts of works not published before 1966. Many of these remain unpublished.

physicist of his era, a rank which was obscured for 200 years but has been re-established by the recent studies of Mr. David Speiser. Euler was the first person to derive an equation of state for a gas from a kinetic-molecular theory. In geometrical optics he invented the achromatic lens. His design for it required glasses of high, distinct, and reproducible quality; attempts to construct lenses according to his prescriptions have been adduced as impulses to the rise of the optical industry in Germany, which was supreme in precision for at least a century. He designed and caused to be built and tried an apparatus for measuring the refractive index of a liquid; it worked, and it remained in use for a century and a half. Euler's hydrodynamics was the first field theory. Perhaps his most important progress in physics other than mechanics is his having taken the observed fact that beams of light pass through each other without interference as justifying use of his linear field theory of acoustic waves to describe waves of light in a luminiferous aether, which he visualized as a subtle fluid.

To study the work of Euler is to survey all the scientific life, and much of the intellectual life generally, of the central half of the eighteenth century. Here I will not even list all the fields of science to which Euler made major additions. The most I attempt is to give some idea what kind of man he was.

Leonard Euler was born in Basel in 1707, the eldest son of a poor pastor who soon moved to a nearby village. The parsonage there had two rooms: the pastor's study and another room, in which the parents and their six children lived. Euler in the brief autobiography he dictated to his eldest son when he was sixty wrote that in his tender age he

The Swiss village, Riehen, where Euler spent his childhood. The bell tower is on the church where Euler's father was pastor.

had been instructed by his father;

> as he had been one of the disciples of the world-famous James Bernoulli, he strove
> at once to put me in possession of the first principles of mathematics, and to this
> end he made use of Christopher Rudolf's *Algebra* with the notes of Michael Stiefel,
> which I studied and worked over with all diligence for several years.

This book, then some 160 years old, only a gifted boy could have used. Soon Euler was
turned over to his grandmother in Basel,

> so as partly by attendance at the gymnasium and partly by private lessons to get a
> foundation in the humanities [i.e., Greek and Latin languages and literatures] and
> at the same time to advance in mathematics.

Documents of the day picture the gymnasium in a lamentable state, with fist-fights in the
classroom and occasional attacks of parents upon teachers. Mathematics was not taught;
Euler was given private lessons by a young university student of theology who was also
a tolerable candidate in mathematics.

At the age of thirteen Euler registered in the faculty of arts of the University of
Basel. There were approximately 100 students and nineteen professors. Instruction was
miserable, and the faculty, underpaid, was mediocre with one exception. The Professor
of Mathematics was John Bernoulli, the younger brother of the great James, by that time
deceased. John Bernoulli, a mighty mathematician and ferocious warrior of the pen, was
universally feared and admired as a geometer second only to the aged and long silent
Newton. Bernoulli had returned, reluctantly, to the backwater of Basel despite brilliant
offers of chairs in the great universities of Holland; he had had to return because of pressure
from his patrician father-in-law. Single-handed, he had made Basel the mathematical
center of Europe. Three of the four principal French mathematicians of the first half of
the century had sought and received instruction from him; his sons and nephews became
mathematicians, some of them outstanding ones. He hated the "English buffoons," as he
called them, and like Horatius at the bridge he had defeated every British champion who
dared challenge him.

Bernoulli discharged his routine lecturing on elementary mathematics at the University
with increasing distaste and decreasing attention. Those few, very few, students whom he
regarded as promising he instructed privately and sometimes gratis. Euler recalled,

> I soon found an opportunity to gain introduction to the famous professor John
> Bernoulli, whose good pleasure it was to advance me further in the mathematical
> sciences. True, because of his business he flatly refused me private lessons, but he
> gave me much wiser advice, namely to get some more difficult mathematical books
> and work through them with all industry, and wherever I should find some check
> or difficulties, he gave me free access to him every Saturday afternoon and was
> so kind as to elucidate all difficulties, which happened with such greatly desired
> advantage that whenever he had obviated one check for me, because of that ten
> others disappeared right away, which is certainly the way to make a happy advance
> in the mathematical sciences.

When he was fifteen, Euler delivered a Latin speech on temperance and received his
prima laurea, first university degree. In the same year he was appointed public opponent

Johann Bernoulli. The frontispiece from his collected works (*Opera Omnia*, 1742)

of claimants for chairs of logic and of the history of law. In the following year he received his master's degree in philosophy, and to the session of 8 June 1724, at which the announcement was made, he gave a public lecture on the philosophies of Descartes and Newton. Meanwhile, he remembered, for the sake of his family

> I had to register in the faculty of theology, and I was to apply myself besides and
> especially to the Greek and Hebrew languages, but not much progress was made,
> for I turned most of my time to mathematical studies, and by my happy fortune
> the Saturday visits to Mr. John Bernoulli continued.

At nineteen Euler published his first mathematical paper, an outgrowth of one of Bernoulli's contests with the English; Euler had found that his teacher's solution of a certain geometrical problem, while indeed better than the English one, could itself be greatly improved, generalized, and shortened. In the case of his own sons, such turns aroused Bernoulli's jealousy and competition, but Euler at once became and remained his favorite disciple.

The next year, at the age of twenty, Euler competed for the Paris prize. These prizes were the principal scientific honors of the century; golden honors they were, too, 2500 livres or even twice or thrice that much, not the empty titles of our time. John Bernoulli himself won the prize twice; his son Daniel, ten times; Euler was to win it twelve times, or about every fourth year of his working life. The assigned topics were usually dull or vague or intricate matters of celestial mechanics, nautics, or physics, never mathematics as such. Often they were directed toward the interests of a specific Frenchman who had something ready and was expected therefore to win, but the competitions were administered fairly, and when an outsider sent in a fine essay, as a rule he was given the prize. The Basler mathematicians had a knack of twisting a promiseless subject into something more fundamental, upon which mathematics could be brought to bear. The prize essays themselves rarely solved the problem announced and usually were works of second class in their authors' total outputs, but the competitions caused the great savants to take up and deepen inquiries they might otherwise never have begun, and so the competitions tended indirectly to broaden the range of mathematical theories of physics. Thus they played, though at a more individual and aristocratic height, a role like that of military support for science in our time. The subject of 1727 was the masting of ships. Euler had never seen a seagoing ship, but his entry received honorable mention and was published forthwith. The winner was Bouguer, for whom the prize had been designed, and who had submitted an entire treatise he had been writing for some years; this treatise immediately became the standard work on the subject. The other two classics of the eighteenth century on naval science, one being much more general and mathematical and profound, and the other being the little handbook to which Turgot referred, were both to be written later by Euler.

In the same year, his twenty-first, Euler on Bernoulli's advice competed for the chair of physics. While he was quickly eliminated as a candidate, he published his specimen essay, *A Physical Dissertation on Sound*. With the clarity and directness that were to become his instantly recognizable signature, in sixteen pages he laid out in order and in simple words, without calculations, all that was then known about the production and propagation of sound, added some details of his own, and listed a number of open problems. This work became a classic at once; it was read and cited for over a hundred years, during which it served as the program for research on acoustics. Euler himself later wrote at least 100 papers directly or indirectly related to the problems set here, and many of these he solved once and for all. The last page lists six annexes. The first denies the principle of pre-established harmony; the second asserts that Newton's Law of gravitation is indeed universal; the fourth affirms that kinetic energy is the true measure of the force of bodies; while the remaining three announce solutions of problems concerning oscillation through a hole in the earth, the rolling of a sphere, and the masting of ships. The professorship was given to a man never heard of again, who in fact was interested primarily in anatomy and botany. Euler at twenty had entered the field of mechanical physics and philosophy as a challenger with firm positions, openly avowed, on every main question then under debate. At the same time, and in equal measure, he was able to announce definite and final solutions to several specific problems. When he died, fifty-five years later, his mastery of all physics as it was then understood, and his ability to solve special problems, were just the same. Indeed, most of the main general advances of the entire century had been made by him, and in addition he had solved many key-problems and hundreds of examples. On the day of his death he had discussed with his disciples the orbit of the planet Uranus, which Herschel had discovered two years before. On his slate was a calculation of the height to which a hot-air balloon could rise. The news of the Montgolfier' first ascent had just reached St. Petersburg, where Euler had been residing for most of his life.

Having had the good luck not to win the chair of physics at Basel, Euler went to Petersburg in 1727. John Bernoulli had been invited but felt himself too old; instead he offered one of his two sons, Daniel and Nicholas, and then adroitly required that neither should go unless the other went too for company and comfort. One was a professor of law and the other was studying medicine in Italy; both were pleased to accept chairs of mathematics or physics. They promised the young Euler the first vacant place, but Russia's thirst for the mathematical sciences was slaked at the moment, and so they suggested he take a position as "Adjunct in Physiology." To this end they advised him to read certain

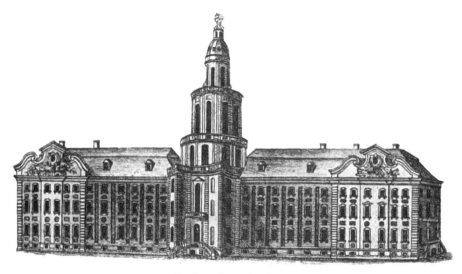

St. Petersburg Academy

books and learn anatomy; accordingly

> I matriculated in the medical faculty of Basel and began to apply myself with all
> industry to the medical course of study ....

Euler arrived in Petersburg on the day the empress died and the Academy fell into

> the greatest consternation, yet I had the pleasure of meeting not only Mr. Daniel
> Bernoulli, whose elder brother Mr. Nicholas had meanwhile died, but also the late
> Professor Hermann, a countryman and also a distant relative of mine, who gave me
> every imaginable assistance. My pay was 300 rubles along with free lodging, heat,
> and light, and since my inclination lay altogether and only toward mathematical
> studies, I was made Adjunct in Higher Mathematics, and the proposal to busy me
> with medicine was dropped. I was given liberty to take part in the meetings of the
> Academy and to present my developments there, which even then were put into
> the *Commentarii* of the Academy.

The academicians were all foreigners—Germans, Swiss, and a Frenchman, not only
the professors but also the students. Thus language was not a problem, but the senior
colleagues were. To a man the chiefs, like university officials today, were tumors, the
only question being whether benign or malignant. The most promising mathematician,
Nicholas II Bernoulli, had died of a fever before Euler arrived. Euler's friends were
Daniel Bernoulli, seven years older and already a famous mathematician and physicist,
and Goldbach, an energetic and intelligent Prussian for whom mathematics was a hobby,
the entire realm of letters an occupation, and espionage a livelihood. The Academy fell on
evil days; its effective director was an Alsatian named Schumacher, whose main interest
lay in the suppression of talent wherever it might rear its inconvenient head. Schumacher
was to play a part in Euler's life for more than a quarter century.

Soon most of the old tumors had been excised by departure or death. So had most
of the capable men. Daniel Bernoulli, after having competed for every vacancy in Basel,
in 1733 finally obtained the chair of anatomy. Once back, he felt himself a new man in
the good Swiss air, but in the rest of his long life he never again reached the level and
the fruitfulness of his eight years in Petersburg, six of which were enlivened by friendly
competition with Euler.

Euler stayed on. For him, these were years of growth as well as production. While
he never lost his love for mechanics and the "higher analysis," he steadily enlarged his
knowledge and power of thought to include all parts of mathematics ever before cultivated
by anyone. He was able to create new synthetic theorems in the Greek style, such as his
magnificent discovery and proof that every rotation has an axis. He sought and read old
books such as Fermat's commentary on Diophantos. On the basis of such antiquarian
studies he recreated the arithmetic theory of numbers, which had been scarcely noticed by
the Bernoullis and Leibniz, in whose school of thought he had been trained. He gave this
subject new life and discovered more major theorems in it than had all mathematicians
before him put together. He was equally at home in the algebra of the seventeenth century,
a field neither easy nor elementary, tightly wed to the theory of numbers. He also probed
new subjects which were to flower only much later. One of these is combinatorial topology,
in which he conjectured but was not able to prove what later became a key-theorem, now

Daniel Bernoulli

called the Euler polyhedron formula.[4] Unifying and subjecting to system the work of many predecessors, he created analytic geometry[5] as we know that discipline today; from

---

[4]Namely, in any simple polyhedron the number of vertices plus the number of faces is greater by two than the number of edges. Euler could not have known that the same assertion lay in an unpublished manuscript of Descartes. Euler did publish a proof, but it is false as it stands; the basic idea of it, nevertheless, is sound and has been applied in countless later researches.

[5]Analytic geometry is ordinarily attributed to Descartes and Fermat. Of course, like any other mathematical innovation, it was neither without antecedents nor beyond improvement. The reader who doubts my statement should draw his own conclusion by comparing Descartes' *La Géométrie*, Volume 2 of Euler's *Introductio in analysin infinitorum*, and a textbook of the 1930s.

Euler's development of analytic geometry is described by C. B. Boyer on pages 180–181 of his *History of Analytic Geometry*, New York, Scripta Mathematica, 1956. Of Euler's *Introductio in analysin infinitorum* Boyer writes

The *Introductio* of Euler is referred to frequently by historians, but its significance generally is underestimated. This book is probably the most influential textbook of modern times. It is the work which made the function concept basic in mathematics. It popularized the definition of logarithms as exponents and the definitions of the trigonometric functions as ratios. It crystallized the distinction between algebraic and transcendental functions and between elementary and higher functions. It developed the use of polar coordinates and of the parametric representation of curves. Many of our commonplace notations are derived from it. In a word, the *Introductio* did for elementary analysis what the *Elements*

his textbook, and from others based upon it, and still others based on them, and so on, students of mathematics learned the subject from 1748 until the 1930s, when it was largely superseded by the rise of modern linear algebra. Students of natural science even today learn it in essentially Euler's way. Euler was the first man to publish a paper on partial differential equations, and the world has learnt most of the elementary calculus of partial derivatives from his books, although some of the rules had been known to Newton and Leibniz but not published by them. It was mainly in his first Petersburg years that Euler developed his taste for pure mathematics, which has remained forever after, in a tradition deriving from him and unbroken by the most violent political changes, a Russian specialty. About one-third of his total product was regarded as "pure" mathematics in his own day; in the classification of our time, this term would apply to only about one-fifth of it; but that small fraction includes many of his deepest and most permanent contributions. One of these is the concept of real function: namely, a rule assigning to each real number in some interval another real number. In his earlier years Euler, like his predecessors, had used a concept of function both narrow and vague, but his own discoveries in the theory of partial-differential equations and wave propagation had shown him the clear way,[6] which every mathematician since 1850 has followed. Other great discoveries were the law of quadratic reciprocity[7] in number theory and the addition theorem for elliptic functions,[8] but these came later than the time of which I am now speaking.

What Euler did for mechanics blanks superlatives. The contents of any one of the two dozen volumes of his *Opera* that concern mechanics primarily would have sufficed to earn its author a place at or near the summit of the field. There is no aspect of it as it stood before his day that he did not change essentially; he solved problems set by his predecessors, applied existing theories to important new instances, simplified ideas while making them more general, unified domains that before him had seemed separate. He created new concepts and new disciplines to embrace phenomena of nature that previously were not understood. Sometimes he worked with the most abstruse mathematics known in his day; he was equally ready to explain his results and their applications by simple rules

---

of Euclid did for geometry. It is, moreover, one of the earliest textbooks on college level mathematics which a modern student can study with ease and enjoyment, with few of the anachronisms which perplex and annoy the reader of many a classical treatise.

Boyer states that Euler's "treatment of the linear equation is characteristic for its generality, but it is startlingly abbreviated." By the standards of modern textbooks for freshmen Euler's book is rather advanced. For example, he stated "the geometry of the straight line is well known."

Finally, writes Boyer,

The Introductio closes with a long and systematic appendix on solid analytic geometry. This is perhaps the most original contribution of Euler to Cartesian geometry, for it represents in a sense the first textbook of algebraic geometry in three dimensions.

By "Cartesian geometry" Boyer refers more or less to what is usually called "analytic geometry"; by "algebraic geometry," to what is usually called "co-ordinate geometry."

[6] The "clear way" is commonly attributed to Dirichlet or other mathematicians of the nineteenth century.

[7] That is in the notation of Gauss, of the two congruences $x^2 \equiv q \pmod{p}$ and $x^2 \equiv p \pmod{q}$, p and q being prime numbers, either both are soluble or neither is except if $p \equiv q \equiv 3 \pmod 4$, in which case one is soluble and the other is not.

[8] That is, in the notation of Jacobi,

$$\mathrm{sn}(u + v) = \frac{(\mathrm{sn}u)(\mathrm{cn}v)(\mathrm{dn}v) + (\mathrm{cn}u)(\mathrm{sn}v)(\mathrm{dn}u)}{1 - k^2(\mathrm{sn}^2 u)(\mathrm{sn}^2 v)}$$

and related formulæ.

of practice; he regularly furnished numerical methods and worked-out instances. Above all, he sought and achieved clarity.

Analysis was the key to mechanics, and in turn mechanics suggested most of the problems of analysis that mathematicians of the eighteenth century attacked. Astronomy and physics were mainly applications of mechanics. Over half of the pages Euler published were expressly devoted to mechanics or closely connected with it.

Nonetheless, there is no evidence that Euler preferred anyone part of mathematics to the rest.[9] The only sure conclusion we can draw from his prodigious output is that he sought to enlarge the domain of mathematics and its applications with a dedication as eager as that which led Don Giovanni to seduce even ugly girls *pel piacer di porle in lista*, but Euler's outposts, even those ridiculed by some of his contemporaries, have been bridgheads to future and permanent, total conquests.

The first Petersburg years brought Euler success, instruction in the facts of life, and misfortune.

> In 1730, when Professors Hermann and Bülfinger returned to their native land, I was named to replace the latter as Professor of Physics, and I made a new contract for four years, granting me 400 rubles for each of the first two and 600 for the next two, along with 60 rubles for lodging, wood, and light.

Then Euler had the experience, not uncommon in the Enlightenment, of being unable to collect all of his contracted salary. In 1731 there was a matter of promotion: Four little men, who up to that time had been receiving less than he, were set equal to him. In a formal protest Euler wrote,

> That we shall each be treated on the same footing is something I can't get through my head at all.... It is true that I have never applied myself so much to physics as to mathematics, but nevertheless I doubt much that you can get from the outside such a person as I for any 400 rubles. In the matter of mathematics, I think the number of those who have carried it as far as I is pretty small in the whole of Europe, and none of those will come for 1000 rubles.

(We should take note of Euler's estimated difference of salaries: 400 for a physicist, 1000 for a mathematician. In those days physics was a speculative or experimental science, not a mathematical one.) [10] Bülfinger, whose talent was modest at best and for mathematics naught, had been Professor of Physics; Daniel Bernoulli, whose lifelong passion was what he himself called physics, was Professor of Higher Mathematics. Schumacher advised the

---

[9] In his beautiful book *Fermat's Last Theorem*, New York etc., Springer-Verlag, H. M. Edwards writes as follows:

> It is a measure of Euler's greatness that when one is studying number theory one has the impression that Euler was primarily interested in number theory, but when one studies divergent series one feels that divergent series were his main interest, when one studies differential equations one imagines that actually differential equations was his favorite subject, and so forth.... Whether or not number theory was a favorite subject of Euler's, it is one in which he showed a lifelong interest and his contributions to number theory alone would suffice to establish a lasting reputation in the annals of mathematics.

[10] This difference in their predecessors is recognized by both mathematicians and physicists today, since the latter are wont to say that the greatest discoveries in mathematics were made by (theoretical) physicists, while the former often remark that most of the major discoveries in theoretical physics were made by mathematicians (until very recently). Usually they are speaking of the same persons, e.g., Huygens and Newton and Euler and Lagrange and Cauchy and Fourier.

President of the Academy not to grant Euler the least concession, since otherwise he would straightway grow impudent. Euler learned a lifelong lesson from this experience: It is futile to argue with administrators but easy to outwork and forget them.

In 1733, Euler states,

> when Professor Daniel Bernoulli, too, went back to his native land, I was given the professorship of Higher Mathematics, and soon thereafter the directing senate ordered me to take over the Department of Geography, on which occasion my salary was increased to 1200 rubles.

Earlier in the same year, even before this splendid increase in his salary, Euler had married, of course choosing a Swiss wife, the daughter of a court artist; in this way he continued the tradition of the Bernoullis, all of whom were either professors or painters, and his younger brother also became a painter. The first of Euler's many children was born the next year. In 1738 a violent fever destroyed the sight of one of Euler's eyes. The work in the geographical department strained his eyesight severely, but he was really interested in constructing a good general map of Russia, and he succeeded in doing so. He wrote to order a school arithmetic text and a great treatise on naval science, receiving for this latter 1200 rubles, in this way doubling his salary one year. Euler's precise recollection of the dates and salaries of his early appointments reflects his Swiss talent for making and saving money. On at least one occasion even Tyche smiled upon him: In the spring of 1749 he wrote to Goldbach that he had received 600 Reichsthaler from a lucky ticket in a lottery, "which was just as good as if I had won a Paris prize this year."

In 1740 Euler was requested to cast the horoscope of the new Czar, who was only a few weeks old. While such a task would have been normal a century earlier, for the Enlightenment it was *retardataire*. Euler smoothly passed the honor on to the Professor of Astronomy. The contents of the horoscope is not known, but in less than a year the child Czar was deposed and hidden; twenty-four years later, still in prison, he died.

In 1740 Frederick II ascended the throne of Prussia. This eccentric and semi-educated general, flute player, and homosexual lay under the spell of France and French men. He wished to create in Berlin a mingled French Académie des Sciences and Académie Française. Voltaire was his Apollo, and Voltaire recommended as director a trifling but extremely eminent French scientist named Maupertuis, whom he dubbed "Le Grand Aplatisseur" for his having led an expedition to Lapland to measure the length of one degree of a meridian, whence he had concluded that the earth was flatter at the poles than at the equator. For Voltaire, who endorsed mathematical philosophy but did not understand it, this proved Descartes wrong and Newton right about everything. The later *philosophes* followed his judgment; the British gleefully followed them; and somehow this minor and precarious if not puerile side issue has assumed in the folklore of science an importance it never for a moment deserved or enjoyed among those who knew what was what in rational mechanics. In addition to being an argonaut, Maupertuis was an *héros de salon* and a *causeur*, a fit table companion for the king; notwithstanding that, he had been a disciple of John Bernoulli, and though no geometer himself, he knew mathematics when he saw it. He proposed to bring all the Bernoullis and Euler to Berlin.

Only Euler was seduced, and at that only because, as he put it, in the regency following the death of Empress Anna "things began to look rather awkward." That the prospect in

Russia was bad indeed, is proved by Euler's consenting to move at no increase in pay. Even so, the Prussian king did not feel himself compelled to discharge his promise in full. After his return to Petersburg, Euler's dictated summary of his twenty-five years in Berlin was "What I encountered there, is well known."

No sooner did Euler arrive in Berlin but the king's wars overturned everything and endangered Maupertuis, who withdrew from Prussia until he was sure Frederick's seat was firm. Euler, meanwhile, was writing mathematical papers. Every associate member of the Academy was required to compose for publication at least one memoir per year; every pensioner, at least two; Euler never presented fewer than ten.

The keys to the treasure house of learning in the eighteenth century—I should be tempted to say also today, were it not that any such statement would be empty because "learning" has been taken off the gold standard—were the Latin language and the infinites-imal calculus. Frederick II understood neither; he detested both. He ordered his Academy to speak and publish only in French, and he encouraged it to cultivate the sciences useful in promotion of trades and manufactures, in the restraint of savage passions, and in the development of a subject's duties. Euler, despite his thoroughly Classical training and his consummate mastery of the new "analysis of curves," easily accepted these conditions. He continued his connection with the Academy of Petersburg, not only sending it a stream of papers, mainly on pure mathematics, but also serving as editor of its publications; in addition, he conveyed to Schumacher information of all sorts regarding the scientific life of the West. In return, of course, he received a salary. These relations continued even through the Seven Years' War, during which Russia joined the alliance against Prussia and at one time overran Berlin. When a farm belonging to Euler[11] was pillaged by the Russians, their commander, General Totleben, saying he did not make war upon the sci-ences, indemnified Euler for more than the damage sustained, and the Empress Elizabeth added a further gift, finally turning the loss into a handsome profit. Euler also lodged and boarded in his house Russian students sent by the Petersburg Academy, one of these being Rasumovski, hetman of the Cossacks, who later became president of the Academy. Euler gave these students instruction in mathematics, this being as close as he ever came to what is called "teaching" in American universities. Euler taught mathematics and physics to the whole world, and down to the present time his influence on instruction in the exact sciences has been second only to Euclid's. In person, had he held a chair in a university, he might have reached a few hundred students at most; like Euclid, by writing Euler has taught mathematics to millions.

By no means all of Euler's books were popular ones. Until about fifteen years ago unopened copies of his more advanced works turned up at low prices on the book market. At least five of these were the first treatises ever published on their subjects, and while easy for a dedicated reader to study, they seemed abstruse to the laity. Few as were the copies sold in Euler's own day,[12] they fell into the right hands. His treatises on

---

[11] The episode has come down to us only through Condorcet's *Eloge*; we do not know whether Euler had more than one farm.

[12] Euler's correspondence with Karsten shows that the printing of his book on the motion of rigid bodies, an acknowledged masterpiece of mechanics, was delayed four years for lack of interest. The publisher demanded subscriptions for 100 copies, but after waiting eighteen months he had received only thirty Euler finally waived royalties; instead, he requested twenty free copies but said he would be satisfied with twelve It seems this latter number was what he did in the end receive. Twenty-five years later, and after Euler's death, the same publisher

rigid-body dynamics, infinite series, differential and integral calculus, and the calculus of variations were mother's milk to three or four generations of mathematicians and theoretical physicists, including the great Frenchmen of the Napoleonic revival, as well as the less eminent but equally influential German and Italian professors of the same period; from the teaching of these three schools the basic core of Euler's work has passed into the common tradition of the mathematical sciences.[13] While it is a rare young Doctor of Philosophy in America today who can decipher a page of Johnson's *London* without a dictionary if not a crib or coach, and while in another academic generation we can confidently expect that *Robinson Crusoe* will have to be translated into "modern English," even the mediocre juniors in engineering the world over have learnt and are able to use a dozen of Euler's discoveries. With the music of the same period, the contrast is more striking. For example, in the eighteenth century no one outside Hamburg can have heard Telemann's *Der Tag des Gerichtes*; few can have been those who heard even some part of Bach's *Messe in H-moll*, and no one, certainly, had heard the whole of it or any part at all of *Die Kunst der Fuge*. While these works seem to us now to stand at the summit of the Enlightenment, even their authors had in their own day merely national or local reputations. Not so with Euler, who was famous far, far beyond the tiny though international circle of those who could understand what he wrote. He was one of those favored few who achieved even from their own contemporaries the respect of which posterity has judged them worthy. Euler won his later fame by the usual method: merciless trials by the fire and water of time. In his own day, from his twenty fifth year onward, he was a senior academician, and he used well the advantages his position gave him.

The academies of the eighteenth century, although few in number, dominated its science, which had become professional. While in the earlier Baroque period there had been many savants, mainly private, who had contributed in some degree to the spring tide of the new natural philosophy, by the time of the Enlightenment science had become a serious business, valued and rewarded though little understood. The high positions were paid well. Euler's initial salary in Berlin was 1600 talers; Maupertuis received nearly twice as much; the junior members, about 300. Paid positions were few, and they were hotly sought. A senior professor in Basel and "the Archimedes of his age," as he justly regarded himself, old John Bernoulli at the end of his life received only about as much as a "student" or "adjunct" in one of the great academies. It is difficult to estimate equivalents in modern currency, but in terms of goods and services in 1982 I think the value of Euler's 1600 talers was around $80,000, tax-free. For example, in 1742 he bought a fine house with a large garden for 2000 taler, one and one quarter year's salary, while the wages of a professor today for the same length of time, after income taxes, would fall well below the price of a run-down row house. Nonetheless we must not be misled by today's social-democratic guilt syndrome, which dictates that the greatest genius of the age must not be paid more than twice the wages of idleness for a congenital fool. The Enlightenment, as its name might suggest, was a period of economic variety, in which Euler found himself further from the top than from the bottom. It would have cost a whole Paris

found it worthwhile to issue the work in a second edition, adding some of Euler's major papers on the subject as an appendix.

[13] It is well known that the British school of the mid-nineteenth century, the greatest representatives of which were Green, Stokes, Kelvin, and Maxwell, learnt mathematics and mathematical physics primarily from French books.

prize to buy a Savonnerie carpet fourteen feet square, had the royal monopoly let any be sold. This is one point where comparisons might be thought simple, since the factory still exists. A carpet of the same size, presumably one of the garish sprays of splotches now regarded as art, cost $36,000 in 1972; as in the Enlightenment, today the total product of the factory is reserved, though no longer for the splendid galleries of kings and their pretty mistresses' bedrooms but rather for the upper beaches washed by the flux and reflux of interchangeable functionaries of the Nth Republic.

Euler practised the thrift for which the Swiss are justly famous. In 1753 he bought a farm for 6000 talers, and with its produce of hay, grain, vegetables, and fruit he cut his household expenses in half. He lodged there his widowed mother, his younger children, and their private tutors. A portrait of the time, a fine pastel by Handmann, shows him in an elegant nightcap and a dressing gown of light and dark blue strips of satin, presumably his working clothes. In this portrait, his blind right eye is turned aside from the beholder. The somewhat confused expression of the mouth is due only to damage to the pastel and does not reflect the ease and decision visible in other portraits of Euler. One of these, unpublished, shows him in a scarlet velvet morning coat. In another he sits on a curvilinear chair with vasiform splat, writing at a carved and gilt table in rococo style.

To learn what an Academy of the eighteenth century was, we may begin with Gulliver's third voyage, published the year before Euler first went to Petersburg. Swift had the Royal Society of London in mind, but the glove fits the more formal academies of the Continent almost as well. First there was the mathematical class:

> ... a race of mortals ... singular in their shapes, habits, and countenances. Their heads were all reclined either to the right or the left; one of their eyes turned inward, and the other directly up to the zenith. Their outward garments were adorned with the figures of suns, moons, and stars, interwoven with those of fiddles, flutes, harps, trumpets, guitars, harpsichords, and many other instruments of music....

> At last we entered the palace, and proceeded into the chamber of presence, where I saw the King seated on his throne, attended on each side by persons of prime quality. Before the throne was a large table filled with globes and spheres, and mathematical instruments of all kinds. His Majesty took not the least notice of us, although our entrance was not without sufficient noise, by the concourse of all persons belonging to the court. But he was then deep in a problem, and we attended at least an hour, before he could solve it.... My dinner was brought.... In the first course there was a shoulder of mutton, cut into an equilateral triangle, a piece of beef into a rhomboides, and a pudding into a cycloid. The second course was two ducks, trussed up into the form of fiddles; sausages and puddings resembling flutes and hautboys, and a breast of veal in the shape of a harp. The servants cut our bread into cones, cylinders, parallelograms, and several other mathematical figures....

> The knowledge I had in mathematics gave me great assistance in acquiring their phraseology, which depended much upon that science and music; and in the latter I was not unskilled. Their ideas are perpetually conversant in lines and figures. If they would, for example, praise the beauty of a woman, or any other animal, they describe it by rhombs, circles, parallelograms, ellipses, and other geometrical terms, or by words of art drawn from music, needless here to repeat.

(We remark that the mathematicians of the Enlightenment shared the common passion for music. Euler himself wrote a major treatise on harmony, which as far as it goes has never been superseded; he projected a treatise on composition; and he published some short papers concerning the function of dissonances. D'Alembert likewise wrote a treatise on music. Some musicians returned the compliment: Rameau wrote,

> Music is a science which should have secure rules; these rules should be drawn from an evident principle, and this principle can scarcely be known to us without the aid of mathematics. Thus I must admit that despite all the experience I could get from music in practising it for so long a time, nevertheless it is only by the help of mathematics that my ideas have grown clear.

Whatever Rameau's study of mathematics may have been, no sign of it may be detected in his book, in which even the experimental facts of acoustics as they were then known are partly misrepresented.)

Thus we see that "relevant" studies were subsidized by the governments of the Enlightenment, that they employed large staffs and needed costly apparatus, and that of modern educational tools only television and computerized dating were yet to be discovered. None of the products of these gossamer schemes for human betterment led to anything we now value. On the other hand, the military projects rarely if ever brought any improvement in the arts of warfare, but they did yield as by-products much basic science which every man curious to understand the world around him must learn today, science upon which rests much of our ordinary technology, that ubiquitous and supremely ugly technology whose products the most humanitarian of humanists insist upon having, and at low cost, however much they may despise the kind of learning that has produced them. For example, Euler's treatise on naval science was based largely on assumptions about the inertial and frictional resistances of water and air which were later shown to be false, and so his tediously scrupulous calculations of the efficiency of sails, oars, and paddle wheels, the design of hulls, and the courses of sailing ships, while correct as calculations, can have been little but useless to the Russian navy, yet his book contains also the first analysis of the stability of floating bodies in general and of the motion of rigid bodies about a variable axis. One device based upon Euler's basic theory but not invented until over 150 years after his death is the gyrocompass, which has saved a thousand times the number of lives it has helped to destroy.

Swift did not mention the disputes of the academicians and the precarious finances of the academies. Although by disposition somewhat irascible, Euler was not quarrelsome; he was exceptionally generous, never once making a claim of priority and in some cases actually giving away discoveries that were his own. He was the first to cite the works of others in what is now regarded as the just way, that is, so as to acknowledge their worth. Up to his time citation had been little more than a weapon of attack, to show where predecessors went wrong. Euler's intellectual generosity can hardly be set as an example, any more than a rich man's scale of giving can be imitated by a poor one: Euler was so wealthy in theorems that loss of a dozen more or less would not be noticed.

It was a different matter with religious issues. Euler maintained throughout his life the simple Protestant faith his father had preached. It had no pretensions in science, and science for Euler had no just pretensions in morality and religion. Thus for Euler the atheism or deism or agnosticism of the French *philosophes* was devilish. King Frederick, on the other hand, while regarding organized religion as desirable for the ignorant, upheld the supremacy of the human intellect so long as it impinged only upon God's rights, not those of earthly kings. A Swiss Protestant was ready to bow to his king, but not to the Devil. Euler published anonymously a booklet called *The Rescue of Divine Revelation from the Objections of the Freethinkers.*

In addition, Euler was a philosopher in his own right. Whereas the *philosophes* ridiculed him as naive, Kant later was to derive his own metaphysics from his study of Euler's writings, but he was not able enough in mathematics to understand Euler's major metaphysical paper, *Reflections on Space and Time*. The ridiculously narrow doctrine of the physical universe we are accustomed to associate with Kant and his successors in German philosophy was evolved after Euler's death, and Euler's point of view did not come into its own until the rise of non-Euclidean geometries and relativity, one and two centuries later.[14]

Maupertuis, President of the Berlin Academy, was not precisely a *philosophe*. Euler was loyal to him, and he stood between Euler and the dislike, even contempt, of the king. Maupertuis had sputtered an overriding law of nature, the Law of Least Action, according to which all natural operations rendered something the smallest it could possibly be. Maupertuis' attempt to phrase this law in its application to mechanics was wrong, and ridiculously so. A year earlier Euler had found a correct statement for the motion of a single particle, greatly more special than Maupertuis' pronouncement, but, as far as it went, right. When he heard of Maupertuis' principle, far from claiming any credit, Euler published his own result as being a confirmation of Maupertuis's grand idea, which he praised beyond measure.

Not so the rest of the world. A distinguished nonresident member of the Academy named Koenig, a good mathematician and a friend and former protégé of Maupertuis, had some objections, which he confided to Maupertuis in a private conversation. A break followed, for Maupertuis tolerated no criticism. The next year Koenig published his objections, along with counterexamples, and he mentioned that in any case the idea had been sketched in a letter of Leibniz, long dead, an extract from which he included. A dreadful rumpus ensued in Berlin. Koenig could not produce the letter, which he said he had seen in the possession of his unfortunate friend Henzi, whom the fathers of the Canton of Bern had beheaded because he had accepted their invitation to make some suggestions regarding the government. Euler came to the defense of Least Action and Maupertuis. Having handed over to Maupertuis as a gift his own discovery of the one case in which the principle could then be proved right, he was sure Maupertuis could not have stolen it from Leibniz, and he had shown that something could be done with the principle if properly corrected. Unfortunately he chose to launch a counterattack against Koenig, claiming that the letter was forged.[15]

---

[14] Euler did not anticipate these much later specific theories, but they are in no way contradictory or repugnant to the general conceptions of space and time he formulated.

[15] In Euler's entire life this episode is the only one that has given rise to any *suspicion* of wrongdoing. With the gleeful desire now in fashion to show that *everyone* is as evil as everyone else—or conversely, that nobody is better than anybody—so that no moral or intellectual values can have any but transitory and subjective, and hence meaningless, meaning, every biographical notice on Euler, no matter how meagre or slipshod, manages to mention his unfairness to Koenig.

Meanwhile Voltaire, who after the death of his mistress the Marquise du Châtelet had no agreeable lodging, came to visit King Frederick at Potsdam. Formerly Voltaire had been a great admirer of Maupertuis and had written:

> *Héros de la physique, Argonautes nouveaux*
> *Qui franchissez les monts, qui traversez les eaux,*
> *Dont le travail immense et l'exact mesure*
> *De la terre étonnée ont fixé la figure.*

> Heroes of physics, new Argonauts,
> Who cross the mountains and the seas,
> Whose immense labor and exact measurement
> Have fixed the figure of the astonished earth.

After having sat for a while as the rival of Maupertuis at the king's table, Voltaire changed his mind and republished the quatrain with "hero" replaced by "courier" and with the couplet about immense work and exact measurement replaced by:

> *Ramenez des climats, soumis aux trois couronnes*
> *Vos perches, vos secteurs, et surtout deux Lapones!*

> You bring back from climes subject to the three crowns
> Your poles, your sectors, and above all two Lapp girls.

Indeed Maupertuis had a strange household, which his Lapp mistress had to share with tropical birds, exotic dogs, and a black man, but this was only the beginning. Just at that time Maupertuis published a medley called *Letter on the Progress of the Sciences*, in which he proposed numerous things worthy of the Academy of Lagado: investigations of the Patagonian giants, methods of prolonging life, a college composed of perfectly educated representatives of all nations, vivisection of criminals, a town where only Latin would be spoken, boring a study hole into the earth, use of drugs to allow experiments on the brain, and other metaphysical matters. Voltaire was thus well prepared to regard the treatment of Koenig by Maupertuis as unjust, and Maupertuis' eccentricities and pretensions furnished an immediate subject for a satire: *Dr. Akakia, Physician of the Pope*. The doctor's mission was to cure Maupertuis of his dreadful case of insufferable arrogance.

The king, while presumably amused by the wit displayed, was insulted by the attack on his own President. It must be remembered that the king himself regularly participated in the doings of his Academy by composing essays on moral philosophy for its memoirs. He forbade Voltaire's satire to be printed. Voltaire printed it anyway, using a permit issued for another work. The king, doubly insulted, had the edition burnt by the hangman. The satire was reprinted in Holland, and Berlin was flooded with copies. Voltaire, in increasing disgrace, left town as quickly as he could gain permission to do so On his slow progress to Switzerland he was in fact arrested and detained for a while by the king's officers. Maupertuis, already sick to death with tuberculosis, also left Berlin to take refuge in the home of one of the Bernoullis in Basel, where in a few years he died. Voltaire published a sequel, in which Akakia induced Maupertuis and Koenig to sign a treaty of peace. Article 19 concerns Euler:

> ...our lieutenant general L. Euler hereby through us openly declares

I. That he has never learnt philosophy and honestly repents that by us he has been misled into the opinion that one could understand it without learning it, and that in future he will rest content with the fame of being the mathematician who in a given time has filled more sheets of paper with calculations than any other.. . .

Unfortunately the further sections of this article of the treaty, while equally witty, repeat some of the specific objections of the Englishman Robins about mathematical points, objections which reflect only the inability of Robins to understand the advanced mathematics of his day. In a typical effusion of literary philosophy, Voltaire did no more than blindly copy passages of bad science.

After Maupertuis' departure all the duties of the presidency fell on Euler, but the king would not have a German (for as such he regarded Euler) assume the title, be given the powers, or receive the pay of the office. The Academy had to finance itself from the sale of almanacs, and Euler had to direct their production and marketing. The depression caused by the Seven Years' War was severe. Serious disputes with the king ensued. Meanwhile, the Academy grew smaller from attrition, until besides Euler there was only one other man of any capacity, namely, the lately arrived, self-taught Genevan genius Lambert, whom Frederick regarded as a bear and could only with great difficulty and after long delay be persuaded to accept.

Almost as soon as he had arrived in Berlin, Euler came to realize that in leaving Russia he had made a grave mistake. He found neither the leisure to work, for he was immediately engulfed in the administration of the academy, nor the stimulation from gifted friends and acquaintances he had enjoyed in Petersburg. After having been in Berlin for eight years he wrote

I and all those who have had the good fortune to spend some time in the Imperial Russian Academy must admit that we owe all we are to the advantageous circumstances in which we found ourselves there. For my part, had I not had that splendid opportunity, I should have had to devote myself primarily to some other field of study, in which by all appearances I should have become only a bungler.

Such vehemence of expression may be due to its having been directed to Schumacher, on whose good will Euler's pension depended, yet because all evidence confirms his truthfulness at other times and in other matters, it is unlikely that what he wrote here differed much from what he felt.

While throughout his long life Frederick again and again expressed his contempt for the infinitesimal calculus, the elements of which, it seems, he had tried to learn several times but in vain, he insisted upon having a mathematician as President of his Academy. At the same time this mathematician had to be French, a man of the world, a lion of society. Few indeed have been the mathematicians of this kind, but Frederick found one.

In 1759, when Maupertuis died, there were besides Euler and Lambert only two other major mathematicians in Europe: Daniel Bernoulli and d'Alembert. The former did not fit any of Frederick's qualifications. The latter, a Frenchman ten years younger than Euler, was at the height of his fame; he was Frederick's ideal, being a man of wit, a *philosophe*, a major collaborator on Diderot's *Encyclopédie*, and a light of literature. Even seven years earlier the king had offered him a salary of 12,000 francs, which was seven times what he was receiving in Paris, and also free lodging in the royal chateau and meals at the royal table, but d'Alembert had preferred freedom in poverty to the dangerous vicinity of a king. Moreover, d'Alembert had quarreled with the Berlin Academy over one of its prizes, and for a time he seemed to be a rival of Euler in mechanics and in some parts of analysis. The major scientific dispute of the mid-century, which concerned the tones and motions of the monochord, was at its hottest; the disputants were d'Alembert, Euler, and Daniel Bernoulli, three powerful parties each consisting in just one man, since there was no one else who could understand the mathematics enough to form a founded opinion, let alone take part. Here,[16] as in several other circumstances of science, the eighteenth century is unique: never before had mathematics been so highly regarded by the community of learning, but never before or after were there so few persons able to enter the arena of mathematical research.

D'Alembert came to visit Frederick at Potsdam in 1763. The academicians, most of whom were Swiss, feared the worst. D'Alembert spoke graciously to them and recommended them to the king. In particular, he declined the presidency and recommended Euler for it; the king positively refused, and indeed all along he had spoken contemptuously of Euler, written to him with harsh disrespect, and declined to grant him the least of the requests he had submitted from time to time on behalf of his family and friends. After d'Alembert had returned to Paris, Frederick wrote for his advice on all matters concerning the Academy of Berlin, to the extent that when the academicians wished to suggest something to the king, they found it best to convey the message first to d'Alembert in Paris, who thereupon, if he agreed, offered it to the king as his own idea.

Euler then found the position intolerable. For a long time he had been negotiating intermittently regarding return to the Petersburg Academy. With the accession of a German princess as Catherine II of Russia in 1762, the auspices for the arts and sciences there improved greatly, and Euler succeeded in obtaining an excellent appointment. He tendered his resignation to King Frederick, who brusquely told him to stop petitioning. Euler desisted from taking part in any activity of the Academy. D'Alembert, meanwhile, had found a replacement for him, the young Lagrange, a Piedmontese who had begun in 1760, at the age of twenty-four, to pour forth brilliant research on analysis and mechanics at Euler's own level and speed. Euler had tried to induce him to come to Berlin, but Lagrange, seeing that he had to choose between Euler and d'Alembert, took d'Alembert

---

[16]While it had antecedents going back for over a century, the dispute began with a paper by d'Alembert published in 1749 and continued through d'Alembert's remaining life. Hankins on page 48 of his biography of d'Alembert states that d'Alembert conceded defeat in a final volume of his *Opuscules*, which exists in manuscript but was never published. On the whole, the controversy was not resolved during the lifetimes of any of the main disputants but rather just died out. Euler solved all the central problems concerning a homogenous string correctly and in generality. Daniel Bernoulli's point of view has been used more often subsequently and is susceptible of greater generalization, but he himself was unable to do much on the basis of it, since the mathematical theory essential for exploiting it was not developed until the middle of the next century. Lagrange also took part from 1760 onward, but his work is largely incomplete or incorrect. While it made a great stir in its day and drew high praise from both Euler and d'Alembert, it stands up but ill under critical scrutiny. For a review of the whole matter, see pages 237–300 of my *Rational Mechanics of Flexible or Elastic Bodies, 1638–1788*, Leonhardi Euleri *Opera omnia* (II) 112, 1960. Although various historians of science have protested that my estimates of Lagrange's work in mechanics and analysis (for I have never formed any judgment whatever concerning his work in algebra and number theory) are too harsh, those estimates are induced from detailed examination of the sources, page by page and line by line, and so I will not revise them until such time as I be shown specific errors in my evaluations of specific passages. Anyone who has read older essays on the history of mathematics will be accustomed to sweeping generalities based on a glancing acquaintance with a few of the more elementary parts of works cited, but I see no reason to respect utterances of this kind today.

as his foster father in the politics of science, though in research he always followed tacitly in Euler's footsteps. The choice reflected Lagrange's sagacity. D'Alembert, though not old, had ceased to produce anything worthwhile and had become merely a conniver; he had quarreled with all mathematicians of his own age or older, and he was detested by his fellow academicians in Paris; vain, he badly needed an admirer at the highest echelon of mathematics. Euler was at the summit and plateau of his creative powers, was on excellent terms with everyone except d'Alembert, Koenig, and King Frederick, and needed nothing but money and rank. D'Alembert arranged that Lagrange go to Berlin as Euler's successor.[17] In order to do so, d'Alembert had to tell Frederick a white lie, namely, that Lagrange was a *philosophe* and man of the world. In fact he was neither; he had no interests outside mathematics and a narrow outlook within it, but in society he knew how to keep his mouth shut when not expressing deference to the views of his seniors. In addition, he could pass more or less for a Frenchman, and he later became one,[18] but he never lost his heavy Piedmontese accent.

In all of Euler's vast correspondence there is no mention of politics and little reference to social conditions. Evidently one country, government, or party was the same as another for him, provided it allowed free worship in the Protestant faith his father had taught him and the chance to do a mountain of mathematics for a good salary. Like many other men of the Enlightenment, Euler expressed a general interest in human well-being and in good works such as widows' pensions, charity for orphans and cripples, and common measures for prevention of disease and promotion of trades and manufactures, but his own contribution to these estimable objectives seems to have been confined, beyond a few special mathematical studies, to an exemplary personal life and a miraculously creative and ageless exercise in mathematical science. Again and again he stated that truth of all kinds, knowledge in general, and mathematics in particular led to the betterment of man's condition, and he never showed evidence of seeing any conflict between service to his prince and service to humanity. While obviously neither a Prussian nationalist nor a Russian one, Euler served both countries with the total loyalty which in those days was regarded as the ordinary, moral duty of a servant to his master. The personal failings of Frederick II as a candidate for God's lieutenant on earth must have been more than obvious to Euler, but it was not those that drove him from Berlin. Rather, he sought a social and financial position worthy of himself and, above all, advancement for his children.

Finally Frederick granted Euler leave to depart with most of his family and some of his servants, eighteen persons all told. Euler, then in his sixtieth year, was entertained en route by the King of Poland and the eminent nobility, and upon arrival in Russia was received by the empress. In addition to his salary of 3000 rubles he was given 8800 rubles

---

[17]The relations between Euler and d'Alembert in 1763–1766 are too complicated to trace here. Like most other savants of the period, Euler despised d'Alembert's character, and he did not wish to remain in the Academy if d'Alembert were to become its president. By the time d'Alembert came to decline the presidency, Euler wished only to leave Berlin and feared that d'Alembert's recommendation of him might result in his being retained against his will; and by the time it came to persuading Frederick to accept Lagrange as Euler's successor, d'Alembert's actions were in Euler's best interest, because without a replacement Euler would not have succeeded in getting permission to go.

[18]Lagrange's mother tongue was the Piedmontese dialect; his first publication was in Latin. The errors of language in his earliest papers in French have been silently corrected in the reprints in his *Œuvres Complètes*, the editors of which, unfortunately, for the most part have not taken similar pains with the numerous errors in mathematics.

to buy a good house and 2000 rubles for furniture. He was not burdened with duties; his counsels were requested regularly and often followed. His greatest reward was that good places in the Academy or the imperial service were found for his sons, and marriages into the nobility were arranged for his daughters.

In his last years in Petersburg Euler had more time free for mathematics than ever before. He soon lost the sight of his one remaining eye. Like Bach, he underwent the torment of an operation for cataract, which was unsuccessful and rendered him almost totally blind. If anything, this enforced end to most of the ordinary duties of life left him still freer to work. About half of his 800 publications were written in these, the last seventeen years of his life. In 1766, the year he moved, Euler composed the first general treatise on hydrodynamics; it was to be about 100 years before anyone wrote another. The next year Euler wrote his famous *Complete Introduction to Algebra*. After Euclid's *Elements*, this is the most widely read of all books on mathematics, having been printed at least thirty times in three editions and in six languages; selections were being used as textbooks in the Boston schools in the 1830s. The next year, 1768, Euler wrote his treatise on geometrical optics in three volumes and his tract on the motion of the moon; both of these are filled with colossal calculations, and the latter contains a single table 144 pages long, calculated under Euler's direction "by the tireless labor" of his son, Krafft, and Lexell, all of them academicians. In 1770 he wrote a monograph on the difficult orbit of a comet which had appeared the year before.

Euler's total blindness put an end to composition of such long treatises, and the great increase in the annual number of his publications reflects the change in his method of work. In the middle of his study he had a large table with a slate top. Being barely able to distinguish white from black, he could write a few large equations. Every morning a young Swiss assistant read him the post, the newspaper, and some mathematical literature. Euler then explained some problem he had been sleeping on and proposed a method of attacking it. The assistant was usually able to produce the outline for a draught of a short memoir, or part of one, by the next morning. In 1775, for example, Euler composed more than one complete paper per week; these run from ten to fifty pages in length and concern widely different special problems.

Two years before his death Euler presented to the Petersburg academy a pair of papers suggested by Vergil's line

anchora de prora jacitur, stant littore puppes.

The problem is to find the motion of a ship whose prow is anchored. The title of the first paper tells us that the problem is "commonplace enough, but very difficult to solve"; Euler derives the differential equation of motion for a much simplified model and obtains some integrals of the motion but despairs of proceeding further; in the second paper he presents and analyses the general solution. The *Acta* for that year include five further papers by Euler, but his output was become too great for the ordinary channels, and in the year of his death the Academy issued in addition to nineteen memoirs in the *Acta* an extra volume called *Opuscula analytica*, which consists in thirteen of his papers composed and presented to the Academy nine to twelve years earlier.

Euler's memory, always extraordinary, had by then become prodigious. He could still recite the *Æneid* in Latin from beginning to end, remembering also which lines were first and last on each page of the edition from which he had learnt it some sixty or seventy years earlier. Enormous equations and vast tables of numbers were ready on demand for the eye of his mind. He became one of the sights of the town for distinguished visitors, with whom he usually spoke on nonmathematical topics. Amazed by the breadth and immediacy of his knowledge concerning every subject of discourse, they spread fairy tales about what he could do in his last years.

Only recently have we been able, by study of the manuscripts he left behind, to determine the course of Euler's thought. We now know, for example, that many of the manuscript memoirs published in the two volumes of posthumous works in 1862 he wrote while still a student in Basel and himself withheld from publication for a reason—which usually was some hidden error or an unacceptable or unconvincing result. The first page of one of these memoirs is reproduced here as Figure 2. The memoir it opens is the one that served to introduce Euler to Daniel Bernoulli and was important in securing him his first post in Petersburg. There can be only one reason Euler did not publish it: Daniel Bernoulli had obtained the same result at about the same time by somewhat different means, and Euler did not wish to detract from his friend's glory. The result itself, the solution of the problem of efflux of water from a vessel, became known through Daniel Bernoulli's book, published twelve years later.

The manuscript is a typical one. The spots are ink from the other side showing through. There are few corrections in the smooth, easy writing. The manuscripts of the books Euler wrote in later life are much the same, but for some remain one or even two complete earlier manuscripts of the whole, showing many differences from the final one. When Euler wished to revise a work, he wrote it all out afresh, neat and clean. Like Mozart, he revised in his head and did not begin to use paper until the revision was complete.

The most interesting of all Euler's remains is his first notebook, written when he was eighteen or nineteen and still a student of John Bernoulli. It could nearly be described as being all his 800 books and papers in little. Much of what he did in his long life is an outgrowth of the projects he outlined in these years of adolescence. Later, he customarily worked in some four domains of mathematics and physics at once, but he kept changing these from year to year. Typically he would develop something as far as he could, write eight or ten memoirs on various aspects of it, publish most of them, and drop the subject. Coming back to it ten or fifteen years later, he would repeat the pattern but from a deeper point of view, incorporating everything he had done before but presenting it more simply and in a broader conceptual framework. Another ten or fifteen years would see the pattern repeated again. To learn the subject, we need consult only his last works upon it, but to learn his course of thought, we must study the earliest ones, especially those he did not himself publish.

In an age when genius, intellectual ambition, and drive were common, no man surpassed Euler in any one, and none came near him in combination of all three. Nevertheless, histories of the eighteenth century and social or intellectual histories in general rarely mention him. The explanation was written by Fontenelle, before Euler was born:

> We like to regard as useless what we do not know; it is a kind of revenge; and since mathematics and physics are rather generally unknown, they rather generally pass for useless. The source of their misfortune is plain; they are prickly, wild, and hard to reach . . .

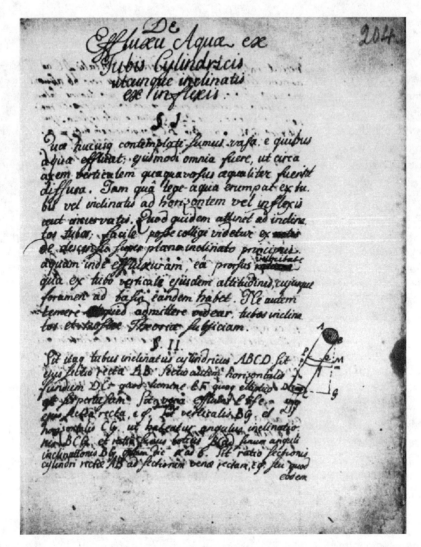

Figure 1. First page of Euler's first paper on fluid mechanics, presented to the Petersburg Academy in 1727, first published 1965. Photograph courtesy of Dr. G. K. Mikhailov.

Such is the destiny of sciences handled by few. The usefulness of their progress is imperceptible to most people, especially if they are practised by professions not particularly illustrious.

Acknowledgment   I am grateful to Dr. Marta Rezler for correction of some details regarding Voltaire.

References

Biography:

Article X, pages 32–60 of *Adumbratio eruditorum Basiliensium meritis apud exteros olim hodieque celebrium,* published as "Adpendix" to *Athenae Rauricae, sive catalogus pro-*

fessorum academiae Basiliensis ab anno 1770 ad annum 1778, cum brevi singulorum biographia, Basileae, 1778. I know this work only through the article by F. Müller, "Uber eine Biographie L. Eulers vom Jahre 1780 and Zusätze zur Euler-Literatur," *Bericht der Deutschen Mathematiker- Vereinigung* 17 (1908): 36–39.

Nicolaus Fuss, *Lobrede auf Herrn Leonhard Euler . . . 23 Octob. 1783 vorgelesen. . . .* Basel, 1786 = pages XLIII–XCV of Leonhardi Euleri *Opera omnia* (I)I, Leipzig & Berlin, Teubner, 1911.

M.-J.-A.-N. Caritat, Marquis de Condorcet, "Eloge de M. Euler," *Histoire de l'Académie Royale des Sciences* (Paris) 1783: 37–68 (1786) = pages 287–310 of Leonhardi Euleri *Opera omnia* (III)12, Zürich, Orell Füssli, 1960.

O. Spiess, *Leonhard Euler, Ein Beitrag zur Geistesgeschichte des XVIII. Jahrhunderts*, Frauenfeld/Leipzig, 1929.

Note: Fuss did not meet Euler until 1773, Euler's sixty-seventh year; Condorcet never met him at all. Neither was competent in more than a small part of the range of science enriched by Euler; both were younger than he by more than thirty years, and neither showed evidence of having studied Euler's early work in detail. Their necrologies of Euler are heavily weighted by hearsay and treat his youth as already legendary. The accounts of Euler's life and work in the general histories of mathematics or collected biographies of mathematicians are mainly if not entirely their authors' personal embroideries upon odds and ends pecked out of the two necrologies. The biography by Spiess, in welcome contrast, is based upon extensive study of unpublished letters and documents as well as all published sources concerning Euler's life. Nevertheless, it is a biography in the literary sense; while Spiess made some attempts to write what is now called intellectual history, his understanding of the contents of Euler's researches was limited not only to what in Spiess's day was called pure mathematics but even to elementary matters such as quadratures, properties of particular curves, explicit sums of series, etc. Thus, inevitably, Euler appears in Spiess's pages as the most dazzling of mathematical jugglers but not as the great creator of concepts and organizer of doctrines he really was. In general, the critical reader who would understand Euler's conceptual frame and intellectual achievement can find today no intermediary between himself and Euler's own writings except the prefaces to some volumes of the *Opera omnia*, for which see below, "Euler's place in the history of science."

A. P. Youschkevitch, article "Euler," *Dictionary of Scientific Biography*, Volume 4, 1971.

Portraits:

Thiersch, "Zur Ikonographie Leonhard und Johann Albrecht Euler's," *Gesellschaft der Wissenschaften zu Göttingen, Nachrichten der Philosophisch-Historischen Classe* 1928: 264–289+4 plates.

H. Thiersch, "Leonhard Euler's `verschollenes' Bildnis und sein Maler," ibid. 1930: 193–217 + Nachtrag + 2 plates.

H. Thiersch, "Weitere Beiträge zur Ikonographie Leonhard und Johann Albrecht Euler's," ibid. 219–249 + 3 plates.

Lists of publications, of manuscripts, and of letters:

G. Eneström, "Verzeichnis der Schriften Leonhard Eulers," Jahresbericht der Deutschen Mathematiker- Vereinigung 4. Ergänzungsband (2 Lieferungen), 388 pages (1910) and 22: 191–205 (1910).

Manuscripta Euleriana Archivi Academiae Scientarum URSS, 1, Moscow & Leningrad, 1962. (This volume describes the scientific manuscripts preserved in Russia. It is reviewed above in Part bI of this essay. According to Eneström, the manuscripts left in the Archives of the Academy in Berlin were once described by Jacobi. I have not seen his description and do not know if it was ever published or if the manuscripts still exist.)

Leonhardi Euleri commercium epistolicum. Descriptio commercii epistolici. Leonhardi Euleri *Opera omnia* (IVA)1, ediderunt A. P. Juskevic, V. I. Smirnov, & W. Habicht, Basel, Birkhäuser, 1975.

Works:

Memoirs, books, and manuscripts, mainly those published at least once before 1911:

Leonhardi Euleri *Opera omnia*, at first Leipzig, then Zürich or other cities of Switzerland, 1911–:
Series I. *Opera mathematica* (complete, 29 volumes issued in 30 parts).
Series II. *Opera mechanica et astronomica* (27 of 32 part-volumes published by the end of 1982).
Series III. *Opera physica et miscellanea* (11 of 12 volumes published by the end of 1982).

Manuscripts not Published before 1911:

Manuscripta Euleriana Archivi Academiae Scientiarum URSS, Volume 2, Moscow & Leningrad, 1965.

Letters:

Leonhardi Euleri *Opera omnia* (IVA)1, the catalogue of the letters, gives references to the some thirty publications in which one or more letters appear. Other volumes in this series are to publish the letters in full. Volume 5 was published in 1980. It includes errata and addenda for Volume 1.

Euler's place in the history of science:

Although it would be hard to find any history of mathematics or physics that does not say something about one or more aspects of Euler's work, and although his name is used as a label for a dozen or more of the commonest and most useful theorems in the mathematical sciences, the bulk and level of his works seem to have discouraged critical study of them. Even volumes of essays devoted to celebrations of Eulerian anniversaries often contain no more than musings by senior scientists who have glanced at a few pages before composing variants of the generalities imparted to them by their teachers in elementary courses half a century earlier. In regard to eighteenth century mathematics and physics the general histories of science or mathematics or physics are grossly unreliable because they are based largely on tale-bearing or caprice or both. Some of the prefaces to individual volumes of Leonhardi Euleri *Opera omnia* explain succinctly some part of Euler's work, especially those in Volumes (I) **4** and **5** (by Fueter), (I) **9** (by A. Speiser), (I) **24** (by Carathéodory), (I) **25** through **29** (by A. Speiser), (II) **3** (by Blanc), (II) **5** (by Fleckenstein), (II) **6**, **7**, and **9** (by Blanc), (II) **11** through **13** (by Truesdell), (II) **14** (by Scherrer), (II) **15** (by Ackeret), (II) **16** and **17** (by Blanc & De Haller), (II) **20** and **21** (by Habicht), (II) **22** (by Courvoisier), (II) **23** (by Fleckenstein), (II) **25** (by Schürer), (II) **28** (by A. Speiser), (II) **29** and **30** (by Courvoisier), (III) **5** (by D. Speiser), (III) **6** (by A. Speiser), (III) **7** (by Habicht), (III) **8** (by Herzberger), (III) **9** (by Habicht), (III) **8** (by D. Speiser), (III) **11** and **12** (by A. Speiser). A few of these also place Euler's work in the setting of its antecedents and its time. For mechanics there is also my book, *Essays in the History of Mechanics*, New York, Springer-Verlag, 1968, and Szabó's, described above in Essay 31; both treat Euler merely incidentally.

The only other occasional yet solid analyses of Euler's work I have found in languages other than Russian are included in Chapter VII of C. R. Boyer's *History of Analytic Geometry*, New York, Scripta Mathematica, 1956, and in six articles in the *Archive for History of Exact Sciences*:

J. E. Hofmann, "Über zahlentheoretische Methoden Fermats und Eulers, ihre Zusammenhänge und ihre Bedeutung," 1(1960/1962): 122–159 (1961).

O. B. Sheynin, "On the mathematical treatment of observations by L. Euler," 9 (1972): 45–56.

H. J. M. Bas, "Differentials, higher-order differentials and the derivative in the Leibnizian calculus," 14 (1974/1975): 1–90 (1974).

R. Calinger, "Euler's 'Letters to a Princess of Germany' as an expression of his mature scientific outlook," 15 (1975/1976): 211–233 (1976).

A. P. Youschkevitsch, "The concept of function up to the middle of the 19th century," 16 (1976/1977): 37–85 (1976).

C. A. Wilson, "Perturbations and solar tables from Lacaille to Delambre: the rapprochement of observation and theory," 22 (1980); 53–304.

Note also the chapter in Edwards' book cited above in Footnote 9.

$\mathcal{E}uler$[1]

André Weil

Until the latter part of the seventeenth century, mathematics had sometimes bestowed high reputation upon its adepts but had seldom provided them with the means to social advancement and honorable employment. Viète had made his living as a lawyer, Fermat as a magistrate; even in Fermat's days, endowed chairs for mathematics were few and far between. In Italy, the University of Bologna ("*lo studio di Bologna*," as it was commonly called), famous throughout Europe, had indeed counted Scipione del Ferro among its professors in the early sixteenth century; but Cardano had been active as a physician; Bombelli was an engineer, and so was Simon Stevin in the Netherlands. Napier, the inventor of logarithms, was a Scottish laird, living in his castle of Merchiston after coming back from the travels of his early youth. Neighboring disciplines did not fare better. Copernicus was an ecclesiastical dignitary. Kepler's teacher Maestlin had been a professor in Tubingen, but Kepler plied his trade as an astrologer and maker of horoscopes. Galileo's genius, coupled with his domineering personality, earned him, first a professorship in Padova, then an enviable position as a protégé of the Grand-Duke of Tuscany, which saved him from the worst consequences of his disastrous conflict with the Church of Rome; his pupil Torricelli succeeded him as "philosopher and mathematician" to the Grand-Duke, while Cavalieri combined the Bologna chair with the priorate of the Gesuati convent in the same city.

Among Fermat's scientific correspondents, few held professorial rank. Roberval, at the Collège de France (then styled Collège Royal), occupied the chair founded in 1572 from a legacy of the scientist and philosopher Pierre de la Ramée. The Savilian chair at Oxford, created for H. Briggs in 1620, was held by Wallis from 1649 until his death in 1703; but his talented younger friend and collaborator William Brouncker, second Viscount, was a nobleman whose career as commissioner of the Navy, and whose amours, are abundantly documented in Pepys's diary. It was only in 1663 that Isaac Barrow became the first Lucasian professor in Cambridge, a position which he relinquished to Newton in 1669 to become preacher to Charles II and achieve high reputation as a divine. In the Netherlands, while Descartes's friend and commentator F. Schooten was a professor in

[1] Reprinted from *Number Theory* (Birkhäuser Boston, 1984), pp. 159–169 with kind permission of Springer Science and Business Media. Also appeared in the *American Mathematical Monthly*, Vol. 91, November, 1984, pp. 537–542.

Leiden, René de Sluse, a mathematician in high esteem among his contemporaries and an attractive personality, was a canon in Liège. Descartes, as he tells us, felt himself, by the grace of God ("*grace à Dieu*": *Discours de la Méthode*, Desc. VI.9), above the need of gainful employment; so were his friends Constantin Huygens and Constantin's son, the great Christian Huygens. Leibniz was in the employ of the Hanoverian court; all his life he preserved his love for mathematics, but his friends marveled sometimes that his occupations left him enough leisure to cultivate them.

Whatever their position, the attitude of such men towards mathematics was often what we can describe as a thoroughly professional one. Whether through the printed word or through their correspondence, they took pains to give proper diffusion to their ideas and results and to keep abreast of contemporary progress; for this they relied largely upon a private network of informants. When they traveled, they looked up foreign scientists. At home they were visited by scientifically inclined travelers, busy bees intent on disseminating the pollen picked up here and there. They eagerly sought correspondents with interests similar to their own; letters passed from hand to hand until they had reached whoever might feel concerned. A private library of reasonable size was almost a necessity. Booksellers had standing orders to supply customers with the latest publications within each one's chosen field. This system, or lack of system, worked fairly well; indeed it subsists down to the present day, supplementing more formalized modes of communication, and its value is undiminished. Nevertheless, even the seventeenth century must have found it increasingly inadequate.

By the time of Euler's birth in 1707, a radical change had taken place; its first signs had become apparent even before Fermat's death. The *Journal des Sçavans* was started in January 1665, just in time to carry an obituary on Fermat ("*ce grand homme*," as he is called there). Louis XIV's far-sighted minister Colbert had attracted Huygens to Paris in 1666, and the astronomer Cassini in 1669, awarding to each a royal pension of the kind hitherto reserved to literati. In 1635 Richelieu had founded the *Académie française*; the more practical-minded Colbert, realizing the value of scientific research (pure no less than applied) for the prosperity of the realm, set up the *Académie des sciences* in 1666 around a nucleus consisting largely of Fermat's former correspondents; Fermat's great friend and former colleague Carcavi was entrusted with its administration and became its first secretary. In England, some degree of political stability had been restored in 1660 by the recall of Charles II; in 1662 the group of amateurs ("*virtuosi*") who had for some time held regular meetings in Gresham College received the charter which made of them the *Royal Society*, with Brouncker as its first president; in 1665 they started the publication of the *Philosophical Transactions*, which has been continued down to the present day. In 1698 the French academy followed suit with a series of yearly volumes, variously entitled *Histoire* and *Mémoires de l'Académie des Sciences*. In 1682 Leibniz was instrumental in creating, not yet an academy, but at least a major scientific journal, the *Acta Eruditorum* of Leipzig, to whose early issues he contributed the articles by which he was giving birth to the infinitesimal calculus.

Soon universities and academies were competing for scientific talent and sparing neither effort nor expense in order to attract it. Jacob Bernoulli had become a professor in his native city of Basel in 1687; as long as he lived, this left little prospect to his younger brother and bitter rival Johann of finding academic employment there; at first he had to teach Leibniz's infinitesimal calculus to a French nobleman, the Marquis de l'Hôpital, even agreeing to a remarkable contract whereby the latter acquired an option upon all of Bernoulli's mathematical discoveries. In 1695, however, Johann Bernoulli became a professor in Groningen, eventually improving his position there by skillfully playing Utrecht against Groningen; finally he settled down in Basel in 1705 after his brother's death. No wonder, then, that in 1741 we find him congratulating Euler on the financial aspects of his Berlin appointment and suggesting at the same time that he would be willing (for a moderate yearly stipend, "*pro modico subsidio annuo*") to enrich the memoirs of the Berlin Academy with regular contributions of his own. In short, scientific life, by the turn of the century, had acquired a structure not too different from what we witness today.

Euler's father, Paul Euler, was a parish priest established in Riehen near Basel; he had studied theology at the university of Basel, while at the same time attending the lectures of Jacob Bernoulli; he had planned a similar career for his son, but placed no obstacle in the way of young Leonhard's inclinations when they became manifest. Clearly, by that time, a bright future was in store for any young man with exceptional scientific talent.

The Euler plaque in Riehen: "Leonhard Euler/1707–1783/Mathematiker, Physiker, Ingenieur, Astronom und Philosoph verbrachte in Riehen seine Jugenjahre. Er war ein grosser Gelehrter und ein gütiger Mensch." (Leonhard Euler, 1707–1783, mathematician, physicist, engineer, astronomer and philosopher, lived in Riehen during his youth. He was a great scientist and a good man.)

Peter the Great

In 1707, when Euler was born, Jacob Bernoulli was dead, and Johann had succeeded him; Johann's two sons Nicolas (born in 1695) and Daniel (born in 1700) were following the family tradition, except that, in contrast to their father and uncle, they loved each other dearly, as they took pains to make known. Euler became their close friend and Johann's favorite disciple; in his old age, he liked to recall how he had visited his teacher every Saturday and laid before him the difficulties he had encountered during the week, and how hard he had worked so as not to bother him with unnecessary questions.

Three monarchs came to play a decisive role in Euler's career: Peter the Great, Frederic the Great, and the Great Catherine. Peter, a truly great czar perhaps, died in 1725; but he had had time to found Saint Petersburg, erect some of its most impressive buildings, and, most important of all for our story, to make plans for an Academy of Sciences modelled on what he had seen in the West; those plans were faithfully carried out by his widow. In 1725 the two younger Bernoullis, Nicolas and Daniel, were called there. Nicolas died the next year, apparently of appendicitis. About the same time an offer went to Euler to join the Petersburg Academy. He was not quite twenty years old; he had just won a prize for an essay on ship-building, never having seen a sea-going ship in his life. He had no early prospects at home. He accepted with alacrity.

From Basel he sailed down the Rhine to Mainz, then traveled to Lubeck, mostly on foot, visiting Christian Wolff on the way; this was a philosopher and follower of Leibniz, banished from Berlin (as he told Euler) by a king with little understanding for philosophy; his hobby-horse was Leibniz's theory of monads, and Euler was clearly not impressed. From Lubeck a ship took the young mathematician to Petersburg.

In those days academies were well-endowed research institutions, provided with ample funds and good libraries. Their members enjoyed considerable freedom; their primary duty was to contribute substantially to the academy's publications and keep high its prestige in the international scientific world. At the same time they were the scientific advisers to the monarch and to state authorities, always on hand for such tasks, congenial or not, as the latter might find fit to assign to them; had it not been so, no state would have undergone the high expense of maintaining such institutions, as Euler once acknowledged to Catherine. In 1758, at the height of his fame, Euler (who had acquired in Petersburg a good command of the Russian language) did not find it beneath him, nor inconsistent with his continuing close relations with the Petersburg Academy, to translate for King Frederic some dispatches seized during military operations against the Russian army.

In 1727, however, the political situation had changed by the time Euler reached Petersburg. Under a new czar all academic appointments were in abeyance. On the strength of his prize essay, Euler was commissioned into the Russian navy, but not for long. Soon he was a salaried member of the academy, at first with the junior rank of "adjunct." When his friend Daniel left for Basel in 1733, he was appointed in his place; thus he could afford to marry, naturally into the local Swiss colony, and to buy for himself a comfortable house. His bride was the daughter of the painter Gsell; in due course she was to give birth to thirteen children, out of whom only three sons survived Euler; little is recorded of her otherwise. The eldest son Johann Albert, born in 1734, was to become one of his father's collaborators, and later a leading member of the academy.

Once Euler was thus well established in Petersburg, his productivity exceeded all expectations, in spite of the comparative isolation in which he had been left by Daniel Bernoulli's departure. It was hardly interrupted by a severe illness in 1735 and the subsequent loss of his right eye. He had beyond doubt become the most valuable member of the Academy, and his reputation had been growing by leaps and bounds, when two events in the higher spheres of European politics brought about a major change in his peaceful life. In Petersburg the death of the czarina in 1740, a regency, and the ensuing turmoils, seemed to threaten the very existence of the Academy. Just at this juncture Frederic II succeeded his father (the same king who had so cavalierly thrown Christian Wolff out of Berlin) on the throne of Prussia; he immediately took steps directed towards the establishment of an academy under his patronage, for which he sought out the most famous names in European science; naturally Euler was on the list. A munificent offer from Frederic, coupled with a fast deteriorating situation in Petersburg, brought Euler to Berlin in July 1741, after a sea-voyage of three weeks on the Baltic during which he alone among his family (or so he claimed) had been free from sea-sickness. In the following year, to his great satisfaction, he was able to purchase an excellent house, well situated, and, by special royal order, exempt from requisition. There he lived for the next 24 years, with apparently the sole interruption of seasonal visits to the country estate he acquired in 1752 in Charlottenburg, and of a family trip to Frankfurt in 1750 to meet his widowed mother who was coming from Basel to live in Berlin with him; his father, who had been disappointed in his hope of getting Euler to return to Basel, had died in 1745.

With Euler's change of residence one might have expected that the steady flow of his publications would be diverted from Petersburg to Berlin; but far from it! He was not only allowed to keep his membership in the Petersburg Academy, but his pension from Petersburg was continued, and he was intent upon giving his former colleagues value for their money. Well might his great-grandson P.-H. Fuss describe Euler's Berlin period as "twenty-five years of prodigious activity." More than 100 memoirs sent to Petersburg, 127 published in Berlin on all possible topics in pure and applied mathematics were the products of those years, side by side with major treatises on analysis, but also on artillery, ship-building, lunar theory; not to mention the prize-winning essays sent to the Paris academy (whose prizes brought substantial cash rewards in addition to high reputation); to which one has to add the *Letters to a German Princess* (one of the most successful popular books on science ever written) and even a defense of Christianity (*Rettung der gottlichen Offenbarung...*) which did nothing to ingratiate its author with the would-be philosopher-king Frederic. At the same time Euler was conducting an increasingly heavy correspondence, scientific, personal and also official since the administrative burdens of the academy tended to fall more and more upon his shoulders.

As years went by, Euler and Frederic became disenchanted with each other. The king was not unaware of the lustre that Euler was throwing upon his academy, but French literati stood far higher in his favor. He was seeking to attract d'Alembert to Berlin and was expected to put him above Euler as head of the Academy. Euler was spared this blow to his self-esteem; d'Alembert, forewarned perhaps by Voltaire's unpleasant experience with Frederic in 1753, enjoyed basking in the king's favor for the time of a short visit but valued his freedom far too highly to alienate it more durably. Nevertheless, as early as 1763, Euler's thoughts started turning again towards Russia.

Fortunately another political upheaval had just taken place there. In 1762 the czar's German wife had seized power as Catherine II after ridding herself and Russia of her husband. One of her first projects was to restore the Petersburg Academy to its former glory. This was almost synonymous with bringing Euler back. Negotiations dragged on for three years. Finally, in 1766, the Russian ambassador in Berlin was instructed to request Euler to write his own contract. Frederic, realizing too late the magnitude of this loss, had tried to put obstacles in the way; he soon found that he could not afford to displease the imperial lady. In the same year Euler was back in Petersburg, after a triumphal journey through Poland where Catherine's former lover, King Stanislas, treated him almost like a fellow-sovereign.

By then Euler was losing his eyesight. He had lost the use of his right eye during his first stay in Petersburg. About the time when he left Berlin, or shortly thereafter, a cataract developed in his left eye. In 1770, in answer to a letter from Lagrange on number theory, he described his condition as follows: "*Je me suis fait lire toutes les opérations que vous avez faites sur la formule $101 = pp - 13qq$ et je suis entièrement convaincu de leur solidité; mais, étant hors d'état de lire ou d'écrire moi-même, je dois vous avouer que mon imagination n'a pas été capable de saisir le fondement de toutes les déductions que vous avez été obligé de faire et encore moins de fixer dans mon esprit la signification de toutes les lettres que vous y avez introduites. Il est bien vrai que de semblables recherches ont fait autrefois mes délices et m'ont coûté bien du temps; mais à présent je ne saurois plus entreprendre que celles que je suis capable de dévellopper dans ma tête et souvent je*

Catherine the Great of Russia

suis obligé de recourir à un ami pour exécuter les calculs que mon imagination projette" ["I have had all your calculations read to me, concerning the equation $101 = p^2 - 13q^2$, and I am fully persuaded of their validity; but, as I am unable to read or write, I must confess that my imagination could not follow the reasons for all the steps you have had to take, nor keep in mind the meaning of all your symbols. It is true that such investigations have formerly been a delight to me and that I have spent much time on them; but now I can only undertake what I can carry out in my head, and often I have to depend upon some friend to do the calculations which I have planned"].

An operation was attempted in 1771 and was successful at first, but the eye soon became infected, and total or near-total blindness ensued. Except for this misfortune, and for a fire which destroyed his house in 1771 among many others in Petersburg, he lived on in comfort, greatly honored and respected; neither old age nor blindness could induce him to take a well-deserved rest. He had assistants, one of whom was his own son; others were sent to him from Basel, with the co-operation of his old friend Daniel Bernoulli; to one of them, N. Fuss, who had come to Petersburg in 1773 and later married a granddaughter of Euler's, we owe a vivid description of Euler's method of work in the last decade of his life. Hundreds of memoirs were written during that period; enough, as Euler had predicted, to fill up the academy publications for many years to come. He died suddenly on 18 September 1783, having preserved excellent general health and his full mental powers until that very day.

A youthful Leonhard Euler
Courtesy of Universitätsbibliothek Basel

Frederick the Great on Mathematics and Mathematicians[1]

Florian Cajori

Frederick William I of Prussia ordered that his son, known later as Frederick the Great, should "learn no Latin"; "let him learn arithmetic, mathematics, artillery,—economy to the very bottom."[2] The old king allowed the Berlin "Society of Sciences," the favorite child of Leibniz, to languish and almost to pass away. His son, Frederick, on the other hand, secretly acquired some Latin, shunned the study of mathematics beyond its rudiments, and brought the Berlin "Academy"[3] to great splendor. Frederick looked upon mathematical study with disfavor. As crown prince, he wrote, on January 26, 1738, to Voltaire his plan of study,[4] "to take up again philosophy, history, poetry, music. As for mathematics, I confess to you that I dislike it; it dries up the mind. We Germans have it only too dry; it is a sterile field which must be cultivated and watered constantly, that it may produce."

Bantering the mathematicians

D'Alembert once wrote to Frederick the Great:[5] "It is the destiny of your majesty to be always at war; in summer with the Austrians, in winter with mathematics." The king himself put it in these words:[6] "I love to wrangle with mathematicians, that I may know

---

[1] Reprinted from the *American Mathematical Monthly*, Vol. 34, March, 1927, pp. 122–130.

[2] Thomas Carlyle, *History of Frederick II of Prussia*, vol. 2, London, 1870, pp. 19, 20.

[3] The institution commonly known as the Berlin Academy was founded by Leibniz in 1700 under the name of *Societas regia scientiarum*. After his death the organization was neglected for about a quarter of a century and it nearly passed out of existence. It received new life under the patronage of Frederick the Great who reorganized it under the name of *Académie royale des sciences et des belles lettres de Berlin*, adopting French, instead of Latin, as the language in which its memoirs should be published. Men of note were invited to the Academy and were given a salary enabling them to devote their time to research. Among the mathematicians thus invited were Maupertuis, Euler, Lambert and Lagrange. At the beginning of the nineteenth century the Academy was again reorganized under the stimulus mainly of Wilhelm von Humboldt, and it was given a strictly German stamp.

[4] *Œuvres de Voltaire*, vol. 53 (Paris, 1831), p. 28.

[5] *Posthumous Works of Frederick II*, Letters between Frederick II and M. D'Alembert. Translated from the French by Thomas Holcroft, vol. 11 (London, 1789), p. 4. Letter of December 23, 1762.

[6] Loc. cit., p. 27. Letter of August 20, 1765.

Frederick the Great

whether, without understanding $xx + y$, it is not possible to be in the right." The king presents an argument why D'Alembert should not decline his invitation to settle in Berlin, ending with the remark:[7] "Such is my refutation. I hold myself to be victorious, and erect a trophy to myself, for having vanquished a great mathematician, wholly to his disgrace."

Another time, he commentated on some essays of D'Alembert:[8] "I read that part of the work in which you condescend to sink the science of *the sublime geometry* to the level of my ignorance." He ventured upon physical subjects:[9] "Is it not true that electricity, and all the prodigies it has hitherto discovered, have only served to excite curiosity? Is it not true, that the doctrine of attraction and gravity, has done nothing more than astonish the imagination? Is it not true that all the operations of chemistry are in the same predicament? But are robbers less numerous or contractors less covetous?" To which D'Alembert replies:[10] "Your majesty is pleased to treat the sublime geometry a little cavalierly. I allow that it frequently is, as your majesty well observes, a luxury in which idle learning indulges; but to prove that it has often been useful we need only recollect the system of the world, the phenomena of which it so well explains." But the king persists:[11] "An algebraist, who lives locked up in his cabinet, sees nothing but numbers, and propositions, which produce no effect in the moral world. The progress of manners is of more worth to society than all the calculations of Newton."

---

[7] Loc. cit., p. 28.

[8] Loc. cit., p. 65. Letter of May 5, 1767.

[9] Loc. cit., p. 79. Letter of January 7, 1768.

[10] Loc. cit., p. 82. Letter of January 29, 1768.

[11] Loc. cit., p. 145. Letter of January 4, 1770.

Nor did he hesitate to make sport of the Germans:[12] "Our geometrician of Berlin is in excellent health, and rather lives in the planet Venus than on this terraqueous globe. The people, who have heard speak of Venus and her passage over the sun's disk (transit of Venus of 1769), have been two nights on the watch, to observe the phenomenon. You will laugh at the expense of my good countrymen; but it is all the wit they have." Once D'Alembert took occasion to observe:[13] "I perceive your majesty has always a lash in store for geometry." Quite natural is the king's dictum relating to the telescopes of Beguelin:[14] "I suppose the calculations according to which they are constructed are admirable, but the fact is that I wished to make use of them, but could not see through them."

The King and Euler

During the first thirteen years at the court of St. Petersburg, Euler gained general recognition throughout Europe as a mathematician of the first rank. Frederick, even before ascending the throne, had determined upon an eager search for men of genius for his academy. How he settled upon the name of Euler is not known. Perhaps through his friend von Suhm who had been in St. Petersburg since 1737, and was buying books for Frederick, including certain memoirs of the St. Petersburg academy. On June 14, 1740, two weeks after ascending the throne, Frederick wrote von Suhm,[15] "do what you can to engage Mr. Euler, the great algebraist, and if you can, bring him with you. I shall give him 1000 or 1200 écus salary." On July 15, the king repeated:[16] "Bring Euler, if you can." But von Suhm died and the Prussian ambassador took up the matter.[17] Euler accepted and reached Berlin, July 25, 1741. On September 4, the king wrote him from the field of the Silesian war, and set his salary at 1600 Taler.[18]

The war delayed the organization of the new academy, but at its close a volume of *Miscellanea* of the old society came from the press, containing five papers by Euler. Anxious that the new academy should be organized as soon as possible, Euler wrote the king in 1743,[19] suggesting that the revenues of the academy resulting from the sales of the annual almanacs would be sufficiently increased by the sale in the newly acquired territory of Silesia, "to support an academy of sciences of rank equal to that of St. Petersburg or Paris." Two days later came the king's inappreciative reply:[20] "But I believe that, being accustomed to the abstractions of magnitudes in algebra, you have sinned against the ordinary rules of calculation. Otherwise you would not have imagined such a large revenue from the sale of almanacs in Silesia." Euler instantly replied[21] that he had been prompted by a desire "to be worthy of the favors which the king had bestowed upon him." The king's delay was due, apparently, to his plan of making Maupertuis the ruling

---

[12]Loc. cit., p. 127. Letter of July 2, 1769.

[13]Loc. cit., p. 235. Letter of March 6, 1771.

[14]Loc. cit., vol. 12, p. 166. Letter of January 29, 1779.

[15]*Œuvres de Frederic Le Grand*, vol. 16 (Berlin, 1850), p. 391.

[16]Loc. cit., p. 394.

[17]G. Valentin, *Festschrift zur Feier des 200. Geburtstages Leonhard Eulers*, (Leipzig und Berlin, 1907), p. 4.

[18]*Œuvres de Frederic Le Grand*, vol. 20 (Berlin, 1852), p. 199.

[19]Loc. cit., pp. 199, 200. Letter of January 19, 1743.

[20]Loc. cit., p. 200.

[21]Loc. cit, p. 201.

head of the academy, which was delayed by war conditions until 1746. Meanwhile Euler
and high officials in Berlin organized informally a scientific society which for six months
held weekly meetings, and then fused with the old society into a new academy which held
its first meeting on January 24, 1744, the king's birthday.[22] This provisional organization
comprised four classes: Physics, mathematics, speculative philosophy and philology. Euler
became director of the mathematics class, and retained this place until his return to St.
Petersburg in 1766.

The king's inquiry for the best book on artillery led to Euler's translation into German
in 1745, of the treatise by Benjamin Robins. Euler's important commentaries thereon
were translated into French and English.[23] In a practical way Euler was of service to the
king also for information on lotteries and aid in planning a canal connecting the Oder
and Havel.[24] The king was aware of the high esteem in which Euler was held for his
researches. Before 1753 Euler had won seven prizes offered by the Paris academy.[25] He
was elected foreign associate of that academy. In 1775 the king mentioned Euler as having
been an ornament of the Berlin academy.[26] Nevertheless, he did not appreciate and admire
Euler. At one time, when commenting on the lack of ability of mathematicians in literature
and art, the king said:[27] "A certain geometer who has lost one eye in calculating presumed
to compose a minuet by *a* plus *b*. Were it to be played before Apollo, the poor geometer
would run the risk of being flayed alive, as was Marsyas." In a letter to Voltaire, the king
speaks[28] of a "huge cyclops of mathematician" (*un gros cyclope de géomètre*). The king
had no knowledge of mathematics and no appreciation of its value in civilization. "*Euler
en vains calculs met sa philosophie*," he wrote.[29]

Moreover, he missed in Euler the power of light, witty, brilliant conversation and
correspondence, in which his French correspondents, Voltaire, D'Alembert, Maupertuis,
and D'Argens so eminently excelled. When Maupertuis left Berlin in 1756, broken in
health, no regular successor was appointed, not even after his death. In 1763 the king
offered the presidency to D'Alembert, but he could not be persuaded to leave France.
Thereupon the king made himself head, but D'Alembert in Paris was the secret president,[30]
the one upon whose judgment the king depended almost entirely in the selection of new
members to the academy.[31] The real burden of routine administration fell for the ten years
following 1756 upon Euler who, according to Harnack,[32] "was conscientious, economical,
but hardly less violent and self-willed than the old president (Maupertuis); just he was
to be sure, but not without prejudice." When the presidency was offered to D'Alembert,

[22]G. Valentin, loc. cit., p. 8.

[23]Nicolaus Fuss, *Lobrede auf Leonhard Euler* (Basel, 1797), p. 47.

[24]Op. cit., p. 48.

[25]P. H. Fuss, *Correspondance mathématique et physique de quelques célèbres géomètres du XVIIIème siècle*,
vol. 1 (St. Petersburg, 1843), p. 608.

[26]*Œuvres*, vol. 3, p. 25.

[27]Loc. cit., vol. 9, p. 64.

[28]Loc. cit., vol. 11, p. 128.

[29]Loc. cit., vol. 10, p. 138.

[30]Adolf Harnack, *Geschichte der Königlichen preussischen Akademie der Wissenschaften zu Berlin*, vol. 1
(1900), p. 359.

[31]After the death of D'Alembert, it was Condorcet who for 16 months acted as adviser to Frederick the Great
in matters relating to the Berlin academy. See Loc. cit., p. 390.

[32]Loc. cit., pp. 351, 352.

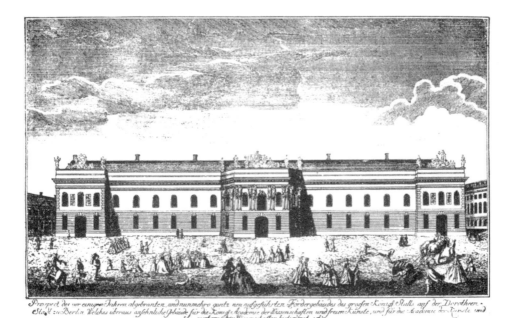

Berlin Academy

Euler felt that he was not appreciated and wrote Goldbach[33] that if the plan persisted to make this a French academy, he might feel constrained to move elsewhere.

On December 2, 1763, Euler asked[34] the king's consent to the marriage of one of his daughters to a cornet in the army, which was promptly denied because of the low rank of the officer. There arose another source of irritation. The king and some members of the academy criticized Euler for allowing an official, Köhler, excessive profits from the sales of the calendar upon which the academy was financially dependent. A commission of investigation was appointed, of which Euler was a member. But, disregarding the commission, Euler made proposals to the king directly. The latter replied:[35] "I cannot compute any curves, but I know quite well that 16000 Taler are more than 13000 Taler." Euler addressed a second letter to the king and received a sharp reply.[36]

These events caused Euler to ask the king for permission to leave Berlin, but not till the third request did the king write a gracious note:[37] Euler "would do him pleasure if he would desist from request for leave and would not revert to this again." But Euler persisted and on May 2, 1766, received leave in a note of two lines. With derision, the king expressed himself to D'Alembert:[38] "Mr. Euler, who is in love even to madness with the great and little bear, has travelled northward to observe them more at his ease. The ship which bore him and his xz and yy has been wrecked. All is lost, which is a pity, for there were materials enough to have formed six volumes, in folio, of memoirs in figures

[33] P. H. Fuss, op. cit., vol. 1, p. 667.

[34] Œuvres, vol. 20, p. 208.

[35] Loc. cit., vol. 20, p. 209.

[36] Adolf Harnack, op. cit., vol. 1, p. 364.

[37] Œuvres, vol. 20, p. 210. Letter of March 17, 1766.

[38] Posthumous Works (Ed. Th. Holcroft), vol. 11, p. 43.

from beginning to end: and Europe, in all probability, will be deprived of the agreeable amusement which such a course of reading would have afforded."

Thus parted the two great men in discord. But ten years softened the feelings. When both were in the autumn of their lives, Euler sent to the king two papers relating to widow's pensions, which the king was introducing, and received in reply gracious acknowledgments. The king also expressed appreciation of his election to honorary membership of the Petersburg academy.[39]

---

[39] *Œuvres*, vol. 20, p. 211; A. Harnack, op. cit., p. 365.

The Euler-Diderot Anecdote[1]

B. H. Brown

No anecdote with regard to a mathematician is better known than the story of the discomfiture of Diderot by Euler. The story was first told by Thiébault [1]; it was later retold, with additions, by De Morgan [2]. Since then a great many authors, all following the more highly colored version of De Morgan, have repeated the story. It is the purpose of this note to show that one addition by De Morgan—the addition which really gives point to the story—is manifestly absurd, and that the credibility of the original story by Thiébault is open to suspicion.

It will be sufficient to give the De Morgan version, noting what he added: "The following story is told by Thiébault, in his *Souvenirs de vingt ans de séjour à Berlin*, published in his old age, about 1804. This volume was fully received as trustworthy; and Marshall Mollendorff told the Duc de Bassano in 1807 that it was the most veracious of books written by the most honest of men. Thiébault says that he has no personal knowledge of the truth of the story, but that it was believed throughout the whole of the north of Europe. Diderot paid a visit to the Russian court at the invitation of the Empress. He conversed very freely, and gave the younger members of the court circle a good deal of lively atheism. The Empress was much amused, but some of the councillors suggested that it might be desirable to check these expositions of doctrine. The Empress did not like to put a direct muzzle on her guest's tongue, so the following plot was contrived. Diderot was informed that a learned mathematician was in possession of an algebraical demonstration of the existence of God, and would give it to him before all the court, if he desired to hear it. Diderot gladly consented; though the name of the mathematician is not given, it was Euler. He advanced toward Diderot, and said gravely and in a tone of perfect conviction: *Monsieur, $(a + b^n)/n = x$, donc Dieu existe; répondez*! Diderot, to whom algebra was Hebrew, was embarrassed and disconcerted; while peals of laughter rose on all sides. He asked permission to return to France at once, which was granted."

This differs from the Thiébault account in three respects:

(1) The formula is slightly different; this affects neither the validity of the proof nor the credibility of the story.

(2) It identifies the mathematician as Euler.

[1] Reprinted from the *American Mathematical Monthly*, Vol. 49, May, 1942, pp. 302–303.

Denis Diderot, oil painting by Louis-Michel van Loo, 1767

(3) The expression "to whom algebra was Hebrew" is an addition. Thiébault says: *"Diderot, voulant prouver la nullité et l'ineptie de cette prétendue preuve, mais ressentant malgré lui, l'embarras où l'on est d'abord lorsqu'on découvre chez les autres, le dessein de nous jour, n'avoit pu échapper aux plaisanteries dont on étoit prêt à l'assaillir."*

Since then the story has, as I have said, been repeated many times. To cite only two instances—both of them by authors of popular books—we find Hogben beginning his *Mathematics for the Million* with this story, but with the substitution of "Arabic" for "Hebrew," and Bell in his *Men of Mathematics* giving a modified version, but with "Chinese" for "Hebrew."

That is the story, and it is a very good story, except that it isn't true. To Diderot algebra was neither Hebrew, nor Arabic, nor even Chinese. Diderot was a very good mathematician, and prior to his Russian trip, was the author of five creditable memoirs on mathematics [3]. To mention only one of these, in the second memoir, *Examen de la développante du cercle*, Diderot shows that if, instead of the Euclidean tools of ruler and compass, we assume a circle and its involute, which last is easily constructed mechanically, then the classical problems, trisection of the angle, duplication of the cube, and quadrature of the circle, may be easily and neatly solved. In proving this, Diderot shows a complete mastery of algebra, geometry, and the calculus.

The anecdote as told by De Morgan and by all who have followed him is thus seen to be absurd. But it may be noted that Thiébault's story is not so unskillful as to aver that Diderot could not reply; it merely says that he sensed the hostility of the audience.

The Thiébault story may have been true; and the mathematician may have been Euler, who was in Russia at that time. What evidence is there for the original story? Thiébault, writing many years later, says merely that the story was believed throughout the north of Europe. No one else tells the story, in particular there is no known Russian source for the story. On the other hand it is known that Frederick the Great, King of Prussia, was a bitter enemy of Diderot. Several sources indicating that stories about Diderot at Saint Petersburg emanated from Berlin are cited and summarized by Tourneux [4]. The alternatives seem to be, first, a rather pointless incident as told by Thiébault; second, and more probable, a canard, inspired by Frederick the Great or by his courtiers.

References

1. *Mes souvenirs de vingt ans de séjour à Berlin*; Paris, 1801, 5 vols; there were several later editions.

2. *A Budget of Paradoxes*, 1872; 2nd ed. edited by David Eugene Smith, 1915.

3. *Œuvres*; Paris, 20 vols., 1875–77. The memoirs are in volume 9.

4. *Diderot et Catherine II*; Paris, 1899. I am indebted to Dr. Arthur M. Wilson, Professor of Biography at Dartmouth College, for this last reference, and for his advance of the second alternative mentioned above.

LEONHARD EULER

Ars Expositionis:
Euler as Writer and Teacher[1]

G. L. Alexanderson

In a recent article by Harold M. Edwards, we are exhorted to "read the masters!" [2]. While we can all agree that in general this is good advice, we may still be reluctant to go to Newton's *Principia* in search of a lucid treatment of the calculus or to Gauss' *Disquisitiones Arithmeticæ* for an account of some famous theorems of number theory and how they came to be proved. One need have no such reluctance, however, about reading Euler. Among the masters he stands out for the clarity of his writing, his willingness to show how he came to his discoveries, and his open admission of failure to prove a conjecture for which he had convincing evidence. George Pólya sums it up when he writes (italics ours):

> A master of inductive research in mathematics, he [Euler] made important discoveries (on infinite series, in the Theory of Numbers, and in other branches of mathematics) by *induction*, that is, by *observation*, *daring guess*, and *shrewd verification*. In this respect, however, Euler is not unique; other mathematicians, great and small, used induction extensively in their work.
>
> Yet Euler seems to me almost unique in one respect: he takes pains to present the relevant inductive evidence carefully, in detail, in good order. He presents it convincingly but honestly, as a genuine scientist should do. His presentation is "the candid exposition of the ideas that led him to those discoveries" and has a distinctive charm. Naturally enough, as any other author, he tries to impress his readers, but, as a really good author, he tries to impress his readers only by such things as have genuinely impressed himself [10, p. 90].

Even among his contemporaries, Euler was known as someone whose writings were particularly clear and elegant. Nicholas Fuss, his assistant in St. Petersburg and a fellow citizen of Basel, wrote in his eulogy on the death of Euler in 1783 [7, p. 14] about the clarity of his ideas, the precision in their statement and the order in which they were arranged. The son of Nicholas Fuss, Paul-Heinrich Fuss, in his preface to the collection

---

[1] Reprinted from *Mathematics Magazine*, Vol. 56, November, 1983, pp. 274–278.

Title page from *Vollständige Anleitung zur Algebra*, 1770

of letters he edited in 1843 [8, p. xli] wrote of Euler's "remarkably simple and lucid exposition of his profound research and his skillful choice of examples."

Euler's career did not involve the teaching that many mathematicians have done routinely over the years, since he spent his professional life at the Imperial Academy of Sciences in St. Petersburg and the Royal Academy of Sciences in Berlin. He nevertheless seems to have had remarkable success at teaching on those occasions when he took on students. N. Fuss tells the story of the young tailor's apprentice Euler brought back to St. Petersburg with him from Berlin in the role of a domestic servant and "who had no smattering of mathematics" but who was the writer to whom Euler dictated his textbook *Vollständige Anleitung zur Algebra*, "as generally admired for the circumstances in which it was composed as for the supreme degree of clarity and of method that prevails throughout. The creative spirit reveals itself even in this purely elementary work" [7, p. 50]. Du Pasquier tells that Euler's son, Johann Albrecht Euler, claimed that by having the text of the algebra dictated to the young servant the boy "became capable of solving by himself even difficult algebraic problems, without need of any help!" [9, p. 113]. The translator of the algebra text into English wrote that

> Here, all is luminous, easy, and obvious. In giving the most difficult demonstrations, and in illustrating the most abstruse subjects, the different steps of the rationale are so many axioms; and it was Euler's great talent to render their order and dependence, in their progress through the mind, clear and evident to the meanest capacity [3, pp. v–vi].

How could one help but learn?

Of course, Euler wrote other texts, the well-known text on the differential calculus (*Institutiones calculi differentialis*) and later his three volume set on the integral calculus (*Institutionum calculi integralis*). These classics set the topics for calculus texts for many years and, again, are known for their clarity and the "choice of examples, as numerous as they are instructive" [9, p. 114]. Perhaps his most famous teaching book is, however, the *Lettres à une Princesse d'Allemagne*. This book, written for the instruction of the 15 year-old future Princess of Anhalt-Dessau, took up topics in mechanics, astronomy, physics, optics, and acoustics "with a marvelous clarity" [9, p. 59]. This set of volumes which appeared in 1768 in St. Petersburg was an immediate success throughout Europe, though it may have been written at a level quite beyond that of a 15 year-old. It was translated into Russian, and appeared in four editions. In Paris, in Leipzig, and in Bern, du Pasquier tells us, there were French editions, twelve in all. They were issued in English nine times, in German six times, in Dutch twice, in Swedish twice, and there were also translations into Italian, Danish and Spanish. The first translator of the *Lettres* into English, Henry Hunter, wrote in his preface

> It was long a matter of surprise to me, that a work so well known, and so justly esteemed, over the whole European Continent, as Euler's *Letters to a German Princess*, should never have made its way into our Island, in the language of the Country. While Petersburg, Berlin, Paris, nay the capital of every petty German principality, was profiting by the ingenious labors of this amiable man, and acute philosopher, the name of Euler was a sound unknown to the ear of youth in the British metropolis. I was mortified to reflect that the specious and seductive productions of a *Rousseau*, and the poisonous effusions of a *Voltaire*, should be in the hands of so many young men, not to say young women, to the perversion of the understanding, and the corruption of the moral principle, while the simple and useful instructions of the virtuous Euler were hardly mentioned.

> As soon as Providence had bestowed on me the blessing of children, I felt it to be my duty to charge myself with their instruction. How I have succeeded it becomes not me to say: but every day I live, the importance of early and proper culture is more deeply impressed on my mind....

> The subjects of these letters, and the author's method of treating them, seem to me much adapted to this purpose. With the assistance of a very moderate apparatus, they might conduct youth of both sexes, with equal delight and emolument, to a very competent knowledge of natural philosophy: very little previous elementary knowledge is necessary to a profitable perusal of them, and that little may be very easily acquired.

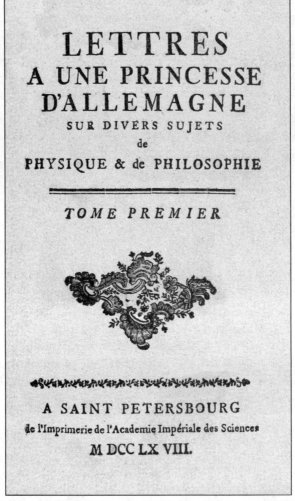

Title page from *Lettres à une Princesse d'Allemagne*, 1768

A considerable part of our common school education, it is well known, consists of the study of the elegant and amusing poetical fictions of antiquity. Without meaning to decry this, may I not be permitted to hint, that it might be of importance frequently to recall young minds from an ideal world, and its ideal inhabitants, to the real world, of which they are a part, and of which it is a shame to be ignorant. [4; xiii–xvi]

As in the other writings of Euler that were intended to instruct, there was much greater content than one often sees in textbooks, and the content of the *Lettres* is still being discussed today [1].

At least one other book of Euler's might be considered a book for instruction, the *Introductio in analysin infinitorum* (1748). Though this is not really a text, it is, nevertheless, a compilation of known work in analysis together with a good bit of new material supplied by Euler. It is one of Euler's most delightful and rewarding works, "as marvellous in its clarity of exposition as for the richness of its contents" [9, p. 113]. The first volume contains a lengthy discussion on the correct definition of a function, but much of the two-volume set is devoted to the solution of wonderful problems. For example, you can find here his argument that the series of the reciprocals of the squares of consecutive integers sums to $\pi^2/6$ and the evaluation of the zeta function for other even integral arguments, the introduction of the theory of partitions, and many properties of logarithms, exponentials and other functions that we now have come to take for granted in courses in classical analysis.

What sets Euler apart from other great masters who wrote mathematics, including textbooks, many of whom even wrote very clearly? As Pólya has noted, part of the answer probably lies in Euler's approach to mathematics and the kind of mathematics that he did so well. He used a great deal of induction in his work, for pattern recognition and discovery of formulas were his forte. Unlike others—Gauss' name comes to mind—Euler did not attempt to hide the origins of his theorems, but, on the contrary, he went out of his way to motivate them by giving many examples. Pólya quotes Condorcet as saying

A bust of Euler in the Bernoullianum at the University of Basel, where several members of the Bernoulli family are commemorated.

that

> He [Euler] preferred instructing his pupils to the little satisfaction of amazing
> them. He would have thought not to have done enough for science if he should
> have failed to add to the discoveries, with which he enriched science, the candid
> exposition of the ideas that led him to those discoveries [10, p. 90].

Euler admits to having trouble proving certain conjectures, but then proceeds to use
unproved results, cautioning the reader that certain steps are not yet completely proved.
In reading Euler we can see a great, creative mind at work. Further, it is encouraging
to us lesser creatures to read of Euler's struggles to discern patterns and to prove his
conjectures. His account of his discovery of what is now termed the pentagonal number
theorem (which for years defied his attempts at proof) was chosen by Pólya as exemplary
of Euler's style of exposition. George Andrews gives a full modern exposition of Euler's
proof of this theorem (this volume, pp. 225–232); here we are interested in Euler's own
early exposition of the result. A full account of Euler's presentation, taken from [4] and
translated by Pólya, can be found in [10, pp. 91–98]. Here we extract a few sections to
demonstrate Euler's style.

In discussing the function $\sigma(n)$, which gives the sum of the divisors of n, he begins

> Till now the mathematicians tried in vain to discover some order in the se-
> quence of the prime numbers and we have every reason to believe that there is
> some mystery which the human mind shall never penetrate. To convince oneself,
> one has only to glance at the tables of the primes, which some people took the
> trouble to compute beyond a hundred thousand, and one perceives that there is
> no order and no rule. This is so much more surprising as the arithmetic gives us
> definite rules with the help of which we can continue the sequence of the primes
> as far as we please, without noticing, however, the least trace of order. I am myself
> certainly far from this goal, but I just happened to discover an extremely strange
> law governing the sums of the divisors of the integers which, at the first glance,
> appear just as irregular as the sequence of the primes, and which, in a certain
> sense, comprise even the latter. This law, which I shall explain in a moment, is,
> in my opinion, so much more remarkable as it is of such a nature that we can
> be assured of its truth without giving it a perfect demonstration. Nevertheless, I
> shall present such evidence for it as might be regarded as almost equivalent to a
> rigorous demonstration.

He proceeds to develop the formula,

$$\sigma(n) = \sum_{j=1}^{\infty} (-1)^{j+1} \sigma(n - n_j),$$

where

$$n_j = \frac{1}{2} j(3j \pm 1) \quad \text{and} \quad \sigma(0) = n,$$

through the use of many examples and gives a heuristic argument which, unfortunately,
depends on the infinite product formula

$$\prod_{k=1}^{\infty} (1 - x^k) = \sum_{n=-\infty}^{\infty} (-1)^n x^{n(3n+1)/2}$$

which he is unable to prove. But at each step he tells what led him to the next discovery and wonders at the beautiful patterns. He says at one point,

> The examples that I have just developed will undoubtedly dispel any qualms which we might have had about the truth of my formula. Now, this beautiful property of the numbers is so much more surprising as we do not perceive any intelligible connection between the structure of my formula and the nature of the divisors with the sum of which we are here concerned.

Later, he writes,

> I confess that I did not hit on this discovery by mere chance, but another proposition opened the path to this beautiful property—another proposition of the same nature which must be accepted as true although I am unable to prove it. And although we consider here the nature of integers to which the Infinitesimal Calculus does not seem to apply, nevertheless I reached my conclusion by differentiations and other devices. I wish that somebody would find a shorter and more natural way, in which the consideration of the path that I followed might be of some help, perhaps.

In reading Euler's exposition, one cannot help but agree with Pólya that from it we can learn "a great deal about mathematics, or the psychology of invention, or inductive reasoning" [10, p. 99]. His techniques as well as his results are a bountiful source of ideas for modern researchers. Several authors in this volume indicate by their discussion of some aspect of Euler's work how his exposition is extraordinarily rich in ideas. As a final example, we note that Euler uses inductive reasoning similar to that described above and a daring, though incorrect, use of analogy to evaluate $\zeta(2)$. In doing so, he seems to be talking to the reader, explaining, sometimes apologizing for the lack of rigor, but always giving insights into the process of discovery [6], [10, pp. 17–21]. A recent paper [11] focuses on Euler's method used on $\zeta(2)$ and shows its fruitfulness.

George Pólya has long been an advocate of Euler's style of presenting mathematics. The author wishes to thank Professor Pólya for his helpful suggestions in the preparation of this note.

References

1. R. Calinger, Euler's letters to a princess of Germany as an expression of his mature scientific outlook, *Arch. Hist. Exact Sci.*, 15 (1975/76) no. 3, 211–233.

2. H. M. Edwards, Read the masters!, *Mathematics Tomorrow*, L. A. Steen, ed., Springer, New York, 1981, 105–110.

3. L. Euler, *Elements of Algebra*, London, 1797.

4. ——, *Letters of Euler on different subjects in physics and philosophy. Addressed to a German princess*, Henry Hunter translator, London, 1795, 1802.

5. ——, *Opera Omnia*, (1) 2, 241–253.

6. ——, *Opera Omnia*, (1) 14, 73–86, 138–155, 156–186.

7. N. Fuss, *Eloge de Monsieur Léonard Euler Lu à l'Académie Impériale des Sciences*, St. Petersburg, 1783.

8. P.-H. Fuss, *Correspondance Mathématique et Physique de Quelques Célèbres Géomètres du XVIIIéme Siècle*, St. Petersburg, 1843.

9. L.-G. du Pasquier, *Léonard Euler et Ses Amis*, Hermann, Paris, 1927.

10. G. Pólya, *Mathematics & Plausible Reasoning, Induction and Analogy in Mathematics*, vol. 1, Princeton, 1954.

11. P. Štulić, A discovery of Euler and some of its consequences, *Matematika* (Belgrade), 5 (1976) no. 2, 84–93.

The Foremost Textbook of Modern Times[1]

C. B. Boyer

The most influential mathematics textbook of ancient times (or, for that matter, of all times) is easily named. The *Elements* of Euclid, appearing in over a thousand editions,[2] has set the pattern for the teaching of elementary geometry ever since it was composed more than two and a quarter millennia ago.

The medieval textbook which most strongly influenced mathematical development is not so easily selected; but a good case can be made out for *Al jabr wa'l muquabala* of Al-Khowarizmi,[3] just about half as old as the *Elements*. From this Arabic work algebra took its name and, to a great extent, its origin.

Is it possible to indicate a modern textbook of comparable influence and prestige? Some would mention the *Géométrie* of Descartes or the *Principia* of Newton or the *Disquisitiones* of Gauss; but in pedagogical significance these classics fell far short of a work less well known. The *Géométrie* was not strictly a textbook, and hence most mathematicians learned their analytic geometry from the works of other authors, such as Schooten, De Witt, Sluze, and Lahire. The *Principia*, the greatest of all works in the field of science, affected the course of pure mathematics only indirectly; few readers understood the elements of the calculus which it contained, and the effective teachers of the differential calculus were Leibniz, L'Hospital, and the Bernoullis. The *Disquisitiones*, a work of great profundity, was too specialized to make its influence widely felt except among ardent number-theorists. It is perhaps significant that none of these three modern works appeared in what may be considered the greatest age of textbooks in recent times.

The eighteenth century often is characterized as a prosy age in the history of mathematics, for it contributed no single discovery which captured the imagination as had analytic geometry and the calculus. And yet the century was of capital importance in the consolidation of earlier work, a task which was facilitated by the appearance of out-

---

[1] Presented at the International Congress of Mathematicians, Cambridge, Mass., September 1, 1950. Reprinted from the *American Mathematical Monthly*, Vol. 58, April, 1951, pp. 223–226.

[2] An excellent English edition is *The thirteen books of the Elements*, translated from the text of Heiberg with introduction and commentary by T. L. Heath, 3 vols., Cambridge, 1908.

[3] See *Robert of Chester's Latin translation of the Algebra of Al-Khowarizmi*, with an introduction, critical notes, and an English version by L. C. Karpinski, New York, 1915.

The frontispiece from *Introductio in analysin infinitorum* (1748)

standing textbook writers. At the opening of the century one finds the texts of L'Hospital dominating the fields of analytical conics and the calculus; at the close there were the textbooks of Lacroix which covered the whole elementary field and which appeared in dozens of editions,[4] not to mention the Legendre *Euclid* of the same time.

But over these well-known textbooks there towers another, a work which appeared in the very middle of the great textbook age and to which virtually all later writers admitted indebtedness. This was the *Introductio in analysin infinitorum* of Euler, published in two volumes in 1748.[5] Here in effect Euler accomplished for analysis what Euclid and Al-Khowarizmi had done for synthetic geometry and elementary algebra, respectively. The function concept and infinite processes had arisen by the seventeenth century, yet it was Euler's *Introductio* which fashioned these into the third member of the mathematical triumvirate comprising geometry, algebra, and analysis. From the point of view of leading textbooks, then, one might refer (with, of course, some oversimplification) to geometry as ancient, algebra as medieval, and analysis as modern.

Euler was not the first to use the word "analysis," or even to incorporate it into the title of a book; but he did give the word a new emphasis. Plato's analysis had reference to the logical order of steps in geometrical reasoning, and the analytic art of Viète was akin to our algebra; but the analysis of Euler comes close to the modern orthodox discipline, the study of functions by means of infinite processes, especially through infinite series. It is in this newer sense that Euler, especially after 1748, used the word; and it is for this reason that he has been referred to as "analysis incarnate" [1]. The word, analysis, took on a new lease of life. Euler himself used it in the titles of dozens of his published papers; and soon others were publishing books on "analytic optics" and "analytical mechanics," on "analytical trigonometry" and "analytic geometry." Euler avoided the phrase "analytic geometry," probably to obviate confusion with the older Platonic usage; yet the second volume of the *Introductio* has been referred to, not inappropriately, as "the first text on analytic geometry" [2], and this appeared more than a century after the publication of *La géométrie*! It is the earliest systematic graphical study of functions, not only of a single variable, but of two as well; and Euler's analysis is transcendental as well as algebraic.

The notion of "elementary function" stems largely from the *Introductio*, but the book contains also such higher curves as $y = x^{\sqrt{2}}$ and $y^x = x^y$. Polar coordinates had appeared in at least half a dozen earlier works, including Newton's *Method of fluxions*; but the clarity and generality of Euler's treatment in the *Introductio* were such that most subsequent writers traced the use of polar coordinates back to this book. Here, in fact, the spiral of Archimedes appears in its dual form, probably for the first time. The use of parametric equations, implicit even in the work of Descartes, was first systemized in the *Introductio*; and here also one finds formalized for the first time the equations for the transformation of coordinates for two and three dimensions, the latter in a form still referred to as "Euler's equations." The *Introductio* was the first textbook to recognize the five proper general quadric surfaces as members of a single family, a century and a half after Kepler had done the same for the conics, and the names proposed were very similar

---

[4] In 1848, for example, there appeared at Paris the 20th edition of his *Traité élémentaire d'arithmétique* and the 16th edition of his *Éléments de géomètrie*; and ten years later his *Éléments d'algèbre* appeared in a 20th edition.

[5] Published at Lausanne.

INTRODUCTIO

IN ANALYSIN

INFINITORUM.

AUCTORE

LEONHARDO EULERO,

Professore Regio BEROLINENSI, *& Academiæ Imperialis Scientiarum* PETROPOLITANÆ *Socio.*

TOMUS PRIMUS

LAUSANNÆ,

Apud MARCUM-MICHAELEM BOUSQUET & Socios.

MDCCXLVIII.

The title page from *Introductio in analysin infinitorum* (1748)

to those now adopted.[6] In this same book Euler also did for plane quartic curves what Newton had done for cubics—he ordered them according to genus and species.

The word "first" is a hazardous one to use in the history of mathematics, and yet it has been applied freely to Euler's *Introductio*. The cases already cited by no means exhaust the respects in which this textbook was first. It contains the earliest algorithmic treatment of logarithms as exponents and of the trigonometric functions as numerical ratios. It was the first textbook to list systematically the multiple-angle formulas, calling attention to the periodicities of the functions; and it included the first general analytic treatment of these as infinite products, as well as their expansion into infinite series. The well-known "Euler identities," relating the trigonometric functions to imaginary exponentials, are also found here. The first volume contains as well an exposition of continued fractions and some excellent work of the zeta function and number theory [3].

In scope alone the *Introductio* ranks among the greatest of textbooks, for it is doubtful that any other essentially didactic work includes as large a portion of original material which survives in the college courses of today. Yet the book is outstanding also for its pedagogical lucidity. The immortal Gauss, a man not given to exaggerated expressions of flattery, held that "The study of Euler's works will remain the best school for the various fields of mathematics, and nothing can replace it" [4]. Written as it was more than two hundred years ago (a letter from Euler to D'Alembert indicates that it was completed by 1745), the *Introductio in analysin infinitorum* nevertheless can be read with comparative ease by the modern student—unlike the *Géométrie*, the *Principia*, or the *Disquisitiones*. Not only is the viewpoint quite similar to that of today; even the terminology and notation are almost modern—or perhaps, as Struik has well written, "we should better say that our notation is almost Euler!" [5].

Under the circumstances one should expect that a textbook exhibiting the qualities of the *Introductio* would boast an impressive list of editions and translations; but the facts belie this. The work was not reprinted until the time of the author's death, thirty-five years later; and, including reprintings and incomplete translations, the fewer than a dozen editions are about equally distributed among the three languages Latin, French, and German.[7] No English translation has appeared, and a partial Russian translation apparently has not been published. [Ed. Note: A two-volume English version, *Introduction to Analysis of the Infinite*, translated by John Blanton, was published by Springer-Verlag in 1988 and 1990.] However, that the worth of a textbook is not necessarily measured by the number of its editions is conclusively evidenced by the *Introductio*, the influence of which was unusually pervasive. Almost without exception the authors of the ubiquitous compendia of the second half of the eighteenth century refer to Euler as the source of their analysis.

---

[6]His names were elliptoides, elliptico-hyperbolicae, hyperbolico-hyperbolicae, elliptico-parabolicae, and parabolico-hyperbolicae.

[7]Latin editions appeared in 1748, 1783, 1797, and in volumes 9 and 10 of Euler's *Opera*, series 1; French editions were published in 1785, 1796, and 1835; German editions appeared in 1788, 1835, and 1885. For bibliographic details on Euler's multitudinous works see Gustav Eneström, "Verzeichnis der Schriften Leonhard Eulers," *Jahresbericht der Deutschen Mathematiker-Vereinigung, Ergänzungsband* IV, Leipzig, 1910. Cf. also P. H. Fuss, *Correspondance mathématique et physique de quelques célèbres géomètres du XVIIIème siècle. Précédée d'une notice sur les travaux de Léonard Euler* (2 vols., St. Pétersbourg, 1843); and Felix Müller, "Über bahnbrechende Arbeiten Leonhard Eulers aus der reinen Mathematik," *Abhandlungen sur Geschichte der mathematischen Wissenschaften*, XXV (1907), 61–116.

The *Introductio* became, in a sense, the prototype of modern textbooks. Is not imitation the sincerest form of flattery?

References

1. See, e.g., E. T. Bell, *Men of mathematics* (New York, 1937), chapter 9.

2. D. J. Struik, *A concise history of mathematics* (2 vols., New York, 1948), II, 169.

3. An extensive account of this work is found in the preface, by Andreas Speiser, to series 1, volume 9, of Euler's *Opera omnia* (Geneva, 1945).

4. Ibid., p. viii. Speiser adds to the words of Gauss: "The Introductio in this connection may stand in first place."

5. Struik, op. cit., II, 174.

Leonhard Euler, 1707–1783[1]

J. J. Burckhardt

Born in 1707, Leonhard Euler grew up in the town of Riehen, near Basel, Switzerland. Encouraged by his father, Paulus, a minister, young Leonhard received very early instruction from Johann I Bernoulli, who immediately recognized Euler's talents. Euler completed his work at the University of Basel at age 15, and at age 19 won a prize in the competition organized by the Academy of Sciences in Paris. His paper discussed the optimal arrangement of masts on sailing ships (*Meditationes super problemate nautico...*). In 1727 Euler attempted unsuccessfully to obtain a professorship of physics in Basel by submitting a dissertation on sound (*Dissertatio physica de sono*); however, this failure, in retrospect, was fortunate. Encouraged by Nicholas and Daniel, sons of his teacher Johann Bernoulli, he went to the St. Petersburg Academy in Russia, a field of action that could accommodate his genius and energy.

In St. Petersburg Euler was met by compatriots Jacob Hermann and Daniel Bernoulli and soon befriended the diplomat and amateur mathematician Christian Goldbach. During the years 1727–1741 spent there, Euler wrote over 100 scientific papers and his fundamental work on mechanics. In 1741, at the invitation of Fredrick the Great, he went to the Akademie in Berlin. During his 25 years in Berlin, his incredible mathematical productivity continued. He created, among other works, the calculus of variations, wrote the *Introductio in analysin infinitorum*, and translated and rewrote the treatise on artillery by Benjamin Robins.

Disputes with the Court led Euler in 1766 to accept a very favorable invitation by Katherine II to return to St. Petersburg. There he was received in a princely manner, and he spent the rest of his life in St. Petersburg. Although totally blind, he wrote, with the help of his students, the famous *Algebra* and over 400 scientific papers; he left many unpublished manuscripts.

In recent decades, numerous important materials concerning Euler have been discovered in the archives in the Academy of Sciences of the USSR. It would seem that there is probably little chance of now discovering an unknown manuscript or something important about his life. Euler himself acknowledged the advantageous circumstances he found at the Academy. Judith Kh. Kopelevic notes, "Euler's tombstone, erected by the Academy;

---

[1]Reprinted from the *Mathematics Magazine*, Vol. 56, November, 1983, pp. 262–273.

his bust in the building of the Presidium of the Academy; the two-centuries-long efforts of the Academy to care for his enormous heritage and publish it—all these show clearly that Euler's encounter with the Petersburg Academy of Sciences was a happy one for both sides."

The legacy of Euler's writings

Euler's productivity is astonishing in its range of content and in the sheer volume of written pages. He wrote landmark books on the subjects of mathematical analysis, analytic and differential geometry, the calculus of variations, mechanics, and algebra. He published over 760 research papers, many of which won awards in competitions, and at his death left hundreds of unpublished works; even today there remain unpublished over 3,000 pages in notebooks. In view of this prodigious collection of written material, it is not surprising that soon after Euler's death the task of surveying and publishing his works encountered extraordinary difficulty.

N. I. Fuss made efforts to publish more writings of the master, but only his son P.-H. Fuss succeeded (with the help of C. G. J. Jacobi) to generate interest among others, including Ostrogradskii. An enterprise in this direction was undertaken in Belgium (1838–1839), but failed after the publication of the fifth volume. In 1844, the Petersburg Academy decided on publication of the manuscripts, but this was not carried out. However, in 1849 the *Commentationes arithmeticae collectae*, edited by P.-H. and N. Fuss, were published; this contains, among others, the important manuscript *Tractatus de doctrina numerorum*.

The centennial of Euler's death in 1883 rekindled interest in Euler's works and in 1896 the most valuable preliminary to any complete publication appeared—the *Index operum Leonhardi Euleri* by J. G. Hagen. As the bicentennial of Euler's birth neared, new life was infused into the project, which was thoroughly discussed by the academies of Petersburg and Berlin in 1903. Although the project was abandoned at this time, the celebrations of the bicentennial of Euler's birth provided the needed impetus for the publication of the *Opera omnia*. The untiring efforts of Ferdinand Rudio led to the decision by the Schweizerische Naturforschende Gesellschaft [Swiss Academy of Sciences] in 1909 to undertake the publication, based on the list of Euler's writings prepared by Gustaf Eneström (1852–1923). He lists 866 papers and books published by then. The financial side appeared assured through gifts and subscriptions. But the first World War led to unforeseen difficulties. We are indebted to Andreas Speiser for his efforts, which made it possible to continue the publication, and who overcame financial and publication difficulties so that at the start of World War II about one half of the project was completed. After the war, Speiser, succeeded by Walter Habicht, completed the series 1 (29 volumes), 2 (31 volumes) and 3 (12 volumes) of the *Opera omnia* except for a few volumes.

In 1947–1948 the manuscripts which had been loaned by the St. Petersburg Academy to the Swiss Academy of Sciences were returned to the archives of the Academy of Sciences of the USSR in Leningrad. Their systematic study was started under the supervision of the Academician V. I. Smirnov, with the goal of publishing a fourth series of the *Opera omnia*. As a first result, there appeared in 1965 a new edition of the correspondence between Euler and Goldbach, edited by A. P. Juškevič and E. Winter. In 1967, the Swiss Academy of Sciences and the Academia Nauk of the USSR formed an International Committee, to which was entrusted the publication of Euler's correspondence in a series 4A, and a critical publication of the remaining manuscripts in a series 4B.

To mark the passage of 200 years since Euler's death, a memorial volume has been produced by the Canton of Basel, *Leonhard Euler 1707–1783, Beiträge zu Leben und Werk*, edited by J. J. Burckhardt, E. A. Fellmann, and W. Habicht (Birkhäuser Verlag, Basel and Boston). From a contemporary point of view, this volume presents the insights of outstanding scientists on various aspects of Euler's achievements and their influence on later works. The complete list of essays and their authors appears at the end of this article. The memorial volume ends with a list, compiled by J. J. Burckhardt, of over 700 papers which are devoted to the work of Euler. It should be stressed that this is certainly an incomplete list, and it is hoped that it will lead to many additional listings which will then be published in an appropriate form. It is hoped that papers little known till now will receive the attention they deserve, and that this effort will lead to an improvement in the collaboration of scientists of all countries.

In the present article, we give a brief overview of the work of Euler. In order to include information from recently discovered work as well as the observations and insights of modern scholars, we draw freely from material found in the memorial volume.

Number Theory

Euler had a passionate lifelong interest in the theory of numbers. Approximately one-sixth of his published work in pure mathematics is in this area; the same is true of the manuscripts left unpublished at his death. Although he had an active correspondence with Goldbach, he complained about the lack of response on the part of other contemporary mathematicians such as Huygens, Clairaut, and Daniel Bernoulli, who considered number theory investigations a waste of time, and were even unaware of Fermat's theorem. (Forty years passed before Euler's investigations into Goldbach's problem were followed up by Lagrange.)

André Weil has commented that if one were to distinguish between "theoretical" and "experimental" researchers, as is done for physicists, then Euler's constant preoccupation with number theory would place him among the former. But in view of his insistence on the "inductive" method of discovery of arithmetic truths, carrying out a wealth of numerical calculations for special cases before tackling the general question, one could equally well call him an "experimental" genius.

At the beginning of the eighteenth century—50 years after Fermat's death—the number theoretical work of Fermat was practically forgotten. In a letter dated December 1, 1727, Christian Goldbach brought to Euler's attention Fermat's assertion that numbers of the form

$$2^{2^{p-1}} + 1, \ p \text{ prime},$$

(i.e., $3, 5, 17, 257, \ldots$) are also prime; this led Euler to a study of Fermat's works. His investigations included Fermat's theorem and its generalizations, representations of numbers as sums of squares of polygonal numbers, and elementary quadratic forms.

In the decade between 1740 and 1750, Euler created the basis of a new theory which, until this day, has not essentially changed its character. The question which motivated this work was posed by Naudé on September 12, 1740, who asked Euler the number of ways in which a given integer can be represented as a sum of integers. For this problem, the "partitio numerorum," as well as for related problems, Euler found solutions by associating with a number-theoretic function its generating function, which can be investigated by analytical methods. Euler clearly understood the importance of his discovery. Although he had not found the proof of several central theorems of his theory, he incorporated the basic ideas and a few elementary but remarkable special results in his fundamental text in analysis, *Introductio in analysin infinitorum*. V. Scharlau comments, "Even today it is hard to imagine a more convincing and interesting introduction to this theory."

Euler used this theory in attempting to find a formula for prime numbers, where he considered the function $\sigma(n)$, the sum of all divisors of n. He obtained the formula

$$\sigma(p^k) = \frac{p^{k+1} - 1}{p - 1}, \quad \text{for } p \text{ prime}$$

from which the computation of $\sigma(n)$ follows. Euler also formulated the recursion rule for $\sigma(n)$,

$$\sigma(n) = \sigma(n-1) + \sigma(n-2) - \sigma(n-5) - \sigma(n-7) + \cdots$$

and observed its similarity to the one for $p(n)$, the number of partitions of n. In 1750, Euler brought these investigations to a conclusion by formulating the identity

$$\sum_{i=1}^{\infty} (1 - x^i) = 1 + \sum_{m=1}^{\infty} (-1)^m (x^{\frac{1}{2}(3m^3 - m)} + x^{\frac{1}{2}(3m^2 + m)})$$

which is a cornerstone for all his related results.

Another interesting application of generating functions can be found in Euler's various investigations of "population dynamics," which probably originated in the years 1750–1755. Scharlau writes:

> From today's point of view it is possibly not surprising that Euler found no additional results on generating functions; indeed it took many decades—almost a century—after the end of his activity before his achievements were substantially surpassed. It is remarkable how little attention was given to Euler's ideas by the mathematicians of the 18th and 19th centuries.... There are very few mathematical theories whose character has changed so little since Euler's time as the theory of generating functions and the partitions of numbers.

Among the unpublished fragments of Euler's work (a total of about 3,000 pages, mainly bound in numbered notebooks) are over 1,000 pages which are devoted to number theory, mostly from the years 1736–1744 and 1767–1783. Euler's technique of investigation emerges clearly from these. After lengthy efforts which at times span many years, he reaches his results based on observations, tables, and empirically established facts.

G. P. Matvievskaja and E. P. Ozigova, who have perused these fragments, note that "the handwritten materials widen our views of Euler's activity in the field of number theory. The same holds for other directions of his research. The manuscripts enable us to recognize the sources of his mathematical discoveries." A few examples serve to illustrate these points. On page 18 in notebook N 131 is the problem of deciding whether a given integer is prime. The same notebook contains an entry about the origin of the zeta function, as well as the first mention of the theorem of four squares, to which Euler returns in notebook N 132 (1740–1744). A particularly interesting entry in notebook N 134 (1752–1755) contains Euler's formulation, a hundred years before Bertrand, of the "Bertrand postulate," that there is at least one prime between any integer n and $2n$.

Analysis

Euler was occupied throughout his life with the concept of function; the treatises he produced in analysis were fundamental to the development of the modern foundations of analysis. As early as 1727 Euler had written a fifteen-page manuscript *Calculus differentialis*; it's interesting to compare this fledgling work with his later treatise *Institutiones calculi differentialis* (1755). Here Euler explains the calculus of finite differences of finite increments and considers calculations with infinitely small quantities. D. Laugwitz, one of the contributors to the modern development of analysis through the adjoining of an infinity symbol Ω, remarks that anyone who reads this work, or Euler's *Introductio in analysin infinitorum* (1748), must be struck by the confidence with which Euler utilizes the calculus of both infinitely large and infinitely small magnitudes. Laugwitz indicates that it is possible to formulate Euler's ideas in the modern setting of nonstandard analysis; hence Euler receives a belated justification of his unorthodox techniques.

The richness and diversity of Euler's work in analysis can be seen by a brief summary of the book *Introductio in analysin infinitorum*. The first chapter discusses the definition of "function" which originated with Johann Bernoulli. In the second, Euler formulates the "fundamental theorem of algebra" and sketches a proof; he presents results on real and complex solutions of algebraic equations, a topic resumed in chapter 12 which deals with the decomposition of rational functions into partial fractions. The third chapter contains the so-called "Euler substitution," and the important replacement of a non-explicit functional dependence by a parametric representation. Particularly remarkable is Euler's strict theory of logarithms, and the consideration of the exponential function in chapter 6. Euler asserts that the logarithms of rationals are either rational or transcendental, a fact which was proved only two hundred years later. Weakly convergent series are considered in chapter 7, as well as the question of convergence of series and the relation between a function and its representation outside the circle of convergence. Subsequent chapters deal with transcendental functions and their representation as series or products. The starting point of Bernhard Riemann's investigation of the distribution of primes is in chapter 15, in the formula

$$\sum_n \frac{1}{n^x} = \prod_p \left(\frac{1}{1 - 1/p^x} \right)$$

in which the summation extends over all positive integers and the product over all primes (see p. 80). In chapter 16 Euler turns to the new topic—rife with algebraic ideas—of *Partitione numerorum*, the additive decomposition of natural numbers (see p. 81). The developments of power series into infinite series found here were continued only by Ramanujan, Hardy and Littlewood. The expressions found here were later called theta

EVOLUTIONE FACTORUM ORTIS.   223

ubi alii numeri non occurrunt, nifi qui ex his duobus 2 & 3 C A P.
per multiplicationem originem trahunt; feu qui alios Divifores X V.
præter 2 & 3 non habent.

273. Si igitur pro α, β, γ, δ, &c., unitas per fingulos omnes numeros primos fcribatur, ac ponatur

$$P = \frac{1}{(1-\frac{1}{2})(1-\frac{1}{3})(1-\frac{1}{5})(1-\frac{1}{7})(1-\frac{1}{11})(1-\frac{1}{13})\,\&c.,}$$

fiet

$$P = 1 + \frac{1}{2} + \frac{1}{3} + \frac{1}{4} + \frac{1}{5} + \frac{1}{6} + \frac{1}{7} + \frac{1}{8} + \frac{1}{9} + \&c.,$$

ubi omnes numeri tam primi, quam qui ex primis per multiplicationem nafcuntur, occurrunt. Cum autem omnes numeri
vel fint ipfi primi, vel ex primis per multiplicationem oriundi,
manifeftum eft, hic omnes omnino numeros integros in deno-
minatoribus adeffe debere.

274. Idem evenit, fi numerorum primorum Poteftates
quæcunque accipiantur : fi enim ponatur

$$P = \frac{1}{(1-\frac{1}{2^n})(1-\frac{1}{3^n})(1-\frac{1}{5^n})(1-\frac{1}{7^n})(1-\frac{1}{11^n})\&c.,}$$

fiet

$$P = 1 + \frac{1}{2^n} + \frac{1}{3^n} + \frac{1}{4^n} + \frac{1}{5^n} + \frac{1}{6^n} + \frac{1}{7^n} + \frac{1}{8^n} + \&c.,$$

ubi omnes numeri naturales nullo excepto occurrunt. Quod
fi autem in Factoribus ubique fignum + ftatuatur, ut fit

$$P = \frac{1}{(1+\frac{1}{2^n})(1+\frac{1}{3^n})(1+\frac{1}{5^n})(1+\frac{1}{7^n})(1+\frac{1}{11^n})\,\&c.,}$$

erit

Euleri *Introduct. in Anal. infin. parv.*   F f $P =$

The Euler product-sum formula from chapter 15 of *Introductio in analysin infinitorum*

functions, and used by Jacobi in the general theory of elliptic functions. The last chapter,
17, deals with the numerical solution of algebraic equations, following Daniel Bernoulli.

A. O. Gelfond, whose essay in the memorial volume contains a deep analysis of the
contents of *Introductio . . .*, interprets Euler's ideas in modern terms and stresses the great
relevance of this work, even to this day.

Euler's interest in the theory of vibrating strings is legendary. In 1747 d'Alembert

C A P U T   X V I.

De Partitione numerorum.

297. $\mathbf{P}$ Ropofita fit ifta expreffio

$$(1 + x^{\alpha}z)(1+x^{\beta}z)(1+x^{\gamma}z)(1+x^{\delta}z)(1+x^{\varepsilon}z)\,\&c.,$$

quæ cujufmodi induat formam, fi per multiplicationem evol-vatur, inquiramus. Ponamus prodire

$$1 + Pz + Qz^2 + Rz^3 + Sz^4 + \&c.,$$

atque manifeftum eft P fore fummam Poteftatum

$x^{\alpha}+ x^{\beta}+ x^{\gamma}+x^{\delta}+ x^{\varepsilon}+ \&c.$. Deinde Q eft fumma Fa-ctorum ex binis Poteftatibus' diverfis, feu Q erit aggregatum plurium Poteftatum ipfius x, quarum Exponentes funt fummæ duorum terminorum diverforum hujus Seriei

$$\alpha, \quad \beta, \quad \gamma, \quad \delta, \quad \varepsilon, \quad \xi, \quad \eta, \quad \&c.$$

Slmili modo R erit aggregatum Poteftatum ipfius x, quarum Exponentes funt fummæ trium terminorum diverforum. At-que S erit aggregatum Poteftatum ipfius x, quarum Expo-nentes funt fummæ quatuor terminorum diverforum ejufdem Seriei, $\alpha, \beta, \gamma, \delta, \varepsilon$, &c., & ita porro.

298. Singulæ hæ Poteftates ipfius x, quæ in valoribus li-terarum P, Q, R, S, &c., infunt, unitatem pro coëffi-ciente habebunt, fi quidem earum Exponentes unico modo ex

I i 3 $\alpha, \beta,$

Chapter 16 title page from *Introductio in analysin infinitorum*

formulated the theory and the corresponding partial differential equation; this prompted Euler in 1750 to develop a solution, although restricted to the case in which the vibrations satisfy certain conditions. Euler's friend Daniel Bernoulli contributed (about 1753) two remarkable articles, and presented the solution in the form of a trigonometric series. The problem is fittingly illuminated by Euler's question "what is the law of the vibrating string if it starts with an arbitrary shape" and d'Alembert's answer "in several cases it is not possible to solve the problem, which transcends the resources of the analysis available at this time."

Euler has sometimes been criticized for seeming to ignore the concept of convergence in his freewheeling calculations. Yet in 1740, Euler gave an incomplete formulation of the criterion of convergence that later received Cauchy's name. Euler's last paper was completed in 1783, the year of his death; it contained the germ of the concept of uniform convergence. His example was utilized by Abel in 1826.

After surveying the rich contributions to analysis made in Euler's time, Pierre Dugac declares, "Euler and d'Alembert were the instigators of the most important work on the foundations of analysis in the nineteenth century."

"Applied" Mathematics (Physics)

Euler's investigations and formulations of basic theory in the areas of optics, electricity and magnetism, mechanics, hydrodynamics and hydraulics are among the most fundamental contributions to the development of physics as we know it today. Euler's views on physics had an immediate influence on the study of physics in Russia; this grew out of his close relationship with the contemporary and most influential Russian scientist, M. V. Lomonosov, his several Russian students, and the publication of a translation (by S. J. Rumovskii) of his very popular "Letters to a German Princess." The "Letters ...," which had originated as lessons to the princess of Anhalt-Dessau, niece of the King of Prussia, during Euler's years in Berlin, served as the first encyclopedia of physics in Russia. A. T. Grigor'jan and V. S. Kirsanov observed that the physicist N. M. Speranskii, a noted statesman and author of a physics book (1797), used to read to his students sections from Euler's "Letters ...."

B. L. van der Waerden, in discussing Euler's justification of the principles of mechanics, has asked, "What did Euler mean by saying that in the computation of the total moment of all forces, the inner forces can be neglected because 'les forces internes se détruisent mutuellement'?" He points out that in order to answer that question it is important to know Euler's concept of solids, fluids, and gases. Are they true continua, or aggregates of small particles? The answer can be found in Euler's letters #69 and #70 to a German princess. He does not consider water, wool and air as true continua, but assumes that they consist of separate particles. However, in hydrodynamics, Euler treats liquids and gases as if they were continua. Euler is well aware that this is only an approximation.

A study of the published works of Daniel and Johann Bernoulli, as well as Euler's unpublished works (in particular, Euler's thick notebook from 1725–1727), by G. K. Mikhailov, gives some new and surprising insights into Euler's contributions to the development of theoretical hydraulics. Mikhailov states:

> It is generally known that the creation of the foundations of modern hydrody-
> namics of ideal fluids is one of the fruits of Euler's scientific activity. Less well
> known is his role in the development of theoretical hydraulics, that is, as usually
> understood, the hydrodynamic theory of fluid motion under a one-dimensional flow
> model. Traditionally—and with good reason—it is assumed that the foundations
> of hydraulics were developed by Daniel and Johann Bernoulli in their works pub-

lished between 1729 and 1743. In fact, during the second quarter of the eighteenth century Euler did not publish even a single paper on the elements of hydraulics. The central theme of most of the recent historical-critical studies on the state of hydraulics in that period is the determination of the respective contributions of Daniel and of Johann Bernoulli. But Euler stood, all this time, just beyond the curtain of the stage on which the action was taking place, although almost no contemporary was aware of that.

Euler's work on the theory of ships culminated in the publication of *Scientia navalis seu tractatus de construendis ac dirigendis navibus*, published in 1749. Walter Habicht notes the fundamental importance of this treatise:

> Following the *Mechanica sive motus scientia analytice exposita* which appeared in 1736, it [the *Scientia navalis* ...] is the second milestone in the development of rational mechanics, and to this day has lost none of its importance. The principles of hydrostatics are presented here, for the first time, in complete clarity; based on them is a scientific foundation of the theory of shipbuilding. In fact, the topics treated here permit insights into all the related developments in mechanics during the eighteenth century.

Although Euler's intense interest in the science of optics appeared before he was 30 and remained with him almost to his death, there is still no monographic evaluation of his contributions to the wide field of physical and geometrical optics. Part of Euler's work is best described by Habicht:

> In the second half of his life, from 1750 on and throughout the sixties, Leonhard Euler worked intensively on problems in geometric optics. His goal was to improve in several ways optical instruments, in particular, telescopes and microscopes. Besides the determination of the enlargement, the light intensity and the field of view, he was primarily interested in the deviations from the point-by-point imaging of objects (caused by the diffraction of light passing through a system of lenses), and also in the even less tractable deviations which arise from the spherical shape of the lenses. To these problems Euler devoted a long series of papers, mainly published by the Berlin academy. He admitted that the computational solution of these problems is very hard. As was his custom, he collected his results in a grandly conceived textbook, the *Dioptricæ* (1769–1771). This book deals with the determination of the path of a ray of light through a system of diffracting spherical surfaces, all of which have their centers on a line, the optical axis of the system. In a first approximation, Euler obtains the familiar formulae of elementary optics. In a second approximation he takes into account the spherical and chromatic aberrations. After passing through a diffracting surface, a pencil of rays issuing from a point on the optical axis is spread out in an interval on the optical axis; this is the so-called "longitudinal aberration." Euler uses the expression "espace de diffusion." If the light passes through several diffracting surfaces, the "espace de diffusion" is determined using a principle of superposition.

> Euler had great expectations for his theory, and believed that using his recipes, the optical instruments could be brought to "the highest degree of perfection."

Unfortunately, the practical realization of his systems of lenses did not yield the hoped-for success. He searched for the causes of failure in the poor quality of the lenses on the one hand, and also in basic errors in the laws of diffraction which were determined experimentally in a manner completely unsatisfactory from a theoretical point of view. Because of the failure of his predictions, Euler's *Dioptrica* is often underrated.

Habicht notes that Euler's theory can be modified to obtain the general imaging theories developed in the nineteenth century. The crucial gap in Euler's treatment consists in neglecting those aberrations which are caused by the distance of the object and its images from the optical axis; with modification it is possible to determine the spherical aberration errors of the third order directly from Euler's formulas.

A responsible evaluation of Euler's contributions to optics will be possible only after Euler's unpublished letters and manuscripts are edited and made generally accessible. E. A. Fellmann provides an example of Euler's method which helps to place Euler's contribution in a historic context. The problem of diffraction in the atmosphere is one which was first seriously considered by Euler:

> He began by deriving a very general differential equation; naturally, it turned out not to be integrable-it would have been a miracle had that not happened. Then he searched for conditions which make a solution possible, and finally he solved the problem in several cases under practically plausible assumptions.
>
> Euler frequently expressed the opinion that the phenomena in optics, electricity and magnetism are closely related (as states of the ether), and that therefore they should receive simultaneous and equal treatment. This prophetic dream of Euler concerning the unity of physics could only be realized after the construction of bridges (experimental as well as theoretical) which were missing in Euler's time. These were later built by Faraday, W. Weber and Maxwell.

Euler was deeply influenced by the work of scientists who preceded him as well as by the work of his contemporaries. This is perhaps best illustrated by his role in the development of potential theory. He acknowledges the influence of the work of Leibniz, the Bernoullis, and Jacob Hermann, whose work he had studied in his days in Basel to 1727. In the decade 1730–1740, the contemporaries Euler, Clairaut and Fontaine all were active in developing the main ideas that would lead to potential theory: the geometry of curves, the calculus of variations, and the study of mechanics. By 1752 Euler's work on fluid mechanics *Principia motus fluidorum* was complete. A summary of his contributions to potential theory is given by Jim Cross:

> He helped, with Fontaine and Clairaut, to develop a logical, well-founded calculus of several variables in a clear notation; he transformed, with Daniel Bernoulli and Clairaut, the Galileo-Leibniz energy equation for a particle falling under gravity, into a general principle applicable to continuous bodies and general forces (the principle of least action with Daniel Bernoulli and Maupertuis forms part of this); and he founded, after the attempts of the Bernoullis, d'Alembert, and especially Clairaut, the modern theory of fluid mechanics on complete differentials for forces and velocities. His work was fruitful: the theories of Lagrange grew from his

writings on extremization, fluids and sound, and mechanics; the work of Laplace followed.

Astronomy

Research by Nina I. Nevskaja based on newly available original documents justifies calling Euler a professional astronomer—and even an observer and experimental scientist. Five hundred books and manuscripts from the private library of Joseph Nicholas Delisle have recently come to light and from these one finds that this scientist found Euler a suitable collaborator and valued his knowledge in spherical trigonometry, analysis and probability.

It was a surprise when the records of observations of the Petersburg observatory during its first 21 years—which were presumed lost—were discovered in 1977 in the Leningrad branch of the archives of the Academy of Sciences of the USSR. For almost ten years, Euler was among those who were regularly taking measurements twice daily. Based on these observations, Delisle and Euler computed the instant of true noon, and the noon correction. Euler's entries were so detailed and numerous that it is possible to deduce from them how he gradually mastered the methods of astronomical observations. Utilizing the insights he obtained, Euler found a simple method of computing tables for the meridional equation of the sun; he presented it in the paper *Methodus computandi aequationem meridiei* (1735).

Euler was fascinated by sunspots; his notes from this period contain enthusiastic comments on his observations. The computation of the trajectories of the sunspots by Delisle's method can be considered the beginning of celestial mechanics. The archives also disclose that Euler helped Delisle by working out analytical methods for the determination of the paths of comets.

A little-noted field of Euler's activities, the theory of motion of celestial bodies, is documented by Otto Volk. Euler's first paper, based on generally formulated differential equations of mechanics, is entitled *Recherches sur le mouvement des corps célestes en général* (1747). Using the tables of planets computed by Thomas Street from the pure Keplerian motion of planets around the sun, Euler discusses in Sections 1 to 17 the observed irregularities. In Section 18 he formulates the differential equations of mechanics, and obtains the solution

$$r = a(1 + e \cos v) = \frac{a(1 - e^2)}{1 - e \cos \phi}$$

in which r is the radius, v is the eccentric anomaly and s is the true anomaly, while e and a are constants. This is a regularization of the so-called inverse problem of Newton. Later, Euler obtains a trigonometric series for ϕ; such Fourier series are the basis of his computation of perturbations. This is the topic treated in detail in the prize proposal to the Paris Academy, *Recherches sur la question des inégalités du mouvement de Saturne et de Jupiter, sujet proposé pour le prix de l'année 1748*. In it Euler uses, for the first time, Newton's laws of gravitation to compute the mutual perturbations of planets.

In his paper *Considerationes de motu corporum coelestium* (1764), Euler is the first to begin considering the three-body problem, under certain restrictions. Euler notes the intractability of the problem:

There is no doubt that Kepler discovered the laws according to which celestial bodies move in their paths, and that Newton proved them—to the greatest advantage of astronomy. But this does not mean that the astronomical theory is at the highest level of perfection. We are able to deal completely with Newton's inverse-square law for two bodies. But if a third body is involved, so that each attracts both other bodies, all the arts of analysis are insufficient .... Since the solution of the general problem of three bodies appears to be beyond the human powers of the author, he tried to solve the restricted problem in which the mass of the third body is negligible compared to the other two. Possibly, starting from special cases, the road to the solution of the general problem may be found. But even in the case of the restricted problem the solution encounters difficulties so great that the author has to admit to have spent much effort in vain attempts at solution.

Euler's investigation of the three-body problem was noted only at a later date; the linear solutions to the equation of the fifth degree were (and sometimes still are) called "Lagrange's solutions," without any mention of Euler. But Euler achieved fame through his theory of perturbations, presented in *Nouvelle méthode de déterminer les dérangemens dans le mouvement des corps célestes, causé par leur action mutuelle*. By iteration he determined, for the first time, the perturbations of the elements of the elliptical paths, and then applied this method to determine the motion of three mutually attracting bodies.

Correspondence

The circle of contemporary scholars who were influenced by and in turn, influenced, Euler's investigations was as wide as one could imagine in the eighteenth century. His voluminous correspondence testifies to the fruitful interaction between scientists through queries, conjectures, critical comments, and praise. Some of the correspondence has been published previously in collected works; a standard reference is the collection *Correspondance Mathématique et Physique*, edited by N. Fuss and published in 1843 by the Imperial Academy of Science, St. Petersburg. New discoveries and more complete information have produced recently published collections. The publication in 1965 of the correspondence between Euler and Christian Goldbach has been mentioned earlier.

It is significant that the first volume, Al, published in the fourth series of Euler's *Opera omnia*, contains a complete list of all existing letters to and from Euler (about 3,000), together with a summary of their contents. Volume A5 of this series (1980), edited by A. P. Juškevič and R. Taton, contains Euler's correspondence with A. C. Clairaut, J. d'Alembert, and J. L. Lagrange.

The correspondence between Euler and Lagrange from 1754 to 1775 gives valuable testimony to the development of personal relations between two of the most important scientists of that time. The letter exchange begins with a letter from the 18-year-old Lagrange, who lived in Turin, containing a query in which he mentions the analogy in the development of the binomial $(a + b)^m$ and the differential $d^m(xy)$. Mathematically isolated, Lagrange expresses his admiration for Euler's work, particularly in mechanics. Especially significant is the second letter to Euler (1755). In it Lagrange announces, without details, his new methods in the calculus of variations; Euler at once notes the advantage of these methods over the ones in his *Methodus inveniendi lineas curvas maximi minimive proprietate gaudentes* (1744), and heartily congratulates Lagrange. In 1756 Lagrange develops the differential calculus for several variables and investigates, for the first time, minimal surfaces. After an interruption of three years, Lagrange continues the correspondence by sending his work *La nature et la propagation du son*, and we find interesting discussions on the problem of vibrating strings, which had been carried on since 1749 between d'Alembert, Euler and Daniel Bernoulli.

After a lengthy pause, Euler resumes the correspondence. The first letter (1765) concerns the discussion with d'Alembert on vibrating strings, and the librations of the moon. In a second, Euler tells Lagrange that he has been granted permission by Friedrich II to return to Petersburg, and is attempting to have Lagrange come there. In later correspondence, the emphasis is on questions in the theory of numbers and in algebra. Pell's equation $x^2 - ay^2 = b$, and in particular $p^2 - 13q^2 = 101$, are discussed. Other topics deal with arithmetic, questions concerning developable surfaces, and the motion of the moon.

In 1770 Lagrange writes of his plan to publish Euler's *Algebra* in French, and to add to it an appendix; the published book is mailed on July 13, 1773. The last of Euler's letters, dated March 23, 1775, is remarkable by the exceptionally warm congratulations for Lagrange's work, especially about elliptic integrals. It may be conjectured that this was not the end of the correspondence, but unfortunately no additional letters have survived.

Postscript

This overview of Euler's life and work touches only a small part of the wealth of material to be found in the scholarly essays in the Basel memorial volume. In addition to careful and detailed analysis of many of Euler's scientific and mathematical achievements, these chapters contain new information on all aspects of Euler's private and academic life, his family, his philosophical and religious views, and the fabric of his life and work at the St. Petersburg Academy. In view of the overwhelming volume and diversity of Euler's work, it may never be possible to produce a comprehensive scientific biography of his genius. It is to be hoped that these newest contributions to the study of his life and work will provide impetus for further study and publication of many of the yet unpublished papers which are the unknown legacy of this mathematical giant.

The author and the editor express deep appreciation to Branko Grünbaum, who translated from the German the author's original manuscript. Doris Schattschneider took the trouble to shorten this manuscript from 38 to 22 pages.

Reference

Leonhard Euler 1707–1783, Beiträge zu Leben und Werk, Gedenkband des Kantons Basel-Stadt, edited by J. J. Burckhardt, E. A. Fellmann, and W. Habicht, Birkhäuser Verlag, Basel, 1983.

in die Analysis des Unendlichen"

André Weil (Princeton, USA) L'œuvre arithmétique d'Euler

Winfried Scharlau (Münster, BRD) Eulers Beiträge zur *partitio numerorum* und zur Theorie der erzeugenden Funktionen

Galina P. Matvievskaja/Helena P. Ožigova (Taškent-Leningrad UdSSR) Eulers Manuskripte zur Zahlentheorie

Adolf P. Juškevič (Moskau, UdSSR) L. Euler's unpublished manuscript *Calculus Differentialis*

Pierre Dugac (Paris, F) Euler, d'Alembert et les fondements de l'analyse

Detlef Laugwitz (Darmstadt, BRD) Die Nichtstandard-Analysis: Eine Wiederaufnahme der Ideen und Methoden von Leibniz und Euler

Isaac J. Schoenberg (Madison, USA) Euler's contribution to cardinal spline interpolation: The exponential Euler splines

David Speiser (Louvain, Belgien) Eulers Schriften zur Optik, zur Elektrizität und zum Magnetismus

Gleb K. Mikhailov (Moskau, UdSSR) Leonhard Euler und die Entwicklung der theoretischen Hydraulik im zweiten Viertel des 18.Jahrhunderts

Walter Habicht (Basel, CH) Einige grundlegende Themen in Leonhard Eulers Schiffstheorie

Bartel L. van der Waerden (Zürich, CH) Eulers Herleitung des Drehimpulssatzes

Walter Habicht (Basel, CH) Betrachtungen zu Eulers Dioptrik

Emil A. Fellmann (Basel, CH) Leonhard Eulers Stellung in der Geschichte der Optik

Jim Cross (Melbourne, Australien) Euler's contributions to Potential Theory 1730–1755

Otto Volk (Würzburg, BRD) Eulers Beiträge zur Theorie der Bewegungen der Himmelskörper

Nina I. Nevskaja (Leningrad, UdSSR) Leonhard Euler und die Astronomie

Judith Kh. Kopelevič (Leningrad, UdSSR) Leonhard Euler und die Petersburger Akademie

Ašot T. Grigor'jan/V. S. Kirsanov (Moskau, UdSSR) Euler's Physics in Russia

Ivor Grattan-Guinness (Barnet, GB) Euler's Mathematics in French Science, 1795–1815

René Taton (Paris, F) Les relations d'Euler et de Lagrange

Pierre Speziali (Geneve, CH) Léonard Euler et Gabriel Cramer

Roger Jaquel (Mulhouse, F) Leonard Euler, son fils Jean-Albrecht et leur ami Jean III Bernoulli

Wolfgang Breidert (Karlsruhe, BRD) Leonhard Euler und die Philosophie

Michael Raith (Riehen bei Basel, CH) Der Vater Paulus Euler. Beiträge zum Verständnis der geistigen Herkunft Leonhard Eulers

René Bernoulli (Basel, CH) Leonhard Eulers Augenkrankheiten

Kurt-Reinhard Biermann (Berlin, DDR) Aus der Vorgeschichte der Euler-Werkausgabe

Johann Jakob Burckhardt (Zürich, CH) Die Eulerkommission der Schweizerischen Naturforschenden Gesellschaft.-Ein Beitrag zur Editionsgeschichte

Johann Jakob Burckhardt (Zürich, CH) Euleriana-Verzeichnis des Schrifttums über Leonhard Euler

Euler's Output, A Historical Note[1]

W. W. R. Ball

Professor D. E. Smith has enriched many of the recent numbers of the *Monthly* with extracts from letters in his possession on details of mathematical history. There is perhaps a similar interest in the following extract from a letter in my possession written by De Morgan and dated 17 October 1858. The first part of the extract only puts in a striking way what is familiar to many students; the allusion in it to 1000 miles in 1000 hours refers of course to the famous bet, on the issue of which about £100,000 was staked, made by Captain Barclay Allardyce in 1809 that he would walk one mile in each of a thousand consecutive hours, covering nearly six weeks. The story in the second part of the extract rests I believe only on tradition, but I have heard it from other sources and have no reason to doubt its truth.

Extract from De Morgan's Letter

Euler's life, dating from 1736, the year in which his productions began to appear with rapidity, is a period of 47 years; during the last 17 years he was totally blind, and throughout the whole of it he suffered from the consequences of a fever which deprived him of the sight of one eye at its commencement. Nor was he secluded from the world: he married a second wife, and was the father of 13 children. His life was not free from such calamities as interrupt the course of study. He saw the deathbeds of ten of his children; his house was set on fire and wholly burnt; and an attempt to restore his sight by couching led to an illness which nearly ended his days. He was fond of conversation, of the society of his family, and of music; and was, throughout the whole of his career, at the orders of a royal or imperial patron. Nevertheless, if his memoirs be counted, and if his separate works (not volumes) be allowed for at the average rate of 20 memoirs each, which is an insufficient rating both as to bulk and matter, the result is as follows: Distribute Euler's work through the whole period—the real and average distributions will not much disagree—and there

---

[1]Reprinted from the *American Mathematical Monthly*, Vol. 31, February, 1924, pp. 83–84.

is for each and every fortnight in 47 years a separate effort of mathematical invention, digested, arranged, written in Latin, and amplified, often to a tedious extent, by corollaries and scholia. Through all this mass, the power of the inventor is almost uniformly distributed, and apparently without effort. There is nothing like this, except this, in the history of science: it is the thousand miles in the thousand hours.

When Euler was at Berlin, and there is no reason to suppose it was otherwise at St. Petersburg, he was in the habit of writing memoir after memoir, and placing each, when finished, at the top of a pile of manuscripts. The secretaries of the academy helped themselves from time to time, by taking papers from the top of the pile, according to their estimate of the bulk of matter likely to be wanted for reading. The consequence was that, as the pile often increased more rapidly than the demands upon it, the memoirs which happened to be at the bottom remained there for a long time. This explains how various memoirs of Euler were published, though considerable extensions and improvements of the matter contained in them had been previously published [under his name].

Discoveries[1]

Marta Sved and Dave Logothetti

> The theorem found by our toiler
> Is hot stuff, a genuine "boiler."
> But, wouldn't you know,
> Alas, what a blow,
> The thing was discovered by Euler.

A street sign in Basel

---

[1] Reprinted from *Mathematics Magazine*, Vol. 62, February, 1989, p. 39.

The "Mathematical" Mt. Rushmore according to E. T. Bell: Archimedes, Newton, and Gauss

Bell's Conjecture[1]

J. D. Memory

For math, the Oscar envelope
(Assured by Price and Waterhouse)
Would list a three-way tie, I'd hope:
Archimedes, Newton, Gauss.

Archimedes' *modern* mind
(Narrowly he bounded pi),
Impelled to seek and swift to find,
Defined the Hellenistic high.

Newton's fluxions formed the frame
That fit the Universal Law.
Even Leibniz spread his fame:
"We know the Lion by his claw."

Many Magi graced the scene
But Gauss was greater than all since.
If Number Theory is the Queen,
Carl Friedrich is its freshest Prince.

[1] Reprinted from the *Mathematics Magazine*, Vol. 70, June, 1997, p. 203.
The title refers to E. T. Bell, *Men of Mathematics*, Simon and Schuster, New York, 1961. This poem is inspired by Bell's famous ranking of Archimedes, Newton, and Gauss as history's three greatest mathematicians.

The complete "Mathematical" Mt. Rushmore: Archimedes, Newton, Euler, and Gauss

A Response to "Bell's Conjecture"[1]

Charlie Marion and William Dunham

Dear Editor:

The poem "Bell's Conjecture" (this volume, p. 93) adopted E. T. Bell's ranking of Archimedes, Newton, and Gauss as the greatest mathematicians of all time. We felt compelled to respond to the omission of Leonhard Euler from such glorious company.

> Before we let you get away,
> Your choices set in stone,
> Consider what we have to say:
> E.T.! O, please! Call home!
>
> Stop the presses! Hold that thought!
> And listen to our voices.
> Ruffled, even overwrought,
> We'll supplement your choices.
>
> Old Archie, Isaac, C. F. Gauss—
> Though each deserves a floor
> In mathematics' honored house,
> Make room for just one more.
>
> Without the Bard of Basel, Bell,
> You've clearly dropped the ball.
> Our votes are cast for Euler, L.
> Whose *Opera* says it all:
>
> Six dozen volumes—what a feat!
> Profound and deep throughout
> Does Leonhard rank with the elite?
> Of this there is no doubt.

---

[1] Reprinted from the *Mathematics Magazine*, Vol. 70, December, 1997, p. 326.

Consider how he summed, in turn
The quite elusive mix
Of one slash n all squared—you'll learn
He got π^2 slash six.

We're shocked we did not see his name
With those you justly sainted.
No Euler in your Hall of Fame?
Your judgment's surely Taine-ted.[2]

It's time to honor one you missed,
To do your duty well.
Add worthy Euler to your list,
And save him by the Bell.

---

[2] Ed. Note: E. T. Bell wrote science fiction under the name "John Taine."

Part II
Mathematics

Introduction to Part II

In the remainder of the book, our authors sample from Euler's vast mathematical output. As should be expected, the content here is more technical in nature, featuring a heavy dose of formulas, series, and integrals. To impose a topical organization, we move from the continuous to the discrete—that is, from infinite series to calculus to number theory to algebra to combinatorics. This seems like a natural flow for articles drawn from multiple journals across multiple decades.

The starting point is Morris Kline's "Euler and Infinite Series" from the 1983 issue of *Mathematics Magazine*. Raymond Ayoub follows with "Euler and the Zeta Function," which received the Lester Ford Award from the MAA in 1975. Then comes E. J. Barbeau's 1979 paper on Euler's approach to divergent series, a look into 18th century analysis as practiced by the master. Jumping back to 1872, we have J. W. L. Glaisher's history of Euler's constant—and add as a postscript Morgan Ward's mnemonic for remembering its first ten digits.

Anthony Ferzola carries readers into the realm of calculus with "Euler and Differentials," which received the MAA's George Pólya Award in 1995. The next two papers are surveys in which Euler figures prominently: Philip J. Davis's history of the gamma function and Victor Katz's account of multiple integrals from Euler to Cartan. The analytic section concludes with Jesper Lützen's discussion of Euler's "general partial differential calculus" from the 1983 *Mathematics Magazine* and Erwin Kreyszig's early history of the calculus of variations, highlighting Euler's contributions to this deep and wonderful subject.

At this point, we switch gears. There are three papers on Euler's number theory, all from the tribute issue of 1983. The first, by Paul Erdős and Underwood Dudley, describes some of Euler's number theoretic triumphs. This is followed by George Andrews's account of the pentagonal number theorem, "one of Euler's most profound discoveries," and Harold Edwards's discussion of another groundbreaking achievement in "Euler and Quadratic Reciprocity."

Gears change again with William Dunham's article "Euler and the Fundamental Theorem of Algebra" (Pólya Award, 1992), after which George Pólya himself puts in an appearance with his "Guessing and Proving" from 1978. Here Pólya imagines how Euler might have discovered the polyhedral formula $V + F = E + 2$. Next come two articles on Euler's discrete mathematics: "The Truth about Königsberg" by Brian Hopkins and

Robin Wilson (Pólya Award, 2005) and "Graeco-Latin Squares and a Mistaken Conjecture of Euler" by Dominic Klyve and Lee Stemkoski from 2006. The book ends with a "Glossary" of mathematical terms, formulas, and equations that carry Euler's name. This was the concluding paper of the 1983 issue of *Mathematics Magazine* and serves the same purpose here.

These articles, of course, barely scratch the surface. But, collectively, they remind us—if reminding is necessary—why Euler's 300th birthday is a cause for mathematical celebration.

Euler and Infinite Series[1]

Morris Kline

The history of mathematics is valuable as an account of the gradual development of the many current branches of mathematics. It is extremely fascinating and instructive to study even the *false* steps made by the greatest minds and in this way reveal their often unsuccessful attempts to formulate correct concepts and proofs, even though they were on the threshold of success. Their efforts to justify their work, which we can now appraise with the advantage of hindsight, often border on the incredible.

These features of history are most conspicuous in the work of Leonhard Euler, the key figure in 18th-century mathematics, and one who should be ranked with Archimedes, Newton, and Gauss. Euler's recorded work on infinite series provides a prime example of the struggles, successes and failures which are an essential part of the creative life of almost all great mathematicians. The few examples discussed in this paper will serve to illustrate how Euler surmounted the difficulties he encountered.

Euler first undertook work on infinite series around 1730, and by that time, John Wallis, Isaac Newton, Gottfried Leibniz, Brook Taylor, and Colin Maclaurin had demonstrated the series calculation of the constants e and π and the use of infinite series to represent functions in order to integrate those that could not be treated in closed form. Hence it is understandable that Euler should have tackled the subject. Like his predecessors, Euler's work lacks rigor, is often *ad hoc*, and contains blunders, but despite this, his calculations reveal an uncanny ability to judge when his methods might lead to correct results. Our discussion will not follow the precise historical order of Euler's investigations of series; he made contributions throughout his lifetime.

To appreciate the first example of Euler's work on series, we must consider some background. A series which caused endless dispute was

$$1 - 1 + 1 - 1 + \cdots . \tag{1}$$

It seemed clear that by writing this series as

$$(1 - 1) + (1 - 1) + (1 - 1) + \cdots$$

[1] Reprinted from the *Mathematics Magazine*, Vol. 56, November, 1983, pp. 307–314.

the sum should be 0. It seemed equally clear, however, that by writing the series as

$$1 - (1 - 1) - (1 - 1) - \cdots .$$

the sum should be 1. But still another sum seemed as reasonable. If S denotes the sum of the series (1), then

$$S = 1 - (1 - 1 + 1 - 1 + \cdots) = 1 - S.$$

Hence $S = \frac{1}{2}$. This value was also supported by the formula for summing a geometric series with common ratio -1.

Guido Grandi (1671–1742), in his little book *Quadratura circula et hyperbolae per infinitas hyperbolas geometrice exhibita* (1703), obtained the third result by a variant of the geometric series argument, using the binomial expansion

$$\frac{1}{1 + x} = 1 - x + x^2 - x^3 + \cdots ,$$

with $x = 1$. (He also argued that since the sum was both 0 and $\frac{1}{2}$, he had proved that the world could be created out of nothing.) In a letter to Christian Wolff published in the *Acta eruditorum* of 1713, Leibniz agreed with Grandi's result but thought that it should be possible to obtain it without resorting to the function $1/(1 + x)$. He argued that, since the successive partial sums are $1, 0, 1, 0, 1, \ldots$, with 1 and 0 equally probable, one should therefore take $\frac{1}{2}$, the arithmetic mean, as the sum. This argument was accepted by James, John and Daniel Bernoulli. Leibniz conceded that his argument was more metaphysical than mathematical, but said that there is more metaphysical truth in mathematics than is generally recognized.

Euler took a hand in this argument. To obtain the sum of the series (1), he argued in a manner similar to Grandi, substituting $x = -1$ in the expansion

$$\frac{1}{1 - x} = 1 + x + x^2 + \cdots$$

and obtained

$$\frac{1}{2} = 1 - 1 + 1 - 1 + \cdots .$$

At this early stage of his work on series, Euler used expansion of functions into series to sum other divergent series. For example, he substituted $x = -1$ in the expansion

$$\frac{1}{(1 + x)^2} = (1 + x)^{-2} = 1 - 2x + 3x^2 - 4x^3 + \cdots$$

and obtained

$$\infty = 1 + 2 + 3 + 4 + \cdots . \tag{2}$$

To Euler, this seemed reasonable; he treated infinity as a number. He then considered the geometric (or binomial) series for $1/(1 - x)$ with $x = 2$ and obtained

$$-1 = 1 + 2 + 4 + 8 + \cdots . \tag{3}$$

Since the terms of series (3) exceed the corresponding terms of series (2), Euler concluded that the sum -1 is larger than infinity. Some of Euler's contemporaries argued that negative numbers larger than infinity are different from those less than 0. Euler objected and argued that infinity separates positive and negative numbers just as 0 does.

In a paper of 1734/35 [7], Euler started with the series

$$y = \sin x = x - \frac{x^3}{3!} + \frac{x^5}{5!} + \cdots \tag{4}$$

and rewrote the equation in the form

$$1 - \frac{x}{y} + \frac{x^3}{3!y} - \frac{x^5}{5!y} + \cdots = 0. \tag{5}$$

He then treated the left side of (5) as an infinite polynomial and argued as follows. (The argument is based on the fact that the sum of the reciprocals of the roots of the polynomial $p(x) = 1 - a_1 x + a_2 x^2 - a_3 x^3 + \cdots + (-1)^k a_k x^k$ is a_1, the sum of the squares of the reciprocals of the roots of $p(x)$ is $a_1^2 - 2a_2$, and so on, for higher roots.) Let $A, B, C, \ldots$ be solutions of equation (5). Then the polynomial can be factored into an infinite product,

$$1 - \frac{x}{y} + \frac{x^3}{3!y} - \frac{x^5}{5!y} + \cdots = \left(1 - \frac{x}{A}\right)\left(1 - \frac{x}{B}\right)\left(1 - \frac{x}{C}\right)\cdots .$$

If A is the smallest value of x whose sine is y, then all other solutions $B, C, \ldots$ are $\pi - A, 2\pi + A, 3\pi - A, \ldots; -\pi - A, -2\pi + A, -3\pi - A, \ldots$. Thus

$$\frac{1}{A} + \frac{1}{\pi - A} + \frac{1}{2\pi + A} + \cdots - \frac{1}{\pi + A} - \frac{1}{2\pi - A} - \cdots = \frac{1}{y} \tag{6}$$

$$\frac{1}{A^2} + \frac{1}{(\pi - A)^2} + \frac{1}{(2\pi + A)^2} + \cdots + \frac{1}{(\pi + A)^2} + \frac{1}{(2\pi - A)^2} + \cdots = \frac{1}{y^2} \tag{7}$$

and so on for higher powers of the reciprocals. If, in equations (4) and (5), we take $y = 1$, then $A = \pi/2$, so that (6) becomes

$$\frac{4}{\pi}\left(1 - \frac{1}{3} + \frac{1}{5} - \frac{1}{7} + \cdots\right) = 1,$$

or

$$1 - \frac{1}{3} + \frac{1}{5} - \frac{1}{7} + \cdots = \frac{\pi}{4} \tag{8}$$

and (7) becomes

$$\frac{8}{\pi^2}\left(1 + \frac{1}{9} + \frac{1}{25} + \frac{1}{49} + \cdots\right) = 1,$$

or

$$1 + \frac{1}{3^2} + \frac{1}{5^2} + \frac{1}{7^2} + \cdots = \frac{\pi^2}{8}. \tag{9}$$

Other series that were "summed" in the same manner are

$$\frac{1}{1^3} - \frac{1}{3^3} + \frac{1}{5^3} - \cdots = \frac{\pi^3}{32},$$

$$\frac{1}{1^4} + \frac{1}{3^4} + \frac{1}{5^4} + \cdots = \frac{\pi^4}{96},$$

$$\frac{1}{1^5} - \frac{1}{3^5} + \frac{1}{5^5} - \cdots = \frac{5\pi^5}{1536},$$

$$\frac{1}{1^6} + \frac{1}{3^6} + \frac{1}{5^6} + \cdots = \frac{\pi^6}{960},$$

and so on. From these series he deduced others. For example, since

$$\left(1 + \frac{1}{2^2} + \frac{1}{3^2} + \frac{1}{4^2} + \cdots\right) - \left(1 + \frac{1}{3^2} + \frac{1}{5^2} + \frac{1}{7^2} + \cdots\right) = \frac{1}{2^2}\left(1 + \frac{1}{2^2} + \frac{1}{3^2} + \frac{1}{4^2} + \cdots\right),$$

one can use (9) to obtain

$$\frac{1}{1^2} + \frac{1}{2^2} + \frac{1}{3^2} + \frac{1}{4^2} + \cdots = \frac{\pi^2/8}{3/4} = \frac{\pi^2}{6},$$

and in a similar manner, obtain

$$\frac{1}{1^4} + \frac{1}{2^4} + \frac{1}{3^4} + \frac{1}{4^4} + \cdots = \frac{\pi^4}{90}$$

and other sums. R. Ayoub [1] discusses Euler's use of (4) to compute such sums, W. F. Eberlein [3] discusses Euler's use of the infinite product for the sine function, and H. H. Goldstine [9, 3.1, 3.2] indicates Euler's expansions of such functions as $\frac{1}{2}(e^x - e^{-x})$ and the use of these expansions in computing sums such as (9).

Euler's attempts to sum the reciprocals of powers of the positive integers were not completely idle. In another paper of the same period [4], Euler made a somewhat bizarre use of infinitesimal calculus to find the difference between the sum of the harmonic series and the logarithm, a difference whose expansion utilizes precisely these series of powers. Let

$$s = 1 + \frac{1}{2} + \frac{1}{3} + \cdots + \frac{1}{n-1}.$$

If we regard n as infinite, then 1 is an infinitesimal and we can write $ds = \frac{1}{n} = \frac{1}{n}\,dn$. An integration yields

$$s = \log n + C.$$

To find C, note that

$$\frac{1}{x} = \log\left(1 + \frac{1}{x}\right) + \frac{1}{2x^2} - \frac{1}{3x^3} + \frac{1}{4x^4} - \frac{1}{5x^5} + \cdots .$$

Setting $x = 1, 2, 3, \ldots, n - 1$, in turn, and adding the $n - 1$ equations yields

$$1 + \frac{1}{2} + \frac{1}{3} + \cdots + \frac{1}{n-1} = \log n + \frac{1}{2}\left(1 + \frac{1}{4} + \frac{1}{9} + \cdots + \frac{1}{(n-1)^2}\right)$$

$$- \frac{1}{3}\left(1 + \frac{1}{8} + \frac{1}{27} + \cdots + \frac{1}{(n-1)^3}\right)$$

$$+ \frac{1}{4}\left(1 + \frac{1}{16} + \frac{1}{81} + \cdots + \frac{1}{(n-1)^4}\right)$$

$$- \cdots$$

The limiting value γ of C as n becomes infinite is today called **Euler's constant**.

In a paper of 1740 [6], Euler obtained one of his finest triumphs, namely,

$$s_{2n} = \sum_{\nu=1}^{\infty} \frac{1}{\nu^{2n}} = (-1)^{n-1} \frac{(2\pi)^{2n}}{2(2n)!} B_{2n},$$

where the B_{2n} are the Bernoulli numbers (see below). The connection with the Bernoulli numbers was actually established a little later in his *Institutiones calculi differentialis* of 1755 [8]. In the 1740 paper he also determined the sum $\sum_{\nu=1}^{\infty}(-1)^{\nu-1}(1/\nu^n)$ for the first few odd values of n.

In *Ars conjectandi* (1713), James Bernoulli, who was treating the subject of probability, had introduced the now widely used Bernoulli numbers. Bernoulli had given the following formula for the sum of powers of consecutive positive integers without demonstration:

$$\sum_{k=1}^{n} k^c = \frac{1}{c+1}n^{c+1} + \frac{1}{2}n^c + \frac{c}{2}B_2 n^{c-1} + \frac{c(c-1)(c-2)}{2\cdot3\cdot4}B_4 n^{c-3}$$

$$+ \frac{c(c-1)(c-2)(c-3)(c-4)}{2\cdot3\cdot4\cdot5\cdot6}B_6 n^{c-5} + \cdots . \tag{10}$$

This series terminates at the last positive power of n, and the B's are the **Bernoulli numbers**:

$$B_2 = \frac{1}{6}, \quad B_4 = -\frac{1}{30}, \quad B_6 = \frac{1}{42}, \quad B_8 = -\frac{1}{30}, \quad B_{10} = \frac{5}{66}, \ldots .$$

Bernoulli also gave the recurrence relation which permits one to calculate these coefficients.

Another famous result of Euler's, the Euler-Maclaurin summation formula, is a generalization of Bernoulli's formula (10). Let $f(x)$ be a real-valued function of the real variable x with $2k + 1$ continuous derivatives on the interval $[0, n]$. Then (in modern notation) Euler's formula is

$$\sum_{i=0}^{n} f(i) = \int_0^n f(x) + \frac{1}{2}[f(n) + f(0)] + \frac{B_2}{2!}[f'(n) - f'(0)]$$

$$+ \frac{B_4}{4!}[f'''(n) - f'''(0)] + \cdots \tag{11}$$

$$+ \frac{B_{2k}}{(2k)!}[f^{(2k-1)}(n) - f^{(2k-1)}(0)] + R_k,$$

where

$$R_k = \int_0^n f^{(2k+1)}(x) P_{2k+1}(x)\, dx.$$

Here n and k are positive integers, and $P_{2k+1}(x)$ is the $(2k+1)$th Bernoulli polynomial (which also appears in Bernoulli's *Ars conjectandi*). It can be represented for $0 \le x \le 1$ by

$$P_{2k+1}(x) = 2(-1)^{k+1} \sum_{m=1}^{\infty} \frac{\sin(2\pi m x)}{(2m\pi)^{2k+l}}.$$

The Bernoulli numbers B_i are related to the Bernoulli polynomials by

$$P_k(x) = \frac{x^k}{k!} + \frac{B_1 x^{k-1}}{1!(k-1)!} + \frac{B_2 x^{k-2}}{2!(k-2)!} + \cdots + \frac{B_k}{k!},$$

where $B_1 = -\frac{1}{2}$, and $B_{2k+1} = 0$ for $k = 1, 2, \dots$. They are often defined today by a relation given later by Euler, namely,

$$t(e^t - 1)^{-1} = \sum_{i=0}^{\infty} B_i \frac{t^i}{i!}.$$

Euler's derivation of formula (11) is interesting in its use of the infinitesimal calculus in treating finite series. He begins by noting that if $s(n) = \sum_{i=0}^n f(i)$, then

$$f(n) = s(n) - s(n-1) = -[s(n-1) - s(n)] = \frac{ds}{dn} - \frac{1}{2!}\frac{d^2 s}{dn^2} + \frac{1}{3!}\frac{d^3 s}{dn^3} - \cdots ; \quad (12)$$

hence (solving for ds/dn and integrating),

$$s = \int f\, dn + \frac{1}{2!}\frac{ds}{dn} - \frac{1}{3!}\frac{d^2 s}{dn^2} + \cdots . \quad (13)$$

In order to express the sum s in terms of f, recursion is used. Differentiating (13) repeatedly gives

$$\frac{df}{dn} = \frac{d^2 s}{dn^2} - \frac{1}{2!}\frac{d^3 s}{dn^3} + \frac{1}{3!}\frac{d^4 s}{dn^4} - \cdots, \quad \text{so} \quad \frac{d^2 s}{dn^2} = \frac{df}{dn} + \frac{1}{2!}\frac{d^3 s}{dn^3} - \cdots .$$

$$\frac{d^2 f}{dn^2} = \frac{d^3 s}{dn^3} - \frac{1}{2!}\frac{d^4 s}{dn^4} + \frac{1}{3!}\frac{d^5 s}{dn^5} - \cdots, \quad \text{so} \quad \frac{d^3 s}{dn^3} = \frac{d^2 f}{dn^2} + \frac{1}{2!}\frac{d^4 s}{dn^4} - \cdots \quad (14)$$

and so on. Substituting these values for ds/dn, ds^2/dn^2, $ds^3/dn^3, \dots$ in (13) gives

$$s = \int f\, dn + \frac{1}{2!}\left[f + \frac{1}{2!}\left(\frac{df}{dn} + \frac{1}{2!}\left(\frac{df^2}{dn^2} + \cdots \right) \right) - \frac{1}{3!}\left(\frac{d^2 f}{dn^2} + \cdots \right) \right]$$

$$- \frac{1}{3!}\left(\frac{df}{dn} + \frac{1}{2!}\left(\frac{d^2 f}{dn^2} + \cdots \right) + \cdots \right) + \frac{1}{4!}\left(\frac{d^2 f}{dn^2} - \cdots \right) - \cdots$$

which is cumbersome, but does show the form in which s can be expressed. Euler wrote

$$s = \int f\, dn + \alpha f + \frac{\beta\, df}{dn} + \frac{\gamma\, d^2 f}{dn^2} + \frac{\delta\, d^3 f}{dn^3} + \cdots$$

227

CAPUT I.

DE TRANSFORMATIONE SERIERUM

I.

Um nobis propofitum fit ufum Calculi differentialis tam in univerfa Analyfi, quam in doctrina de feriebus oftendere; nonnulla fubfidia ex Algebra comuni, quae vulgo tractari non folent, hic erunt repetenda. Quae quamvis maximam partem iam in introductione fumus complexi, tamen quaedam ibi funt praetermifia, vel ftudio quod expediat ea tum demum explicari, quando necefiitas id exigat, vel quia cuncta, quibus opus fit futurum, praevideri non poterant. Huc pertinet transformatio ferierum, cui hoc Caput deftinavimus, qua quaevis feries in innumerabiles alias feries tranfmutatur, quae omnes eandem habeant fummam communem, ita ut, fi feriei propofitae fumma fit cognita, reliquae feries omnes fimul fummari queant. Hoc autem capite praemiffo, eo uberius doctrinam ferierum per calculum differentialem & integralem amplificare poterimus.

2. Confiderabimus autem potiffimum eiufmodi feries, quarum finguli termini per poteftates fuccefivas quantitatis cuiufdam indeterminatae funt multiplicati: quoniam hae latius patent, maioremque utilitatem afferent.

F f 2   Sit

Title page, Chapter 1 of *Institutiones calculi differentialis*, Part II

and substituted s and its derivatives into (13) to obtain recursion relations for the coefficients of f, df/dn, d^2f/dn^2, etc., finally obtaining formula (11). A discussion of Euler's derivation of the Euler-Maclaurin formula as well as some of his interesting applications of it is contained in [**9**, 3.3, 3.4]; a modern summary of Euler's work on the formula is contained in [**1**, p. 1074].

If n is allowed to go to infinity in the Euler-Maclaurin formula (11), the infinite series is divergent for almost all $f(x)$ which occur in applications. Nevertheless, under modest additional hypotheses, the remainder R_k is less than the first term neglected, and so the series and the integral give useful approximations to each other, depending on which is easier to compute.

Independently of Euler, Maclaurin (*Treatise on Fluxions*, 1742) arrived at the same summation formula but by a method a little surer and closer to that which we use today. The remainder was first added and seriously treated by Poisson.

Euler also introduced in his *Institutiones* of 1755 a transformation of series, still known and used [12]. Given a series $\sum_{n=0}^{\infty} b_n$, he wrote it as $\sum_{n=0}^{\infty} (-1)^n a_n$. Then by a number of formal algebraic steps he showed that

$$\sum_{n=0}^{\infty} (-1)^n a_n = \sum_{n=0}^{\infty} (-1)^n \frac{\Delta^n a_0}{2^{n+1}}, \tag{15}$$

where

$$\Delta^0 a_0 = a_0, \ \Delta^1 a_0 = a_1 - a_0, \ \Delta^n a_0 = \Delta^{n-1} a_1 - \Delta^{n-1} a_0 = \sum_{i=0}^{n} (-1)^{n-i} \binom{n}{i} a_i, \quad n \geq 2.$$

His derivation of (15) is as follows. Let $a_n = (-1)^n b_n$ and introduce variables x and y related by $x = y/(1-y) = y + y^2 + y^3 + \cdots$. Then

$$b_0 x + b_1 x^2 + b_2 x^3 + \cdots$$
$$= a_0 x - a_1 x^2 + a_2 x^3 - a_3 x^4 + \cdots$$
$$= a_0(y + y^2 + y^3 + \cdots) - a_1(y^2 + 2y^3 + 3y^4 + 4y^5 + \cdots)$$
$$\quad + a_2(y^3 + 3y^4 + 6y^5 + 10y^6 + \cdots) - a_3(y^4 + 4y^5 + 10y^6 + 20y^7 + \cdots) + \cdots$$
$$= a_0 y - (a_1 - a_0)y^2 + (a_2 - 2a_1 + a_0)y^3 - \cdots .$$

Setting $x = 1$ and $y = \frac{1}{2}$ yields

$$\sum_{n=0}^{\infty} b_n = a_0 - a_1 + a_2 - a_3 + \cdots = \frac{a_0}{2} - \frac{\Delta^1 a_0}{4} + \frac{\Delta^2 a_0}{8} - \cdots ,$$

as required.

The transformation in (15) often converts a convergent series into a more rapidly converging one. However, for Euler, who did not usually distinguish between convergent and divergent series, the transformation could also transform divergent series into convergent ones. For, if one applies (15) to the series (1), which is $\sum_{n=0}^{\infty} (-1)^n$, then since $a_0 = 1$ and $\Delta^n a_0 = 0$ for all $n > 1$, the sum on the right is $1/2$. Likewise for the series

$$1 - 2 + 2^2 - 2^3 + 2^4 - \cdots ,$$

the transformation in (15) gives

$$\sum_{n=0}^{\infty} (-1)^n 2^n = \frac{1}{2}(1) - \frac{1}{4}(1) + \frac{1}{8}(1) - \frac{1}{16}(1) + \cdots = \frac{1}{3}.$$

These results are the same as those Euler got by taking the sum of the series to be the value of the function from which the series is derived.

Euler took up the subject of sums of series in a major paper of 1754/55 entitled "On Divergent Series" [5], in which he recognized the distinction between convergent and

cuius evolutione propofita feries nafcatur.
I. *Sit igitur propofita haec feries Leibnitzii.*
$$S = 1 - 1 + 1 - 1 + 1 - 1 + \&c.$$
in qua cum omnes termini fint aequales, fient omnes diffe-
rentiae $= 0$, ideoque ob $a = 1$, erit $S = \frac{1}{2}$.
II. *Sit propofita ifta feries.*
$$S = 1 - 2 + 3 - 4 + 5 - 6 + \&c.$$
Diff. I. $= 1, 1, 1, 1, 1, \&c.$
Cum ergo fit $a = 1$, $\Delta a = 1$, erit $S = \frac{1}{2} - \frac{1}{4} = \frac{1}{4}$.
III. *Sit propofita haec feries:*
$$S = 1 - 3 + 5 - 7 + 9 - \&c.$$
Diff. I. $= 2, 2, 2, 2, \&c.$
Ob $a = 1$ & $\Delta a = 2$ fit $S = \frac{1}{2} - \frac{2}{4} = 0$.
IV. *Sit propofita haec feries trigonalium numerorum.*
$$S = 1 - 3 + 6 - 10 + 15 - 21 + \&c.$$
Diff. I. $= 2, 3, 4, 5, 6, \&c.$
Diff. II. $= 1, 1, 1, 1, \&c.$
Hic ergo ob $a = 1$, $\Delta a = 2$, & $\Delta \Delta a = 1$; erit
$$S = \frac{1}{2} - \frac{2}{4} + \frac{1}{8} = \frac{1}{8}.$$
V. *Sit propofita feries quadratorum:*
$$S = 1 - 4 + 9 - 16 + 25 - 36 + \&c.$$
Diff. I. $= 3, 5, 7, 9, 11, \&c.$
Diff. II. $= 2, 2, 2, 2, \&c.$

Gg   Ob

Euler demonstrates (*Institutiones*, Part II, Chapter 1) his transformation of series with many examples. Here he "sums" the alternating series of triangular numbers, $\sum_{n=0}^{\infty}(-1)^n (n+1) \times (n+2)/2 = 1/8$, the alternating series of squares, $\sum_{n=0}^{\infty}(-1)^n (n+1)^2 = 0$ and of fourth powers $\sum_{n=0}^{\infty}(-1)^n (n+1)^4 = 0$.

divergent series. Apropos of the former he says that for those series in which by constantly adding terms we approach closer and closer to a fixed number, which happens when the terms continually decrease, the series is said to be convergent and the fixed number is its sum. Series whose terms do not decrease and may even increase are divergent.

On divergent series, Euler says one should not use the term "sum" because this refers to actual addition. Euler then states a general principle which explains what he means by the definite value of a divergent series. He points out that the divergent series comes from finite algebraic expressions and then says that the value of the series is *the value of the algebraic expression from which the series comes.* Euler further states, "Whenever an infinite series is obtained as the development of some closed expression, it may be used in mathematical operations as the equivalent of that expression, even for values of the variable for which the series diverges." He repeats this principle in his *Institutiones* of 1755:

> Let us say, therefore, that the sum of any infinite series is the finite expression, by the expansion of which the series is generated. In this sense the sum of the infinite series $1 - x + x^2 - x^3 + \cdots$ will be $1/(1 + x)$, because the series arises from the expansion of the fraction, whatever number is put in place of x. If this is agreed, the new definition of the word sum coincides with the ordinary meaning when a series converges; and since divergent series have no sum in the proper sense of

the word, no inconvenience can arise from this terminology. Finally, by means of this definition, we can preserve the utility of divergent series and defend their use from all objections.

It is fairly certain that Euler meant to limit this doctrine to power series.

Other 18th-century mathematicians also recognized that a distinction must be made between what we now call convergent and divergent series, though they were not at all clear as to what the distinction should be. They were dealing with a new concept and, like all pioneers, they had to struggle to clear the forest. Certainly the interpretation of series suggested by Newton, and adopted by Leibniz, Euler, and Lagrange, that series are just long polynomials and so belong in the domain of algebra, could not serve as a rigorous foundation for the work with series.

One outstanding characteristic of the 18th-century investigations is that mathematicians trusted the symbols far more than logic. Because infinite series had the same symbolic form for all values of x, the distinction between values of x for which the series converged and values for which they diverged did not seem to demand much attention. And even though it was recognized that some series, such as $1 + 2 + 3 + \cdots$, had infinite sums, mathematicians preferred to try to give meaning to the sums rather than question the applicability of summation. Of course, they were fully aware of the need for some proofs. We have seen that Euler did try to justify his use of divergent series. But the few efforts to achieve rigor, significant because they show that standards of rigor vary with the times [10], did not validate the work of the century, and mathematicians almost willingly took the position that what cannot be cured must be endured.

Though we have only glimpsed some of Euler's work, almost all of the great mathematicians of the 18th century contributed to the subject of infinite series [13]. It is fair to say that in this work the formal view dominated. Aware of the power of formal manipulation, mathematicians either ignored or deferred consideration of any limitations to their techniques, such as the importance of convergence. Their work produced useful results, and they were satisfied with this pragmatic sanction. They exceeded the bounds of what they could justify, but they were at least prudent in their use of divergent series. However, these 18th-century mathematicians were to have the last word. Dimly, they saw in divergent infinite series, ideas which were later to gain acceptance, namely, summability and asymptotic series [2], [11], [13, chapter 47], [14].

I wish to thank Professor Edward J. Barbeau of the University of Toronto for his critique and for supplying some material on Euler's proofs.

References

1. Raymond Ayoub, Euler and the zeta function, *Amer. Math. Monthly*, 81 (1974) 1067–1087.

2. E. J. Barbeau, Euler subdues a very obstreperous series, *Amer. Math. Monthly*, 86 (1979) 356–372.

3. W. F. Eberlein, On Euler's infinite product for the sine, *J. Math. Anal. Appl.*, 58 (1977) 147–151.

4. L. Euler, De progressionibus harmonicis observationes, *Comm. acad. sci. Petrop.*, 7 (1734/35), p. 150–161 = *Opera Omnia*, (1) 14, 87–100.

5. ——, De seriebus divergentibus, *Novi comm. acad. sci. Petrop.*, 5 (1754/55), 1760, pp. 205–237 = *Opera Omnia*, (1) 14, 585–617. An English translation by E. J. Barbeau and P. J. Leah is in *Historia Math.*, 3 (1976) 141–160.

6. ——, De seriebus quibusdam considerationes, *Comm. acad. sci. Petrop.*, 12 (1740), 1750, pp. 53–96 = *Opera Omnia*, (1) 14, 407–462.

7. ——, De summis serierum reciprocarum, *Comm. acad. sci. Petrop.*, 7 (1734/35), 1740, pp. 123–134 = *Opera Omnia*, (1) 14, 73–86.

8. ——, Institutiones calculi differentialis cum lius usu in analysi finitorum ac doctrina serierum, *Acad. Imp. Sci., Petrop.* = Opera Omnia, (1) 10, 309–336.

9. H. H. Goldstine, *A History of Numerical Analysis from the 16th through the 19th Century*, Springer-Verlag, 1977, especially Chapter 3.

10. J. V. Grabiner, Is mathematical truth time-dependent?, *Amer. Math. Monthly*, 81 (1974) 354–365.

11. G. H. Hardy, *Divergent Series*, Oxford University Press, 1949.

12. R. Johnsonbaugh, Summing an alternating series, *Amer. Math. Monthly*, 86 (1979) 637–648.

13. Morris Kline, *Mathematical Thought from Ancient to Modern Times*, Oxford University Press, 1972, especially Chapter 20.

14. John Tucciarone, The development of the theory of summable divergent series, *Arch. Hist. Exact Sci.*, 10 (1973) 1–40.

A street sign in Paris. Photo courtesy of Michael Huber.

Euler and the Zeta Function[1]

Raymond Ayoub

1. Introduction

Mathematics in general appeals to the intellect; great mathematics, however, has, in addition, a kind of perceptual quality which endows it with a beauty comparable to that of great art or music. In this category belongs much of the work of the great 18th century Swiss mathematician, Leonhard Euler (1707–1783).

One of the most enchanting episodes is his work on the zeta function, to which this article is devoted. In anticipation of the later notation of G.F.B. Riemann (1826–1866), let

$$\zeta(s) = \sum_{n=1}^{\infty} \frac{1}{n^s},$$

where for the moment no specification will be made on s. Euler's work on $\zeta(s)$ began about 1730 with approximations to the value of $\zeta(2)$, continued with the evaluation of $\zeta(2n)$, where n is a natural number ≥ 1, and resulted about 1749 in the discovery of the functional equation *almost* 110 *years before Riemann*.

Before beginning the story, we shall give a brief sketch of Euler's life. Most of the facts are taken from a eulogy delivered by Nicholas Fuss (1755–1825), the husband of one of Euler's granddaughters.

The city of Basel in Switzerland was one of many free cities in Europe and by the 17th century had become an important center of trade and commerce. The University became a noted institution, largely through the fame of an extraordinary family—the Bernoullis. This family had come from Antwerp to Basel and the founder of the mathematical dynasty was Nicholas Bernoulli. He had 3 sons, two of whom, James (often referred to as Jacob) (1654–1705), and John (1667–1748), became noted mathematicians. Both were pupils of G. Leibniz (1646–1716) with whom John carried on an extensive correspondence and with whose work both James and John became familiar. James was a professor at Basel

[1] Reprinted from the *American Mathematical Monthly*, Vol. 81, December, 1974, pp. 1067–1086. This article received the Lester Ford Award in 1975.

until his death in 1705. John, who had been a professor at Groningen, replaced him. To give an indication of the mathematical activity of this period, it is worthwhile pointing out that I. Newton (1642–1727) had started his work on the theory of fluxions about 1664, publishing his great work, *Principia Mathematica*, in 1687. On the continent G. Leibniz began his studies on the calculus about 1672 and published much of his work in the journal *Acta Eruditorum*. This was a monthly periodical published in Leipzig and devoted to miscellaneous articles, books and book reviews.

Paul Euler (1670–1745) was a Lutheran Pastor who was mathematically talented and who had studied mathematics with James Bernoulli at the University of Basel. Into this intellectually rich and stimulating environment, Leonhard was born in 1707. He was a precocious child who received much encouragement from his father. He entered the University of Basel and displayed such remarkable talent for mathematics that John Bernoulli gave him special instruction on Saturdays. He graduated with a kind of master's degree in 1724 at the age of 17. He had enrolled in the Faculty of Theology and had written a thesis in Latin on a comparison between Newtonian and Cartesian philosophy. Although Paul expected his son to study theology, he did not discourage Leonhard's interest in mathematics. (Still, mathematics was fine as a hobby, but surely not as a profession!)

At this period there were 3 famous centers of learning, the academies at Berlin, Paris, and St. Petersburg, and it was frequently the case that a young scholar would journey to one of these.

John Bernoulli had 3 sons. Two of them, Nicholas II (1695–1726) and Daniel (1700–1782), were mathematicians who befriended Euler. They both went to St. Petersburg in 1725 and both had a high regard for their younger colleague. After some effort, Daniel wrote to Euler that he had secured for him a stipend in the Academy. The appointment was actually in the physiology section but Euler quickly drifted into the mathematics section. He then left Basel and arrived in St. Petersburg in 1727, remaining there until 1741.

The period had its troubles. Tsarina Catherine I was committed to carrying out the policy of her husband, Peter the Great, in establishing a strong Academy. Unfortunately she died the day Euler set foot in Russia. The throne passed to Peter I's grandson, Peter II, who was only 12 and Russia was ruled by despotic regents who declared that the Academy was very costly and was of little use to the state. Euler despaired of being able to pursue his interests and decided to join the navy. Admiral Sievers saw in him a valuable asset to the navy and offered him a position as lieutenant, with promises of rapid promotion. From the sources available to the author it is not clear to what extent, if any, Euler was active in naval affairs. The death of Peter II brought to an end the despotic regency and the Academy's condition improved, but the despotism had discouraged some foreign scholars, who returned to their homeland. An opportunity arose when Bülfinger left Russia and in 1731 Euler was appointed professor of natural sciences. Two years later, when Daniel decided to return to Basel in 1733, he recommended that Euler be appointed his successor as professor of mathematics. Euler remained in this position until 1741 when he was summoned by Frederick the Great of Prussia to the Berlin Academy. He was in Berlin until 1766. Catherine II, the Great, acceded to the throne of Russia in 1762 and in 1766 summoned Euler back to the Academy in St. Petersburg where he remained until his death in 1783.

Euler did significant work in all areas of mathematics and his work in any one of these would have assured him a place in history. He was a prodigious writer whose collected works run currently to 70 quarto volumes with more to come. In editing Euler's works shortly after his death N. Fuss listed 756 articles distributed in time as follows: 1727–33: 24; 1734–43: 49; 1744–53: 125; 1754–63: 99; 1764–72: 104; 1773–82: 355. The most astonishing feature is the phenomenal number written in the last 10 years of his life, during which years he was blind. Since Fuss's editing activities, numerous additional manuscripts have been found and the total will run to almost 900. In addition to his articles he wrote several books, among the most noted and influential of which was his *Introductio in Analysin Infinitorum*. Some have criticized his writings as being repetitive but it is proper to ignore this kind of pedantry.

Euler's articles were mostly in Latin which is unfortunate in view of our present day ignorance of the classics. On the other hand, the Latin is comparatively simple and, with a rudimentary knowledge, together with a dictionary, the reader will be rewarded for his efforts. It is especially fortunate that the notation is familiar, and where the language is difficult, the mathematics comes to the rescue. It is customary to be surprised at how "modern" his notation is; the truth is that his influence was so profound that we still use much of the notation he helped to establish.

Reading his papers is an exhilarating experience; one is struck by the great imagination and originality. Sometimes a result familiar to the reader will take on an original and illuminating aspect, and one wishes that later writers had not tampered with it.

Euler's personal life, though relatively uneventful, was marred by several tragedies. Though apparently of a strong constitution, he developed a massive infection which resulted in the loss of one eye in 1735. The second eye developed a cataract about 1766 which rendered him blind. He could still distinguish lights and shadows and sometimes wrote mathematics in very large symbols on a blackboard. Despite this handicap, he continued unabated his mathematical activities with the help of young assistants. He once met with J. d'Alembert (1717–1783) who was utterly astonished at Euler's ability to carry out in his head the most complicated analytical calculations.

Euler married Catherine Gsell in 1733. She was the daughter of a well-known artist. She had 13 children 8 of whom tragically died in childhood. Catherine died in 1776, Euler then married her half sister.

His character was that of a kind and gentle man. He had a phenomenal memory, had studied the classics, and is said to have known the Aeneid by heart. Though the recipient of numerous honors during his lifetime, he retained his modesty and humility and it was said of him that he took as much pleasure in the discoveries of others as he did in his own.

He carried on an extensive correspondence with various mathematicians, especially Christian Goldbach (1690–1764). He also wrote a series of letters on various subjects in natural philosophy addressed to a German princess. The quality of all his letters reflects his pleasant personality.

2. Early history of the function $\zeta(s)$

In elementary courses in calculus, one of the first examples of an infinite series is that given by $\zeta(s)$.

The student quickly learns, mainly via the integral test, that

$$\sum_{n=1}^{\infty} \frac{1}{n^s}$$

converges if $s > 1$ and diverges if $s \leq 1$. Some enthusiastic teachers will point out that, in fact,

$$\zeta(2) = \sum_{n=1}^{\infty} \frac{1}{n^2} = \frac{\pi^2}{6}, \tag{1}$$

and perhaps remark that this relation is difficult to prove and that students who go on in mathematics will eventually learn at least one proof. More enthusiastic teachers will further point out that if k is an integer $k \geq 1$, then

$$\zeta(2k) = \frac{(-1)^{k-1} B_{2k} (2\pi)^{2k}}{2(2k)!}, \tag{2}$$

where B_{2k} is a *rational* number, viz. a Bernoulli number, a fact first proved by Euler. The generating function for these numbers is given by

$$\frac{x}{e^x - 1} = 1 - \frac{x}{2} + \sum_{n=2}^{\infty} \frac{B_n}{n!} x^n.$$

However, it is easily seen that $B_{2m+1} = 0$ and that $B_2 = \frac{1}{6}$, $B_4 = -\frac{1}{30}$, $B_6 = \frac{1}{42}, \ldots$. They might further point out that if m is odd, $m = 2k + 1$ ($k \geq 1$), then no such formula is known for $\zeta(m)$, and despite considerable efforts over the years, the arithmetic nature of even $\zeta(3)$ remains an unsolved problem. [Ed. Note: In 1978, Roger Apéry proved that $\zeta(3)$ is irrational; in 2000, Tanguy Rivoal showed that $\zeta(n)$ is irrational for infinitely many odd integers n.]

Before proceeding, it is interesting to note that Euler often worked with

$$\theta(s) = \sum_{n=0}^{\infty} \frac{1}{(2n+1)^s},$$

with

$$\phi(s) = \sum_{n=1}^{\infty} \frac{(-1)^{n+1}}{n^s},$$

and with

$$\psi(s) = \sum_{n=0}^{\infty} \frac{(-1)^n}{(2n+1)^s}.$$

The first two are related to $\zeta(s)$ by

$$\zeta(s) = \theta(s) + \frac{1}{2^s} \zeta(s);$$

hence $\theta(s) = \left(1 - (1/2^s)\right)\zeta(s)$, while

$$\phi(s) = \zeta(s) - \frac{2}{2^s}\zeta(s) = (1 - 2^{1-s})\zeta(s).$$

Thus $\phi(n)$ and $\theta(n)$ can be evaluated if $\zeta(n)$ can be. One important advantage of $\phi(s)$ over $\zeta(s)$ is that the series for $\phi(s)$ converges if $s > 0$, while that for $\zeta(s)$ only for $s > 1$.

By contrast $\psi(s)$ has a superficial resemblance to $\phi(s)$ but although $\psi(2n + 1)$ has been evaluated, $\phi(2n + 1)$ has not. In fact Euler proved that

$$\psi(2n + 1) = (-1)^n \frac{E_{2n}}{2^{2n+2}(2n)!} \pi^{2n+1},$$

where

$$\sec x = \sum_{n=0}^{\infty} (-1)^n \frac{E_{2n}}{2n!} x^{2n}$$

and E_{2n} are called Euler numbers.

Let us begin the story and go back... Infinite series have occurred sporadically in mathematics for centuries—in fact Archimedes (287–212 B.C.), when he derived his famous theorem on the quadrature of the parabola, proved in effect that the series

$$\sum_{n=1}^{\infty} 4^{-n}$$

converges. As far as the harmonic series is concerned, however (despite Plato's interest), the earliest recorded appearance appears to be in the works of Nicholas of Oresme (1323–1382) who proved that the series

$$\sum_{n=1}^{\infty} \frac{1}{n}$$

diverges.

The problem occurs again in 1650 in a book *Novae Quadraturae Arithmeticae* by a professor of mechanics in Bologna named Pietro Mengoli (1625–1686). He related the series to the logarithm and posed the problem of finding the sum of the series

$$\sum_{n=1}^{\infty} \frac{1}{n^2},$$

if it converges.

Whether through the book of Mengoli or (what seems likely) independently, the problem became known in France and England. In fact, the English mathematician John Wallis (1616–1703), professor at Oxford, commented on the problem in 1655 in his book *Arithmetica Infinitorum*. He had computed the value of $\zeta(2)$ to 3 decimal places but it does not appear that he recognized this value, 1.645, as being about $\pi^2/6$.

In a letter to John Bernoulli in 1673, Leibniz wrote: "let

$$dy = \frac{1}{1} + \frac{x}{2} + \frac{x^2}{3} + \cdots$$

then $dy = -\frac{\log(1-x)}{x} dx$, thus

$$x + \frac{x^2}{2^2} + \frac{x^3}{3^3} + \cdots = -\int \frac{\log(1 - x)}{x} dx.$$

As $\log(1 - x)$ is infinite when $x = 1$, consider instead

$$dy = \frac{x}{1} - \frac{x^2}{2} + \frac{x^3}{3} - \cdots$$

and get $y = \int \frac{\log(1+x)}{x} \, dx$."

He now integrates by parts and deduces that the evaluation of the sum

$$\sum_{n=1}^{\infty} \frac{(-1)^{n+1}}{n^2}$$

reduces to the evaluation of integrals of the form $\int x^e (1 + x)^n \, dx$. He continues: "If perhaps it were possible to consider all the cases in order, some light would be shed upon the problem."

In a letter to James Bernoulli in 1691, his brother John wrote, "I see now the route for finding the sum $\frac{1}{1} + \frac{1}{4} + \frac{1}{9} + \frac{1}{16} + \cdots$." No further work, however, was forthcoming from him until 1742 when he published a proof similar to that given by Euler in 1734.

In the St. Petersburg Academy, the members were drawn to the problem and took a great interest in the evaluation of $\zeta(2)$. That it is a tantalizing problem stems in part from the fact that the series has a superficial resemblance to the series

$$\sum_{n=1}^{\infty} \frac{1}{n(n + 1)},$$

whose value is easily seen to be

$$\sum_{n=1}^{\infty} \left(\frac{1}{n} - \frac{1}{n + 1} \right) = 1.$$

This fact was early recognized by the academicians. In 1728, Daniel Bernoulli wrote to Goldbach that he had a method for computing quickly an approximation to $\zeta(2)$ and gave as an approximate value $8/5$. In reply Goldbach wrote that he could show that

$$1\frac{16}{25} = 1.64 < \zeta(2) < 1\frac{2}{3} = 1.66.$$

Neither gave indications of their computations. As noted above, Daniel Bernoulli and Euler both lived in St. Petersburg between 1727 and 1733 and it seems very probable that they discussed the problem together.

3. Euler's early contributions

There the problem lay. Euler's first contribution came in 1731 when he gave an original method for computing $\zeta(2)$. His method appeared in a paper *De summatione innumerabilium progressionum*. He deals with sums of the type

$$\sum_{k=1}^{\infty} \frac{x^k}{(ak + b)^m}.$$

In the special case of $\zeta(2)$ his argument is as follows: Since

$$\frac{\log(1-x)}{x} = -\sum_{n=1}^{\infty} \frac{x^{n-1}}{n},$$

it follows that

$$-\zeta(2) = \int_0^1 \frac{\log(1-x)}{x}\,dx.$$

Replacing $1-x$ by t and splitting the integral, it follows that

$$-\zeta(2) = \int_0^1 \frac{\log t}{1-t}\,dt = \int_0^x \frac{\log t}{1-t}\,dt + \int_x^1 \frac{\log t}{1-t}\,dt = I_1 + I_2.$$

In I_2, put $u = 1-t$, expand in a power series and integrate termwise; then if $y = 1-x$,

$$I_2 = \sum_{n=1}^{\infty} \frac{y^n}{n^2}.$$

On the other hand, in I_1, expand $(1-t)^{-1}$ in a series, and integrate by parts getting

$$I_1 = -\log x \log(1-x) - \sum_{n=1}^{\infty} \frac{x^n}{n^2}.$$

Hence $\zeta(2) = \log x \log(1-x) + \sum_{n=1}^{\infty} x^n/n^2 + \sum_{n=1}^{\infty} (1-x)^n/n^2$. Putting $x = \frac{1}{2}$, we conclude that

$$\zeta(2) = (\log 2)^2 + \sum_{n=1}^{\infty} \frac{1}{2^{n-1} n^2}.$$

What has been achieved by this next argument? The series $\sum_{n=1}^{\infty} 1/2^{n-1} n^2$ converges much more rapidly than does the series for $\zeta(2)$. Knowing that

$$\log 2 = -\log\left(1 - \frac{1}{2}\right) = \sum_{n=1}^{\infty} \frac{1}{n 2^n} \sim .480453,$$

and that

$$\sum_{n=1}^{\infty} \frac{1}{2^{n-1} n^2} \sim 1.164482,$$

Euler concludes that $\zeta(2) \sim 1.644934$.

It should be remarked that in 1730 James Stirling (1692–1770) had computed $\zeta(2)$ to 9 decimal places, of which 8 were correct, but Euler was unaware of these calculations.

Euler's next contribution came in 1732/33 in a paper entitled *Methodus Generalis Summandi Progressiones*. In this he states the "Euler-McLaurin" formula (Colin McLaurin (1698–1746)). In a later paper *Inventio summae cuiusque seriei ex dato Termino generali*, published in 1736, he gives a proof. Although the paper was published in 1736, it is reasonable to assume that the work was done before 1734. We shall give Euler's argument which we modify slightly. Moreover, we shall ignore a few technicalities. Let

$$S(x) = \sum_{n \le x} f(n).$$

The object is to approximate $S(x)$ by an integral. We have

$$f(x) = S(x) - S(x - 1).$$ (3)

Using the Taylor (Brook Taylor, 1685–1731) expansion, it follows that

$$f(x) = \sum_{n=1}^{\infty} \frac{(-1)^{n+1} S^{(n)}(x)}{n!};$$ (4)

(the difficulty, of course, is that in writing (3) we are assuming x to be an integer while in (4), we assume x to be any real number).

Assume now that this series can be inverted; that is, assume there exist constants $b_0, b_1, b_2, \ldots$ such that

$$S^{(1)}(x) = \sum_{n=0}^{\infty} b_n f^{(n)}(x).$$ (5)

Differentiating (4), inserting in (5), and equating coefficients, gives recurrence formulae for the b's, viz.,

$$b_0 = 1, \quad b_1 = \frac{b_0}{2}, \quad b_2 = \frac{b_1}{2!} - \frac{b_0}{3!}, \quad b_3 = \frac{b_2}{2!} - \frac{b_1}{3!} + \frac{b_0}{4!}, \text{ etc.}$$

Hence $S(x) = b_0 \int f(x)\,dx + \sum_{n=1}^{\infty} b_n f^{(n-1)}(x)$. The b's turn out to be essentially the Bernoulli numbers. This fact can be intuitively gleaned from the following argument: let D denote the operator d/dx, then (4) can be written as

$$f(x) = \left(1 - \frac{D}{2!} + \frac{D^2}{3!} - \cdots\right) S^{(1)}(x) = \left(\frac{e^{-D} - 1}{D}\right) S^{(1)}(x),$$

or inverting,

$$S^{(1)}(x) = \left(\frac{D}{e^{-D} - 1}\right) f(x).$$

On the other hand, the generating function for the Bernoulli numbers as noted above, gives

$$\frac{x}{e^x - 1} = 1 - \frac{x}{2} + \frac{B_2 x^2}{2!} + \frac{B_3 x^3}{3!} + \cdots.$$

Hence, replacing x by $-D$ gives the desired result. Euler is evidently excited by this discovery (as which of us would not be!) and proceeds to apply it with great enthusiasm in the paper which appeared in 1736, *Inventio summae cuiusque seriei ex dato Termino generali*, referred to above.

He derives a formula for

$$\sum_{m=1}^{n} m^k \qquad (k \geq 1)$$

and painstakingly computes the necessary constants $B_2, \ldots, B_{16}$ and writes out at length the results for $k = 1, \ldots, 16$. Then he applies it to the harmonic series, showing that

$$\sum_{n \leq x} \frac{1}{n} = \text{const} + \log x + \frac{1}{2x} - \frac{1}{12x^2} + \cdots,$$

and performs calculations for $x = 10^l$ for $l = 1, 2, 3, 4, 5, 6$. Finally among other things, he computes $\zeta(2)$ and $\zeta(3)$ with great accuracy. For $\zeta(2)$, he writes

$$\zeta(2) = \sum_{n=1}^{10} \frac{1}{n^2} + \sum_{n=11}^{\infty} \frac{1}{n^2}.$$

He computes the first term by hand and then estimates the remainder by the formula. His result is that approximately

$$\zeta(2) = 1.64493406684822643647.$$

Still the evaluation of $\zeta(2)$ in closed form eluded him. Needless to say, this method of approximation opened a whole new area of research.

4. First triumph

Euler's first triumph came in 1734. Having previously done work on the roots of polynomials, he conceived the idea of generalizing the factorization of polynomials to transcendental functions. Euler communicated his result to Daniel Bernoulli and, while unfortunately this letter has been lost, the reply does exist. Daniel says: "The theorem on the sum of the series

$$1 + \frac{1}{4} + \frac{1}{9} + \frac{1}{16} + \cdots = \frac{pp}{6} \quad \text{and} \quad 1 + \frac{1}{2^4} + \frac{1}{3^4} + \frac{1}{4^4} + \cdots = \frac{p^4}{90}$$

is very remarkable. You must no doubt have come upon it *a posteriori*. I should very much like to see your solution."

Here is a sketch of it as it appears in *De summis serierum reciprocarum*. Consider the expression $f(x) = 1 - (\sin x / \sin \alpha)$ with α fixed and α not a multiple of π. Leibniz had derived the power series expansion for $\sin x$, so write

$$f(x) = 1 - \frac{x}{\sin \alpha} + \frac{x^3}{3! \sin \alpha} - \cdots.$$

The right-hand side is now viewed as a polynomial of infinite degree. If $a_1, a_2, \ldots, a_n, \ldots$ are the roots, then write

$$f(x) = \left(1 - \frac{x}{a_1}\right)\left(1 - \frac{x}{a_2}\right) \cdots \left(1 - \frac{x}{a_n}\right) \cdots = \prod_{k=1}^{\infty} \left(1 - \frac{x}{a_k}\right).$$

The roots of $f(x)$ however, are evident from the left-hand side, viz.,

$$x = \begin{cases} 2n\pi + \alpha \\ 2n\pi + \pi - \alpha \end{cases} \quad n = 0, \pm 1, \pm 2, \ldots;$$

thus

$$f(x) = \prod_{n=-\infty}^{\infty} \left(1 - \frac{x}{2n\pi + \alpha}\right)\left(1 - \frac{x}{2n\pi + \pi - \alpha}\right)$$

$$= \left(1 - \frac{x}{\alpha}\right) \prod_{n=1}^{\infty} \left(1 - \frac{x}{(2n-1)\pi - \alpha}\right)\left(1 + \frac{x}{(2n-1)\pi + \alpha}\right) \qquad (6)$$

$$\times \left(1 - \frac{x}{2n\pi + \alpha}\right)\left(1 + \frac{x}{2n\pi - \alpha}\right).$$

SERIERVM RECIPROCARVM.　　129

$+\frac{1}{25}+\frac{1}{49}+$ etc. fummam effe $=\frac{q^2}{2}=\frac{p^2}{8}$; denotante p totam circuli peripheriam, cuius diameter eft $=1$. Summa autem huius feriei $1+\frac{1}{9}+\frac{1}{25}+$ etc. pendet a fumma feriei $1+\frac{1}{4}+\frac{1}{9}+\frac{1}{16}+\frac{1}{25}+$etc. quia haec quarta fui parte minuta illam dat. Eft ergo fumma huius feriei aequalis fummae illius cum fui triente. Quamobrem erit $1+\frac{1}{4}+\frac{1}{9}+\frac{1}{16}+\frac{1}{25}+\frac{1}{36}+$ etc. $=\frac{p^2}{6}$, ideoque huius feriei fumma per 6 multiplicata aequalis eft quadrato peripheriae circuli cuius diameter eft 1; quae eft ipfa propofitio cuius initio mentionem feci.

§. 12. Cum igitur cafu quo $y=1$, fit $P=1$ et $Q=1$, erunt reliquarum litterarum R, S, T, V etc. vt fequitur: $R=\frac{1}{2}$; $S=\frac{1}{3}$; $T=\frac{5}{24}$; $V=\frac{2}{15}$; $W=\frac{61}{720}$; $X=\frac{17}{45}$ etc. Cum autem fumma cuborum ipfi $R=\frac{1}{2}$ fit aequalis, erit $\frac{2}{q^3}(1-\frac{1}{3^3}+\frac{1}{5^3}-\frac{1}{7^3}+\frac{1}{9^3}-$ etc.$)=\frac{1}{2}$. Quare erit $1-\frac{1}{3^3}+\frac{1}{5^3}-\frac{1}{7^3}+\frac{1}{9^3}-$ etc. $=\frac{q^3}{4}=\frac{p^3}{32}$. Huius ideo feriei fumma per 32 multiplicata dat cubum peripheriae circuli cuius diameter eft 1. Simili modo fumma biquadratorum, quae eft $\frac{2}{p^4}(1+\frac{1}{3^4}+\frac{1}{5^4}+\frac{1}{7^4}+\frac{1}{9^4}+$ etc.$)$ aequalis effe debet $\frac{1}{3}$, ideoque erit $1+\frac{1}{3^4}+\frac{1}{5^4}+\frac{1}{7^4}+\frac{1}{9^4}+$ etc. $=\frac{q^4}{8}=\frac{p^4}{96}$. Eft vero haec feries per $\frac{16}{15}$ multiplicata aequalis huic $1+\frac{1}{2^4}+\frac{1}{3^4}+\frac{1}{5^4}+\frac{1}{6^4}+$ etc. quare ifta feries aequalis eft $\frac{p^4}{90}$; feu feriei $1+\frac{1}{2^4}+\frac{1}{3^4}+\frac{1}{4^4}+$ etc. fumma per 90 multiplicata dat biquadratum peripheriae circuli cuius diameter eft 1.

§. 13. Simili modo inuenientur fummae fuperiorum poteftatum; prodibit autem vt fequitur $1-\frac{1}{3^5}+\frac{1}{5^5}-\frac{1}{7^5}+\frac{1}{9^5}-$etc. $=\frac{5q^5}{48}=\frac{5p^5}{1536}$; atque $1+\frac{1}{3^6}+\frac{1}{5^6}+\frac{1}{7^6}+\frac{1}{9^6}$

Tom. VII.　　　　　　　　R　　　　　　　　　$+$ etc

The page from *De summis serierum reciprocarum* where Euler states the Basel problem of finding the value of $\sum_{n=1}^{\infty} 1/n^2$

We can now expand the right-hand side in a power series and equate coefficients. The expansion on the right involves the "infinite" elementary symmetric functions and Euler now derived the infinite analogues of Newton's formulae, viz., if $a_1, \ldots, a_n, \ldots$ is a sequence and

$$\sigma_m = \sum_{i_1, \ldots, i_{\ldots}} a_{i_1} \cdots a_{i_m}$$

while $S_m = \sum_{i=1}^{\infty} a_i^m$, then in particular,

$$S_1 = \sigma_1, \quad S_2 = \sigma_1^2 - 2\sigma_2, \quad S_3 = \sigma_1^3 - 3\sigma_1\sigma_2 + 3\sigma_3.$$

The other relations may be similarly derived.

Applying these facts to (6) we get (since $\sigma_2 = 0$),

$$\frac{1}{\alpha} + \sum_{n=1}^{\infty} \left(\frac{1}{(2n-1)\pi - \alpha} - \frac{1}{(2n-1)\pi + \alpha} + \frac{1}{2n\pi + \alpha} - \frac{1}{2n\pi - \alpha} \right) = \frac{1}{\sin\alpha},$$

$$\frac{1}{\alpha^2} + \sum_{n=1}^{\infty} \left(\frac{1}{((2n-1)\pi - \alpha)^2} + \frac{1}{((2n-1)\pi + \alpha)^2} + \frac{1}{(2n\pi + \alpha)^2} + \frac{1}{(2n\pi - \alpha)^2} \right)$$

$$= \frac{1}{\sin^2\alpha},$$

$$\frac{1}{\alpha^3} + \sum_{n=1}^{\infty} \left(\frac{1}{((2n-1)\pi - \alpha)^3} - \frac{1}{((2n-1)\pi + \alpha)^3} + \frac{1}{(2n\pi + \alpha)^3} - \frac{1}{(2n\pi - \alpha)^3} \right)$$

$$= \frac{1}{\sin^3\alpha} - \frac{1}{2\sin\alpha},$$

Putting $\alpha = \pi/2$, the first gives $(4/\pi)(1 - \frac{1}{3} + \frac{1}{5} \cdots) = 1$—a fact already known to James Gregory (1638–1675). The second, however, leads to the long sought after objective, for it gives

$$\frac{8}{\pi^2} \left(1 + \frac{1}{3^2} + \frac{1}{5^2} + \cdots \right) = 1.$$

However, as Euler remarks,

$$\zeta(2) = \left(1 + \frac{1}{3^2} + \frac{1}{5^2} + \cdots \right) + \frac{1}{4}\zeta(2)$$

and this, then, gives $\zeta(2) = \pi^2/6$. Similar arguments give

$$1 - \frac{1}{3^3} + \frac{1}{5^3} - \frac{1}{7^3} + \cdots = \frac{\pi^3}{32},$$

$$1 + \frac{1}{3^4} + \frac{1}{5^4} + \cdots = \frac{\pi^4}{96}.$$

and, hence, $\zeta(4) = \pi^4/90$.

Likewise Euler computes the corresponding series with exponents 5, 6, 7, and 8. If $\alpha = \pi/4$, the first relation gives

$$\frac{\pi}{2\sqrt{2}} = 1 + \frac{1}{3} - \frac{1}{5} - \frac{1}{7} + \frac{1}{9} + \frac{1}{11} - \cdots$$

—a fact he attributes to Newton.

This elegant discovery gave him one of his earliest successes and established him as a mathematician of the first rank.

One is naturally tempted to ask why, if Euler intends to use infinite products, he does not simply use $\sin x$ itself? In fact he does; as a postscript to this paper, he notes that

$$\frac{\sin x}{x} = \prod_{n=1}^{\infty} \left(1 - \frac{x^2}{(n\pi)^2}\right) \tag{7}$$

and deduces more directly, $\zeta(2n)$ for $n = 1, 2, 3, 4, 5, 6$. Equation (7), however, does not give the flexibility of (6) and clearly has no hope of yielding anything about $\zeta(2n + 1)$. One might surmise that he first proved (7) and then the more general result (6).

Two objections were raised to this proof by Daniel Bernoulli. In the first place, one can't compute with infinite series in the same way that one does with polynomials, and in the second place, it is not evident that all the roots of $\sin x = \sin \alpha$ are real. Euler recognizes the second objection as being valid and proceeds to prove that, in fact, all the roots are real. As to the first objection, he rightfully insisted in 1740 that the method is as well founded as any other method and, moreover, it is based upon a principle of which adequate use had not been made. Indeed, it opened up the theory of infinite products and partial fraction decomposition of transcendental functions and its importance goes far beyond the immediate application.

5. Connections with arithmetic

Having achieved his objective of evaluating $\zeta(2)$, Euler now turned to the arithmetic properties of $\zeta(s)$. In 1737 he communicated a paper entitled *Variae Observationes circa series infinitas*.

Here for the first time he proved the famous Euler product decomposition in the form

$$\zeta(s) = \frac{2^s \cdot 3^s \cdot 5^s \cdot 7^s \cdot 11^s \cdots}{(2^s - 1)(3^s - 1)(5^s - 1)(7^s - 1)(11^s - 1) \cdots}.$$

One of his theorems is the statement that

$$\sum_p \frac{1}{p} \sim \log \sum_n \frac{1}{n},$$

where the left-hand side is summed over all p. Nowadays we would insist on writing that as $x \to \infty$

$$\sum_{p \leq x} \frac{1}{p} \sim \log \sum_{n \leq x} \frac{1}{n}.$$

He also "proved" that if $n = p_1^{r_1} \cdots p_l^{r_l}$ and $\lambda(n) = (-1)^{r_1 + r_2 + \cdots + r_l}$, then

$$\sum_{n=1}^{\infty} \frac{\lambda(n)}{n} = 0$$

and the corresponding fact for $\mu(u)$ (what is now called the Möbius function) is stated in his *Introductio*. Regretfully, we have put the word "proved" in quotation marks since the justification of this statement is as deep a result as the prime number theorem itself.

6. Return to $\zeta(s)$

He returned to $\zeta(s)$ in 1740 in a paper entitled *De Seriebus Quibusdam Considerationes*. In this he developed the partial fraction decomposition of various functions. In particular, he proved that

$$\frac{\pi \cos\left(\frac{b-a}{2n}\right)\pi}{n \sin\left(\frac{a+b}{2n}\right)\pi - n \sin\left(\frac{b-a}{2n}\right)\pi} = \frac{1}{a} + \sum_{k=1}^{\infty} \frac{2b}{(2k-1)^2 n^2 - b^2} - \frac{2a}{(2kn)^2 - a^2}.$$

By specializing, once again he deduced the values of $\zeta(2), \zeta(4), \ldots$.

In the meantime what has happened to $\zeta(3)$? In this same paper he computed approximate values of $\zeta(2n+1)$ for $n = 1, 2, 3, 4, 5$ to which he added the known values of $\zeta(2n)$. He wrote these in the form

$$\zeta(n) = N\pi^n.$$

He says that if n is even, then N is rational, while if N is odd then he conjectures that N is a function of $\log 2$.

There is now a slight digression.

Apparently to respond to the earlier criticism concerning his first proof, Euler published a paper in an obscure journal, "Literary Journal of Germany, Switzerland and the North (The Hague)," entitled *Démonstration de la somme de la suite* $1 + \frac{1}{4} + \frac{1}{9} + \cdots$. Here he derived once again the formula for $\zeta(2)$.

Since this method is elementary, and is not generally known, and can be given in an elementary course, we present it an detail. We have

$$\frac{1}{2}(\arcsin x)^2 = \int_0^x \frac{\arcsin t}{\sqrt{1-t^2}}\, dt.$$

If we expand $(1-u^2)^{-1/2}$ by the binomial theorem and integrate termwise, we get

$$\arcsin t = \int_0^t \frac{du}{\sqrt{1-u^2}} = t + \sum_{n=1}^{\infty} \frac{1 \cdot 3 \cdots (2n-1)}{2 \cdot 4 \cdots 2n} \frac{t^{2n+1}}{2n+1}.$$

It follows that

$$\frac{1}{2}(\arcsin x)^2 = \int_0^x \frac{t\, dt}{\sqrt{1-t^2}} + \sum_{n=1}^{\infty} \frac{1 \cdot 3 \cdots (2n-1)}{2 \cdot 4 \cdots 2n} \frac{1}{2n+1} \int_0^x \frac{t^{2n+1}}{\sqrt{1-t^2}}\, dt.$$

Euler then writes, "Since the first term is integrable all the others will also be since the integration of each term reduces to that of the preceding. One can see this clearly if we reflect that in general

$$\int_0^x \frac{t^{n+2}}{\sqrt{1-t^2}}\, dt = \frac{n+1}{n+2} \int_0^x \frac{t^n}{\sqrt{1-t^2}}\, dt - \frac{x^{n+1}}{n+2} \sqrt{1-x^2}."$$

(Apparently the favorite phrases of mathematicians, "clearly etc.," are not of recent origin!) In fact, it takes a few steps to see this "clearly." Let

$$I_n(x) = \int_0^x \frac{t^{n+2}}{\sqrt{1-t^2}}\, dt = \int_0^x \frac{t^{n+1}t\, dt}{\sqrt{1-t^2}}.$$

Integration by parts gives $I_n(x) = -x^{n+1}\sqrt{1-x^2} + (n+1)\int_0^x t^n(\sqrt{1-t^2})\,dt$. Multiplying the integrand by $1 = \sqrt{1-t^2}/\sqrt{1-t^2}$, and splitting into two parts gives

$$I_n(x) = -x^{n+1}\sqrt{1-x^2} + (n+1)I_{n-2}(x) - (n+1)I_n(x),$$

and the result follows.

Thus

$$\int_0^1 \frac{t^{2n+1}}{\sqrt{1-t^2}}\,dt = \frac{2n}{2n+1}\int_0^1 \frac{t^{2n-1}}{\sqrt{1-t^2}}\,dt \tag{8}$$

and as $\int_0^1 t\,dt/\sqrt{1-t^2} = 1$, we conclude that

$$\int_0^1 \frac{t^{2n+1}}{\sqrt{1-t^2}}\,dt = \frac{2n(2n-2)\cdots 2}{(2n+1)(2n-1)\cdots 3}.$$

Therefore $\pi^2/8 = \frac{1}{2}(\arcsin 1)^2 = \sum_{n=0}^{\infty} 1/(2n+1)^2$, which as we know from above is equivalent to $\zeta(2) = \pi^2/6$. The same result may be obtained by first showing that $\frac{1}{2}(\arcsin x)^2$ satisfies the differential equation

$$(1 - x^2)y'' - xy' = 1,$$

then using undetermined coefficients to derive the series for $\frac{1}{2}(\arcsin x)^2$, and finally integrating termwise to get $\frac{1}{6}(\arcsin x)^3$, after using the above result (8). The reader will find it interesting to carry out these steps. The method gives $\zeta(2) = \pi^2/6$ directly. Euler concludes with the remark that despite repeated efforts, he was unable to use this technique to find $\zeta(2n)$ for $n \geq 2$. The reader will note that we have glossed over the mild difficulties associated with the point $x = 1$.

Since the time of Euler, there have been many proofs giving the value of $\zeta(2n)$. The interested reader is urged to consult K. Knopp's book on "Infinite Series."

7. The functional equation and $\zeta(3)$

In the middle of the paper *De Seriebus*... referred to above, Euler began a highly interesting new development. There he states that

$$1 - 3 + 5 - 7 + \cdots = 0$$
$$1 - 3^3 + 5^3 - 7^3 + \cdots = 0,$$

etc., whereas,

$$1 - \tfrac{1}{2} + \tfrac{1}{3} - \tfrac{1}{4} + \cdots = \log 2,$$
$$1 - 2 + 3 - 4 + \cdots = \tfrac{1}{4},$$
$$1 - 2^3 + 3^3 - 4^3 + \cdots = -\tfrac{1}{8},$$
$$1 - 2^5 + 3^5 - 4^5 + \cdots = \tfrac{1}{4},$$
$$1 - 2^7 + 3^7 - 4^7 + \cdots = -\tfrac{17}{16}.$$

On the other hand,

$$1 - 2^2 + 3^2 - 4^2 + \cdots = 0,$$
$$1 - 2^4 + 3^4 - 4^4 + \cdots = 0,$$
$$1 - 2^6 + 3^6 - 4^6 + \cdots = 0.$$

Where do these come from? They are derived as follows. Let

$$f(x) = 1 + x + x^2 + \cdots + x^n + \cdots = \frac{1}{1-x} \quad \text{if } |x| < 1.$$

Euler has no reluctance to put $x = -1$; then $1 - 1 + 1 - 1 + \cdots = \frac{1}{2}$.

To $f(x)$ apply the operator $x(d/dx)$. Then

$$x\frac{d}{dx}f(x) = x + 2x^2 + 3x^3 + \cdots = \frac{x}{(1-x)^2};$$

putting $x = -1$, gives $1 - 2 + 3 - 4 + \cdots = \frac{1}{4}$.

Apply the operator again:

$$x + 2^2x^2 + 3^2x^3 + \cdots = \frac{x(1+x)}{(1-x)^3}.$$

Putting $x = -1$, gives $1 - 2^2 + 3^2 - \cdots = 0$.

As the series converges at each stage of this process for $|x| < 1$, we have Euler anticipating "Abel summability" by some 75 years. Then Euler notes that

$$1 - 2 + 3 - 4 + \cdots = \frac{1}{4} = \frac{2 \cdot 1}{\pi^2}\left(1 + \frac{1}{3^2} + \frac{2}{5^2} + \cdots\right),$$

$$1 - 2^3 + 3^3 - 4^3 + \cdots = -\frac{1}{8} = \frac{-2 \cdot 3!}{\pi^4}\left(1 + \frac{1}{3^4} + \frac{1}{5^4} + \cdots\right),$$

$$1 - 2^5 + 3^5 - 4^5 + \cdots = -\frac{1}{4} = \frac{2 \cdot 5!}{\pi^6}\left(1 + \frac{1}{3^6} + \frac{1}{5^6} + \cdots\right),$$

$$1 - 2^7 + 3^7 - 4^7 + \cdots = -\frac{17}{16} = \frac{-2 \cdot 7!}{\pi^8}\left(1 + \frac{1}{3^8} + \frac{1}{5^8} + \cdots\right),$$

as can be verified by an easy computation using the values of $\theta(2n)$:

As in Section 1, let $\theta(s) = \sum_{n=0}^{\infty} 1/(2n+1)^s$ and $\phi(s) = \sum_{n=1}^{\infty} (-1)^{n-1}/n^s$. These relations can be rephrased as

$$\theta(1-2n) = \frac{(-1)^{n-1}2 \cdot (2n-1)!}{\pi^{2n}}\phi(2n) \quad (n = 1, 2, 3, 4),$$

where, of course, $\theta(m)$, $(m = 0, \pm 1, \pm 2, \ldots)$ is to be understood as

$$\lim_{x \to 1^-} \sum_{n=1}^{\infty} \frac{(-1)^{n-1}x^n}{n^m}.$$

Although he does not explicitly say so, one gets the impression that Euler is trying energetically to develop a technique for evaluating $\zeta(3)$, and this impression is partially confirmed later, as we shall see.

In 1749 he gave a paper to the Berlin Academy entitled *Remarques sur un beau rapport entre les séries des puissances tant directes que réciproques*.

This time he considers

$$\phi(s) = \sum_{n=1}^{\infty} \frac{(-1)^n}{n^s}$$

alone and notes the following relations:

$$\frac{1-2+3-4+5-6+\cdots}{1-\dfrac{1}{2^2}+\dfrac{1}{3^2}-\dfrac{1}{4^2}+\dfrac{1}{5^2}-\dfrac{1}{6^2}+\cdots} = +\frac{1\cdot(2^2-1)}{(2-1)\pi^2},$$

$$\frac{1^2-2^2+3^2-4^2+5^2-6^2+\cdots}{1-\dfrac{1}{2^3}+\dfrac{1}{3^3}-\dfrac{1}{4^3}+\dfrac{1}{5^3}-\dfrac{1}{6^3}+\cdots} = 0,$$

$$\frac{1^3-2^3+3^3-4^3+5^3-6^3+\cdots}{1-\dfrac{1}{2^4}+\dfrac{1}{3^4}-\dfrac{1}{4^4}+\dfrac{1}{5^4}-\dfrac{1}{6^4}+\cdots} = -\frac{1\cdot2\cdot3(2^4-1)}{(2^3-1)\pi^4},$$

$$\frac{1^4-2^4+3^4-4^4+5^4-6^4+\cdots}{1-\dfrac{1}{2^5}+\dfrac{1}{3^5}-\dfrac{1}{4^5}+\dfrac{1}{5^5}-\dfrac{1}{6^5}+\cdots} = 0,$$

or if $n \geq 2$,

$$\frac{\phi(1-n)}{\phi(n)} = \begin{cases} \dfrac{(-1)^{(n/2)+1}(2^n-1)(n-1)!}{(2^{n-1}-1)\pi^n} & \text{if } n \text{ is even,} \\ 0 & \text{if } n \text{ is odd.} \end{cases} \tag{9}$$

These relations are listed for $n = 2, 3, \ldots, 10$. On the other hand, if $n = 1$, we see that

$$\frac{1-1+1-1+\cdots}{1-\frac{1}{2}+\frac{1}{3}-\frac{1}{4}+\cdots} = \frac{1}{2\ln 2},$$

"whose connection with the others is entirely hidden." Equation (9) is now rewritten in the form

$$\frac{\phi(1-n)}{\phi(n)} = \frac{-(n-1)!(2^n-1)}{(2^{n-1}-1)\pi^n}\cos\frac{\pi n}{2},$$

and Euler says "I shall hazard the following conjecture:

$$\frac{\phi(1-s)}{\phi(s)} = \frac{-\Gamma(s)(2^s-1)\cos\pi s/2}{(2^{s-1}-1)\pi^s} \tag{10}$$

is true for all s." Isn't this derivation beautiful!?

Now taking the limit of the right-hand side as $s \to 1$ gives exactly $1/2\ln 2$! Euler continues: "The validity of our conjecture for $s = 1$ (which case first appeared to deviate from the others) is already a strong justification[2] of the truth of our conjecture since it appears unlikely that a false assumption could have upheld the truth of this case. We can therefore regard our conjecture as being solidly based but I shall give other justifications which are equally convincing."

He then checks the formula for $s = \frac{1}{2}, \frac{3}{2}$, and in general $s = (2k+1)/2$.

---

[2] Although Euler uses the word "preuve," the original meaning in English (and presumably also in French) conveys the idea of testing an assumption or statement rather than proving in our sense. Compare, for example, the expression "the exception that 'proves' the rule."

We have seen in Section 1, that

$$\phi(s) = (1 - 2^{1-s})\zeta(s),$$

which leads at once from (10) to

$$\zeta(1-s) = \pi^{-s}2^{1-s}\Gamma(s)\cos\frac{\pi s}{2}\zeta(s),$$

and this is the famous functional equation.[3] It was proved by Riemann in 1859.

It should be noted that Euler could not have used $\zeta(s)$ itself since

$$\lim_{x \to 1^-}\sum_{n=1}^{\infty} n^k x^n$$

does not exist for $k = 0, 1, 2, \ldots$ and therefore he could not have attached a meaning to

$$\sum_{n=1}^{\infty} n^{-(1-s)}$$

for $s = 2, 3, \ldots$.

On the other hand, it can be shown that the series

$$\phi(s) = \sum_{n=1}^{\infty}(-1)^{n+1}n^{-s} = (1 - 2^{1-s})\zeta(s)$$

converges for $s > 0$ (in fact if $s = \sigma + it$, for $\sigma > 0$), but as the pole of $\zeta(s)$ at $s = 1$ has been removed by the factor $(1 - 2^{1-s})$, there remains nothing in the nature of $\phi(s)$ to account for this limitation, and it turns out that

$$\sum_{n=1}^{\infty}(-1)^{n+1}n^{-s}$$

is Abel summable for every value of s.

One is naturally tempted to ask whether Riemann could have seen Euler's work. There is no evidence that he had.[4]

Euler continues: "As far as the sum of the reciprocals of powers (i.e., $\sum_{n=1}^{\infty}(-1)^{n+1}/n^k$) is concerned, I have already observed that their sum can be assigned a value only when k is even and that when k is odd, all my efforts have been useless up to now."

Euler now observes as follows: If $s = 2\lambda + 1$, then

$$\phi(2\lambda + 1) = -\frac{(2^{2\lambda} - 1)\pi^{2\lambda+1}}{\Gamma(2\lambda + 1)(2^{2\lambda+1} - 1)}\frac{\phi(-2\lambda)}{\cos((2\lambda + 1)\pi/2)},$$

and $\phi(-2\lambda)$ as well as $\cos((2\lambda + 1)\pi/2)$ vanish if λ is an integer. Taking the limit as $\lambda \to m$ a positive integer with the help of l'Hôpital's rule, we get

$$\phi(2m + 1) = \frac{2(2^{2m} - 1)\pi^{2m}}{(2m)!(2^{2m+1} - 1)}\frac{\sum_{n=1}^{\infty}(-1)^{n+1}n^{2m}\log n}{\cos \pi m}. \tag{11}$$

---

[3] Since completing this article the author has found that E. Landau has given a rigorous proof of the functional equation in the form (10). See *Bibliotheca Mathematica*, vol. 7 (1906–1907) pp. 69–79.

[4] *Added in proof.* A. Weil remarks that the external evidence supports strongly the view that Riemann was very familiar with Euler's contributions.

"It is necessary therefore to find the value of these sums

$$\sum_{n=1}^{\infty} (-1)^{n+1} n^{2m} \log n.$$

But this research is probably more difficult than the one we have in mind (meaning $\phi(2m+1)$) and I perceive no method whatsoever which could lead us to the proposed objective."

He returned to the question for what appears to be the last time in 1772 in a paper entitled *Exercitationes Analyticae*. Through a striking and elaborate scheme, he proved that

$$1 + \frac{1}{3^3} + \frac{1}{5^3} + \cdots = \frac{\pi^2}{4} \log 2 + 2 \int_0^{\pi/2} x \log \sin x \, dx.$$

Here is a sketch of the proof which invokes the extreme virtuosity of a master:

We know from (11) that

$$1 + \frac{1}{3^3} + \frac{1}{5^3} + \cdots = \frac{\pi^2}{2} Z,$$

where

$$Z = \sum_{n=2}^{\infty} (-1)^n n^2 \log n.$$

This follows from (11) as well as the relations cited in Section 1. Of course we continue to understand that if $\sum_{n=1}^{\infty} a_n$ does not converge but $\sum_{n=1}^{\infty} a_n x^n$ converges for $|x| < 1$, then $\sum_{n=1}^{\infty} a_n$ is defined by

$$\lim_{x \to 1^-} \sum_{n=1}^{\infty} a_n x^n, \quad \text{if this limit exists.}$$

Euler then shows that

$$Z = \sum_{n=1}^{\infty} n^2 \log \frac{(2n)^2}{(2n-1)(2n+1)} - \sum_{n=1}^{\infty} n(n+1) \log \frac{(2n+1)^2}{(2n)(2n+2)}.$$

The expansion of the logarithm is carried out and the series rearranged. Letting $\lambda(s) = \sum_{n=1}^{\infty} 1/(n(n+1))^s$, then

$$Z = \frac{1}{2 \cdot 2^2} + \sum_{n=2}^{\infty} \frac{1}{n 2^{2n}} \left(\zeta(2n-2) + (-1)^n \lambda(n-1) \right).$$

$\lambda(n)$ is then expressed in terms of $\zeta(2k)$ $(k = 1, 2, \ldots, n)$, and if

$$S(n) = \frac{1}{n 2^{2n}} + \sum_{k=1}^{\infty} \frac{(n+k-1)(n+k) \cdots (n+2k-2)}{k!(n+k) 2^{2n+2k}},$$

then

$$Z = -\frac{1}{8} + S(1) + 2 \sum_{n=1}^{\infty} \zeta(2n) \left(\frac{1}{(2n+2) 2^{n+2}} - S(2n+1) \right).$$

He now finds the sum $S(n)$ by showing that

$$S_x(n) = \frac{x^n}{n} + \sum_{k=1}^{\infty} \frac{(n+k-1)(n+k)\cdots(n+2k-2)x^{n+k}}{k!(n+k)}$$

satisfies a difference differential equation and that

$$S_x(1) = \frac{1 + 2x - \sqrt{1-4x}}{4}.$$

This is to be evaluated when $x = \frac{1}{4}$. The result of these intricate details is that

$$S(2n+1) - \frac{1}{(2n+2)2^{2n+2}} = \frac{1}{(2n+1)(2n+2)2^{2n+1}}.$$

$$Z = \frac{1}{2^2} - \sum_{n=1}^{\infty} \frac{\zeta(2n)}{(2n+1)(2n+2)2^{2n}}.$$

We know that $\zeta(2n) = \alpha_{2n}\pi^{2n}$, where α_{2n} is explicitly determined in terms of the Bernoulli numbers.

If

$$f x) = x^2 \sum_{n=1}^{\infty} \frac{\alpha_{2n}x^{2n}}{(2n+1)(2n+2)},$$

then by twice differentiating $f(x)$ we see that it satisfies a differential equation which can be solved in view of the fact that we can evaluate the generating function

$$\sum_{n=1}^{\infty} \alpha_{2n}x^{2n}.$$

Is not this derivation breathtaking, especially in the light of the fact that Euler was now blind and these calculations were performed mentally!

8. Conclusion

So end the main contributions of Euler to the zeta function. He did, however, write a brief paper on the function $\sum_{n=1}^{\infty} x^n/n^2$ toward the end of his life (1779), which was published posthumously. We have given only the highlights of his work on $\zeta(s)$. Scattered throughout his papers on analysis and in his correspondence with Goldbach and the Bernoullis are many results which are related to the problem.

While he did not succeed in every objective he set himself, his triumphs stand like a grand fresco—a monument to his extraordinary imagination and sense of beauty and harmony.

Acknowledgments

In addition to the original papers themselves, the author has found the following sources especially helpful:

1. The preface of Vol. 16 of series 1 of Euler's Collected Works is an article entitled "Übersicht über die Bände 14, 5.16, 16," by Georg Faber. Faber gives a summary of the contents, classified by topics.

2. The article by Paul Stäckel, "Eine vergessene Abhandlung Eulers." This first appeared in the now defunct journal *Bibliotheca Mathematica*, 83 (1907–1908) 37–54. In this, Stäckel discusses the article "Démonstration de la somme..." and gives numerous interesting historical facts. It is reprinted in Euler's Collected Works, Vol. 14, pp. 156–176.

3. Correspondence between Euler and Goldbach published by Deutsche Akademie der Wissenschaften and edited by A. Juškevič and E. Winter. The editors' comments on the letters were very helpful.

4. The paper of Landau referred to in the footnote.

5. The referee kindly suggested stylistic changes and pointed out some errors.

Addendum to:
"Euler and the Zeta Function"[1]

A. G. Howson

Readers of Raymond Ayoub's article (this volume, pp. 113–132) may be amused to learn that one of the questions set to candidates for the first London University Matriculation Examination (in 1838), an examination for students of 19 years or under who wished to enter the university, was

"Find the sums to infinity of the series

$$\frac{1}{1^2} + \frac{1}{2^2} + \frac{1}{3^2} + \cdots, \qquad \frac{1}{1 \cdot 2} + \frac{1}{2 \cdot 3} + \frac{1}{3 \cdot 4} + \cdots."$$

There is no indication how the examiner intended the question to be solved; the examination syllabus, which did not include the calculus, referred only to "arithmetical and geometrical progressions" and "arithmetic and algebra." It can, however, be inferred from a previous question

"Prove that

$$\mathrm{Nap}^n \log x = (x - x^{-1}) - \frac{1}{2(x^2 - x^{-2})} + \frac{1}{3(x^3 - x^{-3})} - \cdots."$$

that a cavalier treatment of infinite series was not only tolerated but actively encouraged.

---

[1]Reprinted from the *American Mathematical Monthly*, Vol. 82, August-September, 1975, p. 737.

Bust of Euler

Euler Subdues a Very Obstreperous Series[1]

E. J. Barbeau

The task of evaluating the infinite series $1 - 1! + 2! - 3! + \cdots$ caused Euler to clarify his ideas on the meaning of assigning a sum to a series, even one which, in modern eyes, is divergent. In this article, we summarize these ideas and outline four ingenious approaches of Euler to evaluate the above series. The consistency of these approaches is discussed, with reference to summability methods, extrapolation, continued fractions, and infinite differential operators.

1. Assigning a value to a divergent series

In the late seventeenth and early eighteenth centuries, mathematicians were busily developing what promised to be a significant body of powerful techniques consequent to the creation of the calculus. However, there was considerable uncertainty about the best formulation of the underlying concepts. Probably no better example of this can be found than in the discussion of the meaning of the sum or value of an infinite series. Since sums of monomials can be easily differentiated and integrated, the discovery by Newton and his contemporaries that a great many functions could be developed as power series meant that calculus had quite wide applicability. Consequently, the question of attaching a sum to a series attracted much interest and controversy.

Although mathematicians of this period were aware intuitively that, for some series, the sum could be regarded as the limit of the partial sums, in their view this did not adequately cover the matter. Even when this limit did not exist, many series nevertheless seemed to possess a natural value. Their attitude was influenced in part by their notion of a function as an analytic expression defined over the widest possible domain, including complex numbers and quantities infinitely great or small. Not being in possession of pathological counterexamples, they considered that two analytic expressions agreeing on a continuous set must agree everywhere. Thus, for example, if $(1 + x)^c$ is synonymous with its binomial expansion for $|x| < 1$, then $(1 + x)^c$ must be the value of that expansion for all x, except possibly for obvious singularities.

[1] Reprinted from the *American Mathematical Monthly*, Vol. 86, May, 1979, pp. 356–372.

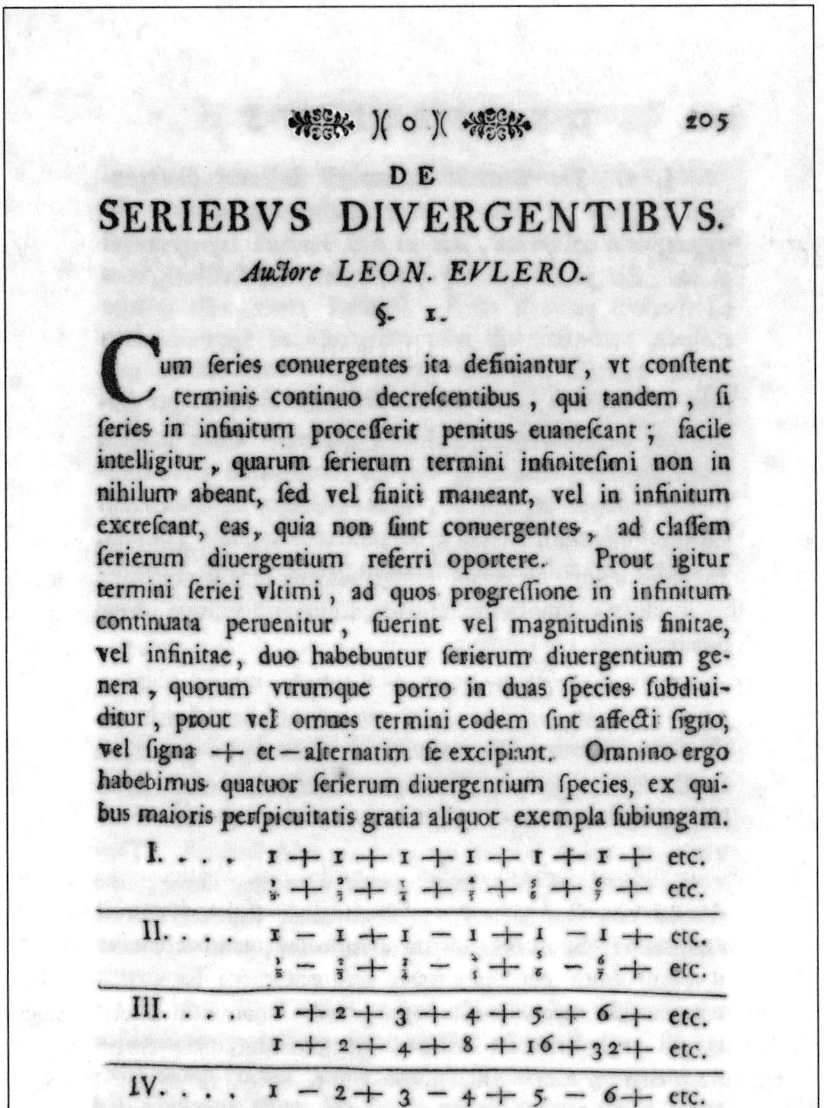

)(o)(205

DE

SERIEBVS DIVERGENTIBVS.

Auctore LEON. EVLERO.

§. 1.

Cum ſeries conuergentes ita definiantur, vt conſtent terminis continuo decreſcentibus, qui tandem, ſi ſeries in infinitum proceſſerit penitus euaneſcant; facile intelligitur, quarum ſerierum termini infiniteſimi non in nihilum abeant, ſed vel finiti maneant, vel in infinitum excreſcant, eas, quia non ſint conuergentes, ad claſſem ſerierum diuergentium referri oportere. Prout igitur termini ſeriei vltimi, ad quos progreſſione in infinitum continuata peruenitur, fuerint vel magnitudinis finitae, vel infinitae, duo habebuntur ſerierum diuergentium genera, quorum vtrumque porro in duas ſpecies ſubdiuiditur, prout vel omnes termini eodem ſint affecti ſigno, vel ſigna + et − alternatim ſe excipiant. Omnino ergo habebimus quatuor ſerierum diuergentium ſpecies, ex quibus maioris perſpicuitatis gratia aliquot exempla ſubiungam.

I. . . . $1 + 1 + 1 + 1 + 1 + 1 +$ etc.
$\frac{1}{2} + \frac{2}{3} + \frac{3}{4} + \frac{4}{5} + \frac{5}{6} + \frac{6}{7} +$ etc.

II. . . . $1 - 1 + 1 - 1 + 1 - 1 +$ etc.
$\frac{1}{2} - \frac{2}{3} + \frac{3}{4} - \frac{4}{5} + \frac{5}{6} - \frac{6}{7} +$ etc.

III. . . . $1 + 2 + 3 + 4 + 5 + 6 +$ etc.
$1 + 2 + 4 + 8 + 16 + 32 +$ etc.

IV. . . . $1 - 2 + 3 - 4 + 5 - 6 +$ etc.
$1 - 2 + 4 - 8 + 16 - 32 +$ etc.

Cc 3 §. 2.

Figure 1. Title page from "De seriebus divergentibus." Taken from *Novi Commentarii* vol. 5, p. 205.

These opinions were buttressed by experience. It was generally found that, where there were several ways of determining the value of an infinite series, they gave the same result. Moreover, in computations, the practice of interchanging an infinite series with its value did not appear to cause trouble.

Euler's paper, "De seriebus divergentibus" [4], published in 1760, illuminates this spirit well. It can be split into two parts. The first subtly treats the question of assigning a value to a series. The second is devoted to evaluating "Wallis' hypergeometric series"

$$1 - 1! + 2! - 3! + 4! - 5! + \cdots .$$

Here we have a somewhat different approach to mathematical acceptability than that of today. Euler's concern is to put his result beyond all reasonable doubt, and this he does by arriving at it by a number of routes. It is consistency, as much as logical argument, which puts its stamp of approval on the mathematics. (See [7] for a wider discussion of this issue.)

Although the modern investigator would quarrel with details of the work of Euler or of his contemporaries, it nevertheless displays a compelling consistency and usually leads to results demonstrably correct according to today's standards of rigour. Consequently, unusual methods of assigning a value to an infinite series have not been disdained during the past century, but rather formalized, studied in detail, compared, and extended. In situations where normal convergence fails, it is possible to find an alternative definition of "sum" which retains many of the properties associated with the usual concept (and, indeed, agrees with it for series convergent in the normal sense) and which will assign to a given infinite series the value of the function which generates it. This can be done for the binomial expansion and other power series beyond the circle of convergence, as well as for Fourier series, witness Fejér's theorem on the Cesàro-summability of the Fourier expansion of a continuous function. A discussion of summability from the modern point of view can be found elsewhere [2, pp. 5–10], [9], [10].

2. Euler's general outlook

The prospectus to his paper [4] declares Euler's intention "to clarify a concept causing up to now the greatest difficulties." While he would not accept that mathematics is necessarily free of controversy, he is confident that mathematical disputes, unlike those in other areas, can be completely resolved once the evidence has been thoroughly weighed. So it is with assigning values to infinite series. Infinite series can be divided into four categories according as the terms are positive or alternating, bounded or unbounded. Examples of the four groups are

$$
\begin{array}{ll}
\text{I.} & 1 + 1 + 1 + 1 + \cdots \\
\text{II.} & 1 - 1 + 1 - 1 + \cdots \\
\text{III.} & 1 + 2 + 4 + 8 + \cdots \\
\text{IV.} & 1 - 2 + 4 - 8 + \cdots
\end{array}
$$

Series in group I present no difficulty. Either they converge to a finite sum in the modern sense, or they diverge to the infinite sum, $a/0$. More contentious are the series of group II. Euler bases his discussion on the expansion $1 - a + a^2 - a^3 + \cdots$ of the fraction $1/(1 + a)$. While no one would deny that these two expressions agree when $|a| < 1$, one might object to assigning the fraction as the sum of the series when $|a| \geq 1$ on the grounds that the remainder term $\pm a^{n+1}/(1 + a)$ in the equation

$$\frac{1}{1+a} = 1 - a + a^2 - a^3 + \cdots \mp a^n \pm \frac{a^{n+1}}{1+a}$$

cannot be neglected. Some of those who support the fraction as sum counter that, for infinite n, the ambiguous sign makes the remainder indeterminate, so that the remainder should be forgotten. In any case, they say, when you sum to infinity, you never reach the place where the remainder has to be inserted. Euler reserves his own position until later.

Those who would assign sums to divergent series appear to be in deep trouble with series in group III. Although it might seem appropriate to assign for these series, as for those of group I, an infinite sum, there occur situations in which the sum indicated by analysis is not only finite but negative. For example, substituting -3 for a in the expansion of $1/(1 + a)$ yields the paradoxical equation

$$-\frac{1}{2} = 1 + 3 + 9 + 27 + 81 + \cdots.$$

Here one is in the absurd position of adding together positive terms to get a negative sum. Nevertheless, explaining this is a mere challenge to the ingenuity! To resolve the paradox (says Euler), some try to distinguish between two types of negative numbers, those that are less than zero and those that are greater than infinity. An example of the first type is the difference between an integer and its successor: $-1 = n - (n + 1)$. An example of the second type is $-1 = 1/-1$, since it fits naturally into the "increasing" sequence

$$\cdots, \frac{1}{3}, \frac{1}{2}, \frac{1}{1}, \frac{1}{0}, \frac{1}{-1}, \frac{1}{-2}, \frac{1}{-3}, \cdots.$$

Euler disapproves of this distinction on the grounds that it "does violence to the certitude of analysis" to have two different concepts of -1. However, he is prepared to accept that "the same quantities which are less than zero can be considered to be greater than infinity."

Series in group IV can sometimes be handled as those in group II, already treated. For example, from the expansion of $1/(1+1)^2$, it is found that $\frac{1}{4} = 1 - 2 + 3 - 4 + 5 - 6 + \cdots$. Euler says very little about this type in general, except to remark that it "is usually burdened with problems of its very own."

Euler affirms that the real justification for assigning a value to a divergent series does not rest in any of the specious arguments given above, but rather in a substitution principle. If an infinite expansion can be replaced in a calculation by the expression which generates it without any ensuing error, then this replacement should be considered valid. One has only to be careful that the rules for doing this are properly investigated. As for the techniques to determine exactly what the value of a given series is, their power can be demonstrated by treating the particularly violent specimen which occupies the rest of the paper.

3. Euler's treatment of Wallis' series

Euler's attribution of the series $1 - 1! + 2! - 3! + \cdots$ to Wallis is a mystery. While Wallis had much to say about summing progressions, I have found no reference in his work to this particular series. His interest in the factorial function lay in interpolating its values for nonintegral arguments. He discusses this question in the Scholium to Proposition 190 in his *Arithmetica infinitorum* (1655) and, again, in a letter to Leibniz dated January 16, 1699 [6,

p. 59], where he seeks a formula for $n!$ which makes sense for nonintegral n comparable to the formula $\frac{1}{2}(n^2 + n)$ for the sum $1 + 2 + 3 + \cdots + n$. The adjective "hypergeometric" simply signifies that each term is obtained from its predecessor by multiplying by a factor which varies (presumably in some regular way) from term to term. This is in contrast to a "geometric" progression in which the multiplying factor remains the same. However, with the great interest in the factorial function, it is likely that the problem of summing "Wallis'" series was formulated in the correspondence of the early eighteenth century. Euler himself discussed it in at least two letters to Nicholas Bernoulli [5, pp. 538, 543, 546] before publishing his findings in the paper under discussion. The series gets brief mention in the books by Kline [3, pp. 451, 1114], Bromwich [1, p. 323] and Hardy [2, pp. 26–29].

Euler evaluates the series by four different methods. In the first he is content to get a rough numerical approximation by exploiting the fact that the series is alternating. To motivate his approach, let us first consider the sum of an alternating series which converges: $1 - \frac{1}{2} + \frac{1}{3} - \frac{1}{4} + \frac{1}{5} - \frac{1}{6} + \cdots = \log 2$. An upper bound for the sum is any partial sum whose last term is positive—for example, $1 - \frac{1}{2} + \frac{1}{3} - \frac{1}{4} + \frac{1}{5} = \frac{47}{60}$; a lower bound is any partial sum whose last term is negative—for example, $1 - \frac{1}{2} + \frac{1}{3} - \frac{1}{4} = \frac{7}{12}$. Apply the same reasoning to Wallis' series. The partial sums are 1, 0, 2, −4, 20, −100, 620,.... The odd partial sums give upper bounds for the value of the series; the even partial sums give lower bounds. However, because the general term does not tend to zero, we do not obtain a very good estimate. Indeed, all that can be said is that the value lies between 0 and 1. Consequently, we would like to transform the series into an equivalent series which is alternating but which is capable of giving a better estimate.

To see how this might be done, notice that the summing of the alternating series $a_1 - a_2 + a_3 - a_4 + a_5 - a_6 + \cdots$ can be achieved by evaluating the power series

$$a_1 x - a_2 x^2 + a_3 x^3 - a_4 x^4 + a_5 x^5 - a_6 x^6 + \cdots$$

at $x = 1$. We effect a change of variables to produce the required transformation. Introduce y by the equation

$$x = y(1 - y)^{-1} = y + y^2 + y^3 + y^4 + \cdots .$$

After substitution and some formal manipulation, the power series becomes

$$a_1 y - (\Delta a_1) y^2 + (\Delta^2 a_1) y^3 - \cdots + (-1)^{k-1}(\Delta^{k-1} a_1) y^k + \cdots$$

where Δ is the forward difference operator defined by

$$\Delta^0 a_i = a_i, \qquad \Delta^1 a_i = \Delta a_i = a_{i+1} - a_i,$$

$$\Delta^k a_i = \Delta^{k-1} a_{i+1} - \Delta^{k-1} a_i = \sum_{j=0}^{k} (-1)^{k-j} \binom{k}{j} a_{i+j}$$

for $i \geq 1, k \geq 2$. Since $x = 1$ corresponds to $y = \frac{1}{2}$, we can evaluate $a_1 - a_2 + a_3 - a_4 + \cdots$ by evaluating the y-series at $y = \frac{1}{2}$:

$$\frac{1}{2} a_1 - \frac{1}{4}(\Delta a_1) + \frac{1}{8}(\Delta^2 a_1) - \frac{1}{16}(\Delta^3 a_1) + \cdots .$$

Euler applies this to obtaining the value

$$A \equiv 1 - 1 + 2 - 6 + 24 - 120 + 720 - 5040 + 40320 - \cdots .$$

First remove the first two terms, $1 - 1$, which cancel, and divide by 2 to get

$$\frac{A}{2} = 1 - 3 + 12 \quad - \quad 60 \quad + \quad 360 \quad - \quad 2520 \quad + \quad 20160 \quad - \quad 181440$$

$$2 \quad 9 \qquad 48 \qquad\quad 300 \qquad\quad 2160 \qquad\quad 17640 \qquad\quad 161280$$

$$7 \quad 39 \qquad 252 \qquad 1860 \qquad\quad 15480 \qquad\quad 143640$$

$$32 \quad 213 \qquad 1608 \qquad 13620 \qquad\quad 128160$$

$$181 \qquad 1395 \qquad 12012 \qquad 114540$$

$$1214 \qquad 10617 \qquad 102528$$

$$9403 \qquad 91911$$

$$82508$$

The rows under the series give, for the absolute values of its terms, differences of the first, second, third, etc., orders, respectively. Applying the transformation, we find that

$$\frac{A}{2} = \frac{1}{2} - \frac{2}{4} + \frac{7}{8} - \frac{32}{16} + \frac{181}{32} - \frac{1214}{64} + \frac{9403}{128} - \frac{82508}{256} + \cdots .$$

Cancelling the first two terms and multiplying by 2 gives

$$A = \frac{7}{8} - \frac{32}{8} + \frac{181}{16} - \frac{1214}{32} + \frac{9403}{64} - \frac{82508}{128} + \cdots .$$

It can be seen that not much progress has been made! However, Euler continues transforming the series, to get at the next turn of the crank,

$$A = \frac{7}{8} - \frac{18}{32} + \frac{81}{128} - \frac{456}{512} + \frac{3123}{2048} - \frac{24894}{8192} + \cdots .$$

Now the second term has smaller magnitude than the first. From the first two partial sums, A must be between $7/8$ and $5/16$. After one more application of the transformation, Euler is prepared to say that A is about 0.580.

The difference operator intervenes also in Euler's second attack on the series. His strategy is to define a sequence whose zeroth term is formally Wallis' series and then to compute this zeroth term numerically. This requires Newton's method of extrapolation, which will be briefly described. For a given sequence, $(a_1, a_2, \ldots, a_n)$, observe that

$$a_{m+1} = a_m + (a_{m+1} - a_m) = a_m + \Delta a_m \equiv (1 + \Delta) a_m$$
$$a_{m+2} = a_{m+1} + \Delta a_{m+1} = (1 + \Delta) a_{m+1} = (1 + \Delta)(1 + \Delta) a_m$$
$$= a_m + 2 \Delta a_m + \Delta^2 a_m$$

and, for any positive integers m and k,

$$a_{m+k} = (1 + \Delta)^k a_m$$

$$\equiv a_m + k \Delta a_m + \frac{k(k-1)}{2} \Delta^2 a_m + \frac{k(k-1)(k-2)}{6} \Delta^3 a_m + \cdots . \qquad (1)$$

For k other than a nonnegative integer, the right side of (2) still makes sense, so that (2) can be used to represent "terms" of the sequence corresponding to indices other than natural numbers.

Euler considers the sequence (P_n) whose terms are given by $P_1 = 1$, $P_2 = 2$, $P_3 = 5$, $P_4 = 16$, $P_5 = 65$, and, generally,

$$P_{n+1} = nP_n + 1 \quad \text{for } n = 2, 3, 4, \ldots. \tag{2}$$

From the fact that $\Delta^i P_i = i!$ ($i = 0, 1, 2, 3, \ldots$), the formula (2) with $m = 1$, $k = n-1$, yields a formula for P_n:

$$P_n = (1 + \Delta)^{n-1} P_1 = P_1 + (n-1)\Delta P_1 + \binom{n-1}{2} \Delta^2 P_1 + \binom{n-1}{3} \Delta^3 P_1 + \cdots$$

$$= 1 + (n-1) + (n-1)(n-2) + (n-1)(n-2)(n-3) + \cdots.$$

Further, substituting 0 for n gives

$$P_0 = 1 - 1! + 2! - 3! + 4! - \cdots.$$

How can a numerical value for P_0 be found?

Euler next applies (2) with $m = 1$, $k = -1$ to the sequences whose general terms are $a_n = 1/P_n$ and $a_n = \log_{10} P_n$. In the first case, the zeroth term is found to be

$$1 - \left(-\frac{1}{2}\right) + \left(\frac{1}{5}\right) - \left(-\frac{3}{80}\right) + \left(-\frac{36}{1040}\right) - \left(\frac{11271}{220376}\right) + \cdots$$

$$= 1 + 0.5 + 0.2 + 0.0375 - 0.0364154 - 0.0511444 + \cdots$$

$$= 1.651740 \quad \text{(Euler's figure)}.$$

Taking 1.651740 as $1/P_0$, we have that $P_0 = 0.60542$. Analysis of the second sequence, $(\log_{10} P_n)$, corroborates this determination of P_0 quite well. The zeroth term of the sequence is

$$0 - 0.3010300 + 0.0969100 - 0.0103000 - 0.0128666 - 0.0053006 + \cdots$$

and this Euler, using the transformation procedure of his first method, computes as 1.7779089. Thereupon, $P_0 = 0.59966$.

This method raises two interesting questions. First, are the series obtained for $1/P_0$ and $\log_{10} P_0$ actually convergent? Second, if the terms of one sequence are a certain function of the corresponding terms of another, how reasonable is it to expect that the functional relation will persist to the extrapolated terms as well? This does not always happen; if, for positive integers n, $a_n = n$, $b_n = f(a_n)$ with $f(z) = \sin \pi z/(\pi z)$, $f(0) = 1$, then Newton's extrapolation procedure yields $a_0 = b_0 = 0$; but $f(a_0) = 1$. One suspects that it is not enough for f to be analytic but that it should have less than exponential growth at infinity as well.

The last two approaches of Euler hinge on finding a closed expression for a power series in x, which, for $x = 1$, produces Wallis' series. In the third method, he observes that the power series

$$s(x) = x - 1x^2 + 2x^3 - 6x^4 + 24x^5 - 120x^6 + \cdots \tag{3}$$

formally satisfies the differential equation

$$s' + \frac{s}{x^2} = \frac{1}{x}. \tag{4}$$

This first order equation can be solved in the usual way; the solution which vanishes for $x = 0$ is

$$s(x) = e^{1/x} \int_0^x \frac{e^{-1/t}}{t} \, dt. \tag{5}$$

Using the substitution $v = \exp(1 - 1/t)$, $t = 1/(1 - \log v)$, $dt/t = dv/v(1 - \log v)$, Euler transforms (6) to

$$s(x) = e^{(1/x-1)} \int_0^{e^{1-1/x}} \frac{dv}{1 - \log v}. \tag{6}$$

For future reference, we record here that, making the substitution $s = \frac{1}{t} - \frac{1}{x}$, the integral can be rendered

$$s(x) = \int_0^\infty \frac{xe^{-s}}{1 + xs} \, ds. \tag{7}$$

These three integrals yield the following alternative forms for the value of Wallis' series:

$$s(1) = e \int_0^1 \frac{e^{-1/t}}{t} \, dt = \int_0^1 \frac{dv}{1 - \log v} = \int_0^\infty \frac{e^{-s}}{1 + s} \, ds, \tag{8}$$

of which Euler computes the approximate values of the first and second by the trapezoidal rule. Euler checks that the second integral of (8) ought to give Wallis' series by substituting $y = 1$ into the expansion (obtained by integrating by parts),

$$\int_0^y \frac{dv}{1 - \log v} = \frac{y}{1 - \log y} - \frac{1 \cdot y}{(1 - \log y)^2} + \frac{1 \cdot 2 \cdot y}{(1 - \log y)^3} - \frac{1 \cdot 2 \cdot 3 \cdot y}{(1 - \log y)^4} + \cdots. \tag{9}$$

This integral allows for an alternative computation of the value of Wallis' series which Euler does not mention in the paper but which he confides in a letter to Bernoulli [5, p. 546]. The left side of (9) is expanded in ascending powers of $(1 - v)$, and the series is integrated term by term. Upon substitution of 1 for y, there results

$$1 - 1 + 2 - 6 + 24 - \cdots = 1 - \frac{1}{2} + \frac{1}{6} - \frac{1}{12} + \frac{1}{30} - \frac{7}{360} + \frac{19}{2520} - \frac{3}{560} + \cdots. \tag{10}$$

Euler's fourth approach is to obtain a "continued fraction" expansion for the power series

$$u(x) = \frac{s(x)}{x} = 1 - x + 2x^2 - 6x^3 + 24x^4 - 120x^5 + \cdots. \tag{11}$$

This has the form $1/(1 + B)$ where

$$B = \frac{x - 2x^2 + 6x^3 - \cdots}{1 - x + 2x^2 - 6x^3 + \cdots}.$$

In turn, B can be put in the form $x/(1 + C)$, with

$$C = \frac{x - 4x^2 + 18x^3 - 96x^4 + \cdots}{1 - 2x + 6x^2 - \cdots}.$$

So far, we have found that

$$u(x) = \cfrac{1}{1 + \cfrac{x}{1 + C}},$$

Carrying on indefinitely, Euler finds

$$u(x) = \cfrac{1}{1 + \cfrac{x}{1 + \cfrac{x}{1 + \cfrac{2x}{1 + \cfrac{2x}{1 + \cfrac{3x}{1 + \cfrac{3x}{1 + \cfrac{4x}{1 + \cfrac{4x}{1 + \cdots}}}}}}}}}. \tag{12}$$

The value of $u(x)$ can be approximated by the convergents obtained by stopping the continued fraction (12) at any point. These are

$$\frac{p_1(x)}{q_1(x)} = \frac{1}{1+x}; \qquad \frac{p_2(x)}{q_2(x)} = \cfrac{1}{1 + \cfrac{x}{1+x}}; \qquad \frac{p_3(x)}{q_3(x)} = \cfrac{1}{1 + \cfrac{x}{1 + \cfrac{x}{1+2x}}}.$$

The nth convergent is $p_n(x)/q_n(x)$, where, for small values of n, $p_n(x)$ and $q_n(x)$ are given in the following table:

n	$p_n(x)$	$q_n(x)$
1	1	$1 + x$
2	$1 + x$	$1 + 2x$
3	$1 + 3x$	$1 + 4x + 2x^2$
4	$1 + 5x + 2x^2$	$1 + 6x + 6x^2$
5	$1 + 8x + 11x^2$	$1 + 9x + 18x^2 + 6x^3$
6	$1 + 11x + 26x^2 + 6x^3$	$1 + 12x + 36x^2 + 24x^3$
7	$1 + 15x + 58x^2 + 50x^3$	$1 + 16x + 72x^2 + 96x^3 + 24x^4.$

In general, for $n \geq 1$, these relations hold:

$$\begin{aligned} p_{2n+1}(x) &= p_{2n}(x) + (n+1)xp_{2n-1}(x) & p_{2n+2}(x) &= p_{2n+1}(x) + (n+1)xp_{2n}(x) \\ q_{2n+1}(x) &= q_{2n}(x) + (n+1)xq_{2n-1}(x) & q_{2n+2}(x) &= q_{2n+1}(x) + (n+1)xq_{2n}(x). \end{aligned} \tag{13}$$

The sum of Wallis' series ought to be $u(1)$. For $x = 1$, the convergents of the continued fraction (12) are

$$\frac{1}{2}, \frac{2}{3}, \frac{4}{7}, \frac{8}{13}, \frac{20}{34}, \frac{44}{73}, \frac{124}{209}, \frac{300}{501}, \dots,$$

forming an apparently convergent sequence.

Euler has another, somewhat curious, way of evaluating the continued fraction at $x = 1$. He writes

$$A = \cfrac{1}{1 + \cfrac{1}{1 + \cfrac{1}{1 + \cfrac{2}{1 + \cfrac{2}{1 + \cfrac{3}{1 + \cfrac{3}{1 + \cfrac{4}{1 + \cfrac{4}{1 + \cfrac{5}{\cfrac{\cdots}{1 + \cfrac{10}{1 + \cfrac{10}{1 + p}}}}}}}}}}}}$$

where

$$p = \cfrac{11}{1 + \cfrac{11}{1 + \cfrac{12}{\cfrac{\cdots}{1 + \cfrac{15}{1 + \cfrac{15}{1 + q}}}}}}, \qquad q = \cfrac{16}{1 + \cfrac{16}{1 + \cfrac{17}{\cfrac{\cdots}{1 + \cfrac{20}{1 + \cfrac{20}{1 + r}}}}}},$$

and where

$$r = \cfrac{21}{1 + \cfrac{21}{1 + \cfrac{22}{1 + \cfrac{22}{1 + \cfrac{23}{1 + \cfrac{23}{1 + \cfrac{\cdots}{}}}}}}}.$$

From this

$$A = \frac{491459820 + 139931620p}{824073141 + 234662231p}; \quad p = \frac{2381951 + 649286q}{887640 + 187440q}; \quad q = \frac{11437136 + 2924816r}{3697925 + 643025r}.$$

The calculation depends on determining r. If, in the definition of r, we replace the numbers $22, 23, 24, 25, \ldots$ all by 21, we obtain the approximate equation $r = 21/(1 + r)$, which is satisfied by

$$r = \frac{1}{2}(\sqrt{85} - 1) = 4.10977\ldots. \tag{14}$$

Euler has a second way of finding r. We have that

$$r = \cfrac{21}{1 + \cfrac{21}{1 + s}} = \frac{21 + 21s}{22 + s},$$

where

$$s = \cfrac{22}{1 + \cfrac{22}{1+t}} = \frac{22 + 22t}{23 + t} \qquad \text{and} \qquad t = \cfrac{23}{1 + \cfrac{23}{1 + \cfrac{24}{1 + \cfrac{24}{1 + \ddots}}}}.$$

Euler assumes that r, s, and t are in arithmetic progression, so that $r + t = 2s$. Since $t = (23s - 22)/(22 - s)$, he finds that

$$r + t = \frac{2s^2 + 925s - 22}{484 - s^2} = 2s,$$

whence

$$2s^3 + 2s^2 - 43s - 22 = 0.$$

This is solved by an approximate method (Newton's) to obtain $s = 4.423$, from which $r = 4.31$, $q = 3.71645446$, $p = 3.0266600163$, $A = 0.5963473621372$ (Euler's accuracy). Euler notes that close rational approximations can be obtained from the convergents of the simple continued fraction expansion of A,

$$\cfrac{1}{1 + \cfrac{1}{1 + \cfrac{1}{2 + \cfrac{1}{10 + \cfrac{1}{1 + \cfrac{1}{1 + \cfrac{1}{4 + \cfrac{1}{2 + \cfrac{1}{2 + \cfrac{1}{13 + \cfrac{1}{4 + \cdots}}}}}}}}}}}.$$

Euler's ingenuity has brought forth a variety of ways of handling the seemingly impossible problem of attaching a value to Wallis' series. If these all lead to the same numerical result, then it will reinforce the conclusion that Wallis' series has a natural value and that, within computational error, we have found it. Let us make the test. Euler's first method is crude, but does give the value 0.580. His second gives the values 0.60542 (from $1/P_0$) and 0.59966 (from $\log P_0$). The trapezoidal role for approximate integration with ten subintervals gives, respectively, 0.59637255 and 0.58734359 for the first and second integrals of (8). Formula (10) gives about 0.59940472. Using the convergents 124/209 and 300/501 of the expansion (12) for $u(1)$ puts the value between 0.5933 and 0.5988. Using (14) for r gives $A = 0.59634738$, and using $r = 4.31$ gives 0.596347362. It can be fairly concluded that these results are consistent. Whatever differences arise seem to reflect the accuracy or efficiency of the method. What should the answer be? Hardy [2, p. 26], by computing (7), obtains

$$1 - 1! + 2! - 3! + \cdots = -e\left(\gamma - 1 + \frac{1}{2 \cdot 2!} - \frac{1}{3 \cdot 3!} + \cdots\right)$$

where $\gamma = 0.577215664901533\ldots$ is Euler's constant. The value obtained is about 0.59635.

4. Closing Remarks

While finding a sum for Wallis' series is hardly of great mathematical significance, there is some fascination attached to the problem. Doubtless, Euler's analysis can be the starting point for a deeper excursion into mathematical interrelationships in a variety of areas—asymptotic expansions, continued fractions, summability, moment problems, factorial series, rational function approximations, infinite differential operators. In this, as in Euler's other investigations, the breadth and ingenuity justify study in something close to the original form by mathematical students.

It could be pointed out that the difficulty of showing that Wallis' series has a value is a result of the field in which we chose to operate. For any prime p, it is clear that Wallis' series converges in the p-adic completion of the rationals, and to an integer, too!

Acknowledgment   I am indebted to Professor Morris Kline for his comments on an earlier draft of this article, and to the referee for his suggestions on presentation and his indication of [8].

References

1. T. J. I'A. Bromwich, *An Introduction to the Theory of Infinite Series*, London, 1931.

2. G. H. Hardy, *Divergent Series*, Oxford, 1949.

3. Morris Kline, *Mathematical Thought from Ancient to Modern Times*, Oxford, New York, 1972.

4. L. Euler, De seriebus divergentibus, *Novi. Comm. Acad. Sci. Petrop.*, 5 (1754/55) 19–23, 205–237 = *Opera Omnia* (1) 14, 585–617. An English translation and paraphrase by E. J. Barbeau and P. J. Leah appears in *Historia Mathematica*, 3 (1976) 141–160.

5. ——, *Opera Postuma, Mathematica et Physica*, 1 (ed. P. H. Fuss and N. Fuss) St. Petersburg, 1862, Six letters to N. Bernoulli, 519–549.

6. C. I. Gerhardt, ed., *G. W. Leibniz Mathematische Schriften 4*, Olms, Hildesheim, New York, 1971.

7. Judith V. Grabiner, Is mathematical truth time-dependent?, *Amer. Math. Monthly*, 81 (1974) 354–365.

8. G. H. Hardy, Note on a divergent series, *Proc. Cambridge Philos. Soc.*, 37 (1941) 1–8.

9. C. N. Moore, Summability of series, *Amer. Math. Monthly*, 39 (1932) 62–71 = *Selected Papers on Calculus* (MAA, 1968) 333–341.

10. John Tucciarone, The development of the theory of summable divergent series from 1880 to 1925, *Arch. History Exact Sci.*, 10 (1973) 1–40.

On the History of Euler's Constant[1]

J. W. L. Glaisher

Of all the mathematical constants π, the ratio of the circumference of a circle to its diameter, and e, the base of the Napierian logarithms, are unquestionably the most remarkable, not only on account of their frequent use, but also from their connection with mathematical history: next however, both in interest and importance must be placed Euler's constant $.57721566\ldots$, usually defined as the limit of $1 + \frac{1}{2} + \cdots + \frac{1}{x} - \log x$, when x is infinite.

The value of π has engaged the attention of many mathematicians and calculators from the time of Archimedes to the present day, and has been computed from so many different formulæ, that a complete account of its calculation would almost amount to a history of mathematics. For every practical purpose π to twenty decimal places is as useful as π to five hundred, but nevertheless, every increase in the number of places has been attended with great interest, not so much because the additional figures were of importance, as because each extension has been in general due to some improvement in the method or formula by which the value was obtained; and thus by examining the different determinations we can trace the transition from the geometrical method of Archimedes to the infinite series of recent times, and it by no means necessarily follows that the 530 places of Shanks and Rutherford in 1853 represent more labour than did the 32 places of Van Ceulen in 1619.[2] Of scarcely inferior interest is the calculation of e, connected as it is with the whole history of logarithms. Euler's constant (which throughout this note will be called γ after Mascheroni, De Morgan, etc.), though of far less celebrity than π or e, has still strong claims to notice; it was introduced at one of the most remarkable periods in mathematical history, and although originally merely connected with the harmonic series, it has since acquired importance, both from its connection with the Gamma Function, and its occurrence in the expansions of the cosine-integral, exponential-integral and logarithm-integral. Additional interest has also been conferred on the history of the constant, by the fact that, owing to the miscalculation of Mascheroni, two values of it have been in existence, and this has occasioned much uncertainty as to its true value, which has only been removed within the last fourteen years by the independent calculations of Lindman and Oettinger.

[1] Reprinted from *Messenger of Mathematics,* Vol. 1, 1872, pp. 25–30.
[2] Prof. de Haan expressly states that the earlier calculations were quite as laborious as those of recent times. *Amsterdam Transactions*, Vol. IV.

The constant is now best known to mathematicians from its occurrence in the formula

$$1 + \frac{1}{2} + \cdots + \frac{1}{x} = \gamma + \log x + \frac{1}{2x} - \frac{B_1}{2x^2} + \frac{B_2}{4x^4} - \frac{B^3}{6x^6} + \cdots \qquad (1)$$

$B_1, B_2, \ldots$, being Bernoulli's numbers, but although its introduction by Euler was sub-sequent to Bernoulli's discussion of the numbers that bear his name, it was not until after the lapse of many years that the latter became sufficiently well known to lead to the summation of the harmonic series in the form (1).

During the first half of the last century the theory of infinite series began to excite great attention, and numerous memoirs of Euler and the Bernoulli's were written on the subject. The series $1 + \frac{1}{2^2} + \frac{1}{3^2} + \cdots$ ('the despair of analysts' as Montucla calls it) for many years exercised in vain the powers of mathematicians, until it was at last successfully summed by Euler in 1748. The harmonic series $1 + \frac{1}{2} + \frac{1}{3} + \cdots$ also, remarkable as being the only divergent series of the group $1 + \frac{1}{2^n} + \frac{1}{3^n} + \cdots$, received no small share of attention about this time.

The discovery that the series, $1 + \frac{1}{2} + \frac{1}{3} + \cdots$ is divergent is attributed by James Bernoulli to his brother (*Ars Conjectandi*, p. 250), but the connection between $1 + \frac{1}{2} + \cdots + \frac{1}{x}$ and $\log x$ was first established by Euler ("De Progressionibus harmonicis observationes," *Comm. Acad. Petropol.* t. VII. for 1734 and 1735, p. 156) as follows:

$$1 = \log 2 \quad + \frac{1}{2} \quad - \frac{1}{3} \quad + \frac{1}{4} \quad - \frac{1}{5} \quad + \cdots$$

$$\frac{1}{2} = \log \frac{3}{2} \quad + \frac{1}{2} \cdot \frac{1}{2^2} \quad - \frac{1}{3} \cdot \frac{1}{2^3} \quad + \frac{1}{4} \cdot \frac{1}{2^4} \quad - \frac{1}{5} \cdot \frac{1}{2^5} \quad + \cdots$$

$$\vdots$$

$$\frac{1}{i} = \log \frac{i+1}{i} \quad + \frac{1}{2} \cdot \frac{1}{i^2} \quad - \frac{1}{3} \cdot \frac{1}{i^3} \quad + \frac{1}{4} \cdot \frac{1}{i^4} \quad - \frac{1}{5} \cdot \frac{1}{i^5} \quad + \cdots,$$

and thence by addition

$$1 + \frac{1}{2} + \cdots + \frac{1}{i} = \log(i+1) + \frac{1}{2}\left(1 + \frac{1}{2^2} + \frac{1}{3^2} + \cdots\right)$$

$$- \frac{1}{3}\left(1 + \frac{1}{2^3} + \frac{1}{3^3} + \cdots\right) + \frac{1}{4}\left(1 + \frac{1}{2^4} + \frac{1}{3^4} + \cdots\right) - \cdots \qquad (2)$$

"Quae series," Euler proceeds, "cum sint convergentes, si proxime summentur prodibit

$$1 + \frac{1}{2} + \cdots + \frac{1}{i} = \log(i+1) + 0.577218.$$

Si summa dicatur s, foret, ut supra fecimus, $ds = di/(i+1)$, ideoque $s = \log(i+1) + C$. Hujus igitur quantitatis constantis C valorem deteximus, quippe est $C = 0.577218$." This is, I believe, the first mention of the constant in mathematics.

In the same transactions, for 1769 (t. XIV., Part I., p. 153) in a paper "De summis serierum numeros Bernoullianos involventium," (which contains Euler's well-known val-ues of the first seventeen Bernoulli's numbers) Euler returns to the harmonic series and

DE PROGRESSIONIBVS HARMONICIS 157

Quae feries, cum fint conuergentes, fi proxime fummentur prodibit $1 + \frac{1}{2} + \frac{1}{3} - - - \frac{1}{i} = l(i+1) + 0,577218$ Si fumma dicatur s, foret, vt fupra fecimus, $ds = \frac{di}{i+1}$, ideoque $s = l(i+1) + C$. Huius igitur quantitatis conftantis c valorem deteximus, quippe eft $C = 0,577218$.

§. 12. Si feries $1 + \frac{1}{2} + \frac{1}{3} - - - - - \frac{1}{i}$ vlterius in infinitum continuetur, et in membra diuidatur, quorum quodvis vt ipfa feries i terminos contineat; erit membrum inter $\frac{1}{i}$ et $\frac{1}{2i}$ contentum $= l2$, fequens $= l\frac{3}{2}$, tertium $= l\frac{4}{3}$, etc. Atque cum ipfius feriei fumma fit log. infiniti, poterit ad analogiam poni $l\frac{1}{0}$. Hocque modo fequens fchema obtinebimus non parum curiofum

Series.	$1 + \frac{1}{2}$	$\frac{3}{2}$	$\frac{1}{2i}$	$\frac{1}{3i}$	$\frac{1}{4i}$	$\frac{1}{5i}$	etc.
Summae	$l\frac{1}{0}$	$l\frac{3}{2}$	$l\frac{5}{3}$	$l\frac{4}{3}$	$l\frac{5}{4}$	

§. 13. Difficile quidem videatur has easdem proprietates progreffionum harmonicarum et logarithmorum expreffiones analytice, eoque modo, quem alibi ad feries fummandas tradidi, inuenire. At rem attentius perpendenti hoc non folum fieri, fed multo generalius etiam fieri poffe deprehenfum eft. Confidero enim non fimplicem progreffionem harmonicam, fed cum geometrica coniunctam, cuiusmodi eft $\frac{cx}{a} + \frac{cx^2}{a+b} + \frac{cx^3}{a+2b} + \frac{cx^4}{a+3b}$

etc. Huius fummam pono s, et vtroque per $bx^{\frac{a-b}{b}}$

multiplicato erit $bx^{\frac{a-b}{b}}s = \frac{bcx^{\frac{a}{b}}}{a} + \frac{bcx^{\frac{a+b}{b}}}{a+b} + \frac{bcx^{\frac{a+2b}{b}}}{a+2b}$

Sumtisque differentialibus habebitur $bD.x^{\frac{a-b}{b}}s = dx(cx^{\frac{a-b}{b}}$

V 3

Page from *De progressionibus harmonicis observationes* showing Euler's constant

gives the formula marked (1) in this note, from which by putting $x = 10$ he calculates $\gamma = .5772156649015325\ldots$. He also gives another formula very similar to (2), viz.

$$\gamma = \frac{1}{2}\left(\frac{1}{2^2} + \frac{1}{3^2} + \cdots\right) + \frac{2}{3}\left(\frac{1}{2^3} + \frac{1}{3^3} + \cdots\right) + \frac{3}{4}\left(\frac{1}{2^4} + \frac{1}{3^4} + \cdots\right) + \cdots, \quad (3)$$

and concludes this portion of his memoir by the remark "Manet ergo quaestio magni momenti, cujusdam indolis sit numerus iste γ, et ad quodnam genus quantitatum sit referendus."

How strongly Euler felt this desire to connect γ with some other known constant, or to discover some property attaching to it, is evident from a memoir, having the discussion of this constant for its sole object, which he communicated to the Petersburg Academy in 1781.[3] Near the commencement of this paper Euler speaks of γ as "magis notatu dignus, quod eum nullo adhuc modo ad quampiam mensuram cognitam revocare mihi quidem licuit." As the formula (1) however when x is very great becomes

$$1 + \frac{1}{2} + \cdots + \frac{1}{x} = \gamma + \log x,$$

he thinks it probable that γ may be the logarithm of some remarkable number, so that if $\gamma = \log N$ we should have $\gamma + \log x = \log(Nx)$. Having met with no success in endeavouring to connect N with any other constant, Euler proceeds to investigate some new formulæ for γ, giving as his reasons the complicated law which Bernoulli's numbers follow and the uncertainty which might attach to the calculation of γ from (1), since that series is ultimately divergent.

The new formulæ obtained in this memoir are

$$1 - \gamma = \frac{1}{2}(s_2 - 1) + \frac{1}{3}(s_3 - 1) + \frac{1}{4}(s_4 - 1) + \cdots, \tag{4}$$

$$2\gamma - 1 = \left(\frac{1}{2} + \frac{1}{3} - \frac{2}{3}s_2\right) + \left(\frac{1}{4} + \frac{1}{5} - \frac{2}{5}s_4\right) + \left(\frac{1}{6} + \frac{1}{7} - \frac{2}{7}s_6\right) + \cdots, \tag{5}$$

$$2 - 2\log 2 - \gamma = \left(\frac{2}{3} \cdot \frac{7}{8}s_2 - \frac{2}{3}\right) + \left(\frac{2}{5} \cdot \frac{31}{32}s_4 - \frac{2}{5}\right) + \left(\frac{2}{7} \cdot \frac{127}{128}s_6 - \frac{2}{7}\right) + \cdots, \tag{6}$$

$$1 - 2\log 2 + \gamma = \left(\frac{1}{2} - \frac{1}{3} \cdot \frac{2}{3 \cdot 2^3}s_2\right) + \left(\frac{1}{4} - \frac{1}{5} - \frac{2}{5 \cdot 2^5}s_4\right) + \cdots, \tag{7}$$

$$\log 2 - \gamma = \frac{1}{3 \cdot 2^2}s_2 + \frac{1}{5 \cdot 2^4}s_4 + \frac{1}{7 \cdot 2^6}s_6 + \cdots, \tag{8}$$

$$1 - \log \frac{3}{2} - \gamma = \frac{1}{3 \cdot 2^2}(s_2 - 1) + \frac{1}{5 \cdot 2^4}(s_4 - 1) + \frac{1}{7 \cdot 2^6}(s_6 - 1) + \cdots, \tag{9}$$

where

$$s_n = 1 + \frac{1}{2^n} + \frac{1}{3^n} + \cdots.$$

From (4) Euler calculates γ to five places correctly and from (9) to twelve places.

Mascheroni, in his *Adnotationes ad Euleri Calculum Integralem*,[4] subsequently extended the calculation to 32 figures as follows:

$$\gamma = .57721\ 56649\ 01532\ 86061\ 81120\ 90082\ 39\ldots$$

In 1809, Soldner published at Munich his "Théorie d'une nouvelle fonction transcendante," the new transcendent being the logarithm-integral

$$li\,x = \int_0^x \frac{dx}{\log x} = \gamma + \log\log x + \log x + \frac{(\log x)^2}{2^2} + \cdots;$$

---

[3] "De numero memorabili in summatione progressionis harmonicae naturalis occurrente," *Acta Petrop.*, t. v., Part II., p, 45.

[4] This work I have not seen, it is referred to by Lacroix (*Calc. Diff. et Int.*, t. III., p. 521).

the value of γ Soldner gives on p. 13 as

$$\gamma = .57721\ 56649\ 01532\ 86060\ 6065\ldots$$

which differs from Mascheroni's value in the twentieth place.

This want of agreement led Gauss,[5] in 1812, to urge F. G. B. Nicolai, "juvenem calculo indefessum," to undertake the calculation and decide which was the true value. Nicolai used formula (1) and obtained γ correctly to forty places, both from $n = 50$ and $x = 100$. This double calculation, which confirmed Soldner's value, set the matter completely at rest; but, unfortunately, the note in Gauss's memoir, which contains the result of Nicolai's calculation, seems to have attracted little notice, and Mascheroni's value has been repeatedly quoted since; thus it is given by Lacroix (*Calc. Diff. et Int.*, t. III., p. 521); by Bretschneider (*Crelle*, t. 17, p. 260); by Bessel (*Königsberger Archiv*, p. 4); and in Grunert's supplement to Klügel's *Wörterbuch*, both values are given, the author having clearly not been aware of Nicolai's calculations.

The occurrence of the two values in Klügel induced Lindman to recalculate the constant, and his values are given in Grunert's *Archiv*[6] for 1857. He used formula (1) and obtained γ correctly to 34 places from $x = 100$ and to 24 places from $x = 20$. Four years later, in ignorance of what Lindman had done, Oettinger[7] applied himself to the investigation of the true value, which he verified by four different calculations; using formula (1) he obtained γ correctly to 18, 25, 34, and 41 places, by giving x successively the values 10, 20, 50, and 100.

Thus it will be seen that Mascheroni's error has led to eight additional calculations of the constant.

It should be mentioned that Bretschneider in giving Mascheroni's value (*Crelle*, t. 19, p. 260) attributes it to Kramp. This is most probably a mistake, as both Lacroix and Gauss refer it to Mascheroni; and Bretschneider himself, in alluding to the error (*Zeitschrift für Mathematik und Physik*, t. 6, p. 131) speaks of it as Mascheroni's calculation; it would, however, have been more satisfactory if he had either corrected or confirmed definitely his assertion about Kramp. I may mention that I have made a cursory examination of all Kramp's papers given in the Royal Society's Catalogue which seemed to have any chance of containing it, as well as his treatise on Refraction, without success.

On account of its occurrence in the formula

$$\log \Gamma(1 + x) = -\gamma x + \frac{1}{2}s_2 x^2 - \frac{1}{3}s_3 x^3 + \frac{1}{4}s_4 x^4 - \cdots$$

$$\log \Gamma(1 + x) = \frac{1}{2}\log\left(\frac{\pi x}{\sin \pi x}\right) - \gamma x - \frac{1}{3}s_3 x^3 - \frac{1}{5}x^5 - \cdots$$

Legendre[8] was naturally concerned with the constant and calculated it to 19 places by (1) from $x = 10$; he also made two other calculations to 15 places, the one from the formula

$$1 - \gamma - \frac{1}{2}\log 2 = \frac{1}{3}(s_3 - 1) + \frac{1}{5}(s_5 - 1) + \cdots \tag{10}$$

---

[5] Disquisitiones generales circa seriem infinitam $1 + \frac{\alpha\beta}{\gamma}x + \cdots$, *Comment. Soc. Reg. Gotting.* t. II., see the note p. 86 of the memoir.

[6] "De vero valore constantis quae in, logarithmo integrali occurrit," t. 29, p. 238.

[7] "Ueber die richtige Werthbestimmung der Constante des Integrallogarithmus," *Crelle*, t. 70, p, 375.

[8] *Traité des Fonctionz Elliptiques*, t. II., p. 434.

and the other from that marked (9) in this note, chiefly with the view of verifying the values of $s_1, s_2, \ldots$.

The most recent determinations of the constant are those of Mr. Shanks (*Proc. Roy. Soc.*, t. 15, p. 429), who has calculated it, from (1) to 21, 28, 39, 46, 54, 59, and 59 places from $x = 10, 20, 50, 100, 200, 500$, and 1000 respectively. The value (after correction of an error in the fiftieth place) to 59 places is

$$\gamma = .57721\ 56649\ 01532\ 86060\ 65120\ 90082\ 40243\ 10421\ 59335\ 93992\ 35988\ 0577\ldots.$$

It is clearly convenient that the constant should generally be denoted by the same letter. Euler used C and O for it; Legendre, Lindman, &c., C; De Haan A; and Mascheroni, De Morgan, Boole, &c., have written it γ, which is clearly the most suitable, if it is to have a distinctive letter assigned to it. It has sometimes (as in *Crelle*, t. 57, p. 128) been quoted as Mascheroni's constant, but it is evident that Euler's labours have abundantly justified his claim to its being named after him.

A Mnemonic for Euler's Constant[1]

Morgan Ward

The first ten digits for Euler's constant may be remembered by aid of the following mnemonic:

These numbers proceed to a limit Euler's subtle mind discerned.

5   7   7   2 1 5   6   6   4   9

Since the next significant figure is a zero, the sentence comes to a very natural stopping place.

[1] Reprinted from the *American Mathematical Monthly*, Vol. 38, November, 1931, p. 522.

Leonhard Euler in a portrait by E. Handmann (1753)

Euler and Differentials[1]

Anthony P. Ferzola

Two recent articles by Dunham [5] and Flusser [10] have presented examples of Leonhard Euler's work in algebra. Both papers are a joy to read; watching Euler manipulate and calculate with incredible facility is a pleasure. A modern mathematician can see the logical flaws in some of the arguments, yet at the same time be aware that the mind behind it all is that of a unique master.

These two articles reminded me how much fun it is to read Euler. In researching the evolution of the differential a few years ago, I found the work of Euler refreshingly different from that of other seventeenth- and eighteenth-century mathematicians. One can read about Euler's use and misuse of infinite series in most histories of mathematics (e.g., [2, pp. 486–490]). This paper offers a glimpse at how Euler used infinitesimals and infinite series to compute differentials for the elementary functions encountered in a typical undergraduate calculus sequence. I hope the reader of this brief survey of Euler's work with differentials will seek out original sources such as [8] and [9]. As Harold Edwards [7] has cogently argued, we have much to learn from reading the masters.

Euler and the 18th Century

Euler (1707–1783) was the most prolific and one of the most influential mathematicians who ever lived. He made major contributions to both pure and applied mathematics and his collected works amount to over 70 volumes. So strong was his influence that historians like Boyer [2] and Edwards [6] refer to the eighteenth century as the Age of Euler.

Euler made the function concept fundamental in analysis. He saw a function as both any quantity depending on variables and also as any algebraic combination of constants and variables (including infinite sums or products). This is obviously not a modern definition of a function. Still, Euler used his function concept to maximal advantage. As we examine some of Euler's computations, keep in mind the immense insight and unity he achieved with the function approach—a point of view we now take for granted.

---

[1]Reprinted from the *College Mathematics Journal*, Vol. 25, March, 1994, pp. 102–111. This article received the George Pólya Award in 1995.

In his *Introducio in analysin infinitorum* (1748), one sees the first systematic inter-pretation of logarithms as exponents. Prior to Euler, logarithms were typically viewed as terms of an arithmetic series in one-to-one correspondence with terms of a geometric series [3]. Euler viewed trigonometric functions as numerical ratios rather than as ratios of line segments. He also studied properties of the elementary transcendental functions by the frequent use of their infinite series expansions [6, p. 270]. Euler often used infinite series indiscriminately, without regard to questions of convergence.

Euler's understanding and use of differentials within the framework of functions is the focus of this paper. Before presenting his work, a word about the differential before Euler.

For Leibniz (1646–1716) the differentials dx and dy were, as the name suggests, (infinitesimal) differences in the abscissa x and the ordinate y, respectively [4, pp. 70–76]. The infinitesimal was considered to be a number smaller than any positive number. The omission of the "even smaller" higher-order infinitesimals such as $(dx)^2$ or $dx\, dy$, which were deemed negligible relative to dx and dy, was basic to his methods. So powerful were the notation and methods that the differential calculus was truly a *differential* calculus for nearly one and a half centuries: The differential (and not the derivative) was the main object of study.

Leibniz gave other interpretations of the differential, but the mathematicians working in the early eighteenth century tended to favor Leibniz's formulation of a differential as an infinitesimal. It appears in the work of Johann Bernoulli (1667–1748) and in the first calculus textbook, *Analyse des infiniment petits pour l'intelligence des lignes courbes* (1696), which was written by L'Hôpital and which made free use of Bernoulli's ideas (see [18, p. 315]). Euler was one of Bernoulli's pupils.

Many of Euler's results and infinite series discussed below were known to Newton, Leibniz, Bernoulli, and others. Euler's work with differentials is unique, however, in his definition of infinitesimals as absolute zeros and in his heavy reliance on infinite series to *develop* his differential calculus.

Differentials as Absolute Zeros

In his *Institutiones calculi differentialis* (1755), Euler stated: "To those who ask what the infinitely small quantity in mathematics is, we answer it is actually equal to zero" [18, p. 384]. Euler felt that the view of the infinitesimal as zero adequately removed the mystery and ambiguity of statements such as "The infinitesimal is smaller than any given quantity" or the postulate of Johann Bernoulli that "Adding an infinitesimal to a quantity leaves the quantity unchanged."

Euler then said that the quotient $0/0$ can actually take on any value because

$$n \cdot 0 = 0$$

for all real n and therefore, he concluded,

$$\frac{n}{1} = \frac{0}{0}. \tag{1}$$

He noted that if two zeros can have an arbitrary ratio, then different symbols should be used for the zero in the numerator and the zero in the denominator of the fraction on the right-hand side of equation (1). It is here that Euler introduced the Leibnizian notation of differentials.

Euler denoted an infinitely small quantity by dx. Here $dx = 0$ and $a\,dx = 0$ for any finite quantity a. But for Euler these two zeros are different zeros that cannot be confused when the ratio $a\,dx/dx = a$ is investigated [18, p. 385]. In a similar way dy/dx can denote a finite ratio even though dx and dy are zero. "Thus for Euler the calculus was simply the determination of the ratios of evanescent increments—a heuristic procedure for finding the value of the expression $0/0$" [1].

The neglect of higher-order infinitesimals was also explained employing quotients. Noting that $dx = 0$ and $(dx)^2 = 0$, where $(dx)^2$ is a zero (or infinitesimal) of second order, Euler reasoned that

$$dx + (dx)^2 = dx$$

because

$$\frac{dx + (dx)^2}{dx} = 1 + dx = 1.$$

By the same reasoning, Euler established that

$$dx + (dx)^{n+1} = dx$$

for all $n > 0$. The omission of higher-order differentials was frequently utilized by Euler in finding the differential dy, where y is a function of x.

Computations with Elementary Functions

The computations discussed in this section are all found in Euler's *Institutiones calculi differentialis*. Their most noteworthy feature is the use of power series expressions for functions from the outset, with no mention of questions of convergence. Thus, whereas in modern textbooks the justification of such infinite series expansions is an advanced topic in differential calculus, for Euler they were the foundation for the calculation of derivatives.

To find dy if $y = x^n$ (n any real number), Euler used the binomial expansion [9, p. 99]. If x is increased by an infinitesimal amount dx, then y experiences a change of dy where

$$\begin{aligned}
dy &= (x + dx)^n - x^n \\
&= nx^{n-1}\,dx + \frac{n(n-1)}{1}x^{n-2}(dx)^2 + \cdots \\
&= nx^{n-1}\,dx
\end{aligned}$$

upon the omission of the higher-order infinitesimals $(dx)^2$, etc. Newton and Leibniz did similar computations for finding the derivative of $y = x^n$, Leibniz using a comparable differential argument while Newton worked with fluxions [6, p. 192]. Within the rigorous context of

$$\lim_{\Delta x \to 0} \frac{(x + \Delta x)^n - x^n}{\Delta x}$$

we all use the essence of this computation (for positive integer powers of x) in our first semester calculus courses.

Euler derived the product rule as follows:

$$d(pq) = (p + dp)(q + dq) - pq$$
$$= p\,dq + q\,dp + dp\,dq$$
$$= p\,dq + q\,dp$$

where the last step is due to the omission of the higher-order infinitesimal $dp\,dq$. Similar computations were done by Leibniz [4, p. 143]. This argument is analogous to the proof of the product rule still found in a few present-day textbooks (e.g., [12]).

Euler's derivation of the quotient rule is unique in its use of a geometric series [9, p. 103]:

$$\frac{1}{q + dq} = \frac{1}{q}\left(\frac{1}{1 + dq/q}\right)$$
$$= \frac{1}{q}\left(1 - \frac{dq}{q} + \frac{dq^2}{q^2} - \cdots\right)$$
$$= \frac{1}{q} - \frac{dq}{q^2}.$$

Then

$$d\left(\frac{p}{q}\right) = \frac{p + dp}{q + dq} - \frac{p}{q}$$
$$= (p + dp)\frac{1}{q + dq} - \frac{p}{q}$$
$$= (p + dp)\left(\frac{1}{q} - \frac{dq}{q^2}\right) - \frac{p}{q}$$
$$= \frac{dp}{q} - \frac{p\,dq}{q^2}$$
$$= \frac{q\,dp - p\,dq}{q^2}.$$

In chapter 6, Euler found the differentials of transcendental functions. For computing the differential of the natural logarithm (which he denoted by the single letter "ℓ" but which we will denote by the usual "log"), Euler used Mercator's series [9, p. 122]:

$$\log(1 + z) = z - \frac{z^2}{2} + \frac{z^3}{3} - \cdots.$$

Given $y = \log(x)$ then

$$dy = \log(x + dx) - \log(x)$$
$$= \log\left(1 + \frac{dx}{x}\right)$$
$$= \frac{dx}{x} - \frac{(dx)^2}{2x^2} + \frac{(dx)^3}{3x^3} - \cdots$$
$$= \frac{dx}{x}.$$

To illustrate the chain rule, Euler did many examples. For instance, if $y = \log(x^n)$ then letting $p = x^n$ yields $y = \log(p)$, which implies that $dy = dp/p$ where $dp = nx^{n-1}\, dx$. Thus $dy = n\, dx/x$.

Euler's computation of dy for $y = \log(x)$ can be found in a modern nonstandard analysis text [15, p. 65]. This may seem unremarkable since nonstandard analysis was developed by Abraham Robinson in the mid-twentieth century to place the notion of infinitesimals and their manipulation on solid logical ground. In fact, it is rare to find nonstandard analysis arguments that are exactly like Euler's, because nonstandard analysis arguments are rarely done in the context of infinite series (see [11] and [16]).

As an example of Euler's work with trigonometric functions, consider the computation of dy for $y = \sin x$ [9, p. 132]. For this purpose he explicitly used the sine and cosine series

$$\sin x = x - \frac{x^3}{3!} + \frac{x^5}{5!} - \cdots \tag{2}$$

$$\cos x = 1 - \frac{x^2}{2!} + \frac{x^4}{4!} - \cdots \tag{3}$$

to show that $\sin(dx) = dx$ and $\cos(dx) = 1$. He obtained these results by substituting dx into (2) and (3) and ignoring higher-order differentials. He also employed the trigonometric identity

$$\sin(a + b) = \sin a \cos b + \sin b \cos a. \tag{4}$$

Thus, using (4):

$$\begin{aligned}
dy &= \sin(x + dx) - \sin x \\
&= \sin x \cos dx + \sin dx \cos x - \sin x \\
&= \sin x + \cos x\, dx - \sin x \\
&= \cos x\, dx.
\end{aligned}$$

This is the most beautifully efficient computation of all those presented, especially when compared to the usual limit computation of the derivative of $y = \sin x$. There one needs to work as follows:

$$\lim_{\Delta x \to 0} \frac{\sin(x + \Delta x) - \sin x}{\Delta x} = \lim_{\Delta x \to 0} \frac{\sin x \cos(\Delta x) + \sin(\Delta x) \cos x - \sin x}{\Delta x}$$

$$= \cos x \lim_{\Delta x \to 0} \frac{\sin(\Delta x)}{\Delta x} + \sin x \lim_{\Delta x \to 0} \frac{\cos(\Delta x) - 1}{\Delta x}$$

Then $y' = \cos x$ is obtained using two limits (which must be proven):

$$\lim_{x \to 0} \frac{\sin x}{x} = 1 \tag{5}$$

and

$$\lim_{x \to 0} \frac{\cos x - 1}{x} = 0.$$

The first of these limits is captured in Euler's equation $\sin(dx) = dx$. The second limit is comparable to Euler's equation $\cos(dx) = 1$ or $\cos(dx) - 1 = 0$. Although

Euler's derivation is computationally more compact than the standard modern approach, the latter is logically sound. Any method for differentiating the sine function must deal in particular with (5). This is proven geometrically, since in the standard modern approach one defines at the outset the geometric meaning of the trigonometric functions (i.e., cosine and sine parametrize the unit circle). The proof of (5) is relatively easy when compared to the difficulty involved in showing the geometric meaning of the functions Euler defined (without regard to questions of convergence) as the sums of the power series (2) and (3).

In Euler's three-volume *Institutiones calculi integralis* (1768–1770), he defined integration, like Leibniz and Johann Bernoulli, as the formal inverse of the differential. He used the integral symbol and wrote, for example,

$$\int n x^{n-1}\, dx = x^n$$

$$\int \frac{dx}{x} = \log x$$

$$\int \cos x\, dx = \sin x$$

all plus or minus an appropriate constant. The first volume of this work reads like a modern calculus textbook chapter on techniques of integration. Integration by substitution, by parts, by partial fractions, and by trigonometric substitution are all illustrated in a logical and systematic way. Undoubtedly, Euler's well-organized and all-encompassing use of differentials in a function context did much to solidify the popularity of the differential and integral notations on the continent.

The Total Differential

Euler's *Institutiones calculi differentialis* was the first systematic exposition of the calculus of functions of several variables. He understood a function of n variables to be any finite or infinite expression involving these variables. As soon as he introduced these functions, Euler addressed the question of the relationship among the differentials of all the variables involved.

He obtained the result that if

$$V = f(x, y, z)$$

then

$$dV = p\, dx + q\, dy + r\, dz,$$

where p, q, and r are all functions of x, y, and z [9, pp. 144–145]. He arrived at this formula in an interesting way. If X is a function of x alone and is increased by an infinitesimal amount dx, then

$$dX = P\, dx$$

by the usual one-variable argument. Similarly, if Y and Z are functions of y alone and z alone respectively, then

$$dY = Q\, dy$$

99

CAPUT V.

DE DIFFERENTIATIONE FUNCTIONUM
ALGEBRAICARUM UNICAM
VARIABILEM INVOLVENTIUM.

152.

Quia quantitatis variabilis x differentiale est $= dx$ erit x in proximum promovendo $x^1 = x + dx$. Quare si fuerit y quaecunque functio ipsius x, si in ea loco x ponatur $x + dx$, ea abibit in y^1, atque differentia $y^1 - y$ dabit differentiale ipsius y. Si igitur ponamus $y = x^n$ fiet

$$y^1 = (x + dx)^n = x^n + nx^{n-1} dx + \frac{n(n-1)}{1 \cdot 2} x^{n-2} dx^2 + \&c.$$

eritque ergo

$$dy = y^1 - y = nx^{n-1} dx + \frac{n(n-1)}{1 \cdot 2} x^{n-2} dx^2 + \&c.$$

At in hac expressione terminus secundus cum reliquis sequentibus prae primo evanescit, eritque idcirco $nx^{n-1} dx$ differentiale ipsius x^n, seu $d.x^n = nx^{n-1} dx$. Unde si a sit numerus seu quantitas constans, erit quoque $d.ax^n = nax^{n-1} dx$. Cuiuscunque ergo ipsius x potestatis differentiale invenitur, multiplicando eam per exponentem, dividendo per x, & reliquum per dx multiplicando, quae regula facile memoria retinetur.

153. Cognito differentiali primo ipsius x^n, ex eo facile differentiale secundum reperitur, dummodo, ut hic constanter assumemus, differentiale dx constans statuatur. Cum enim in differentiali $nx^{n-1} dx$ factor ndx sit constans, alterius factoris x^{n-1} differentiale sumi debet, quod proinde erit $(n-1)x^{n-2} dx$. Hoc ergo per ndx multiplicatum dabit differen-

N 2   ren-

Title page, Chapter 5 of *Institutiones calculi differentialis*

and

$$dZ = R\,dz.$$

If $V = X + Y + Z$ (i.e., a special function of three variables) then

$$dV = dX + dY + dZ$$
$$= P\,dx + Q\,dy + R\,dz.$$

If $V = XYZ$, then

$$dV = (X + P\,dx)(Y + Q\,dy)(Z + R\,dz) - XYZ.$$

This simplifies (upon omission of higher-order differential terms such as $ZPQ\,dx\,dy$) to

$$dV = YZP\,dx + XZQ\,dy + XYR\,dz.$$

From these two examples, Euler expected that any algebraic expression of x, y, and z has differential

$$dV = p\,dx + q\,dy + r\,dz \tag{6}$$

because a function of three variables can be thought of as a sum of products of these variables. He generalized the result for any number of variables [9, p. 146].

Later in the same work, he addressed the concept of partial differentiation [9, pp. 156–157]. If y and z are held constant, then by equation (6)

$$dV = p\,dx$$

as there is no change in y or z. (Notice how, for Euler, no change in y is not the same as saying dy is the infinitesimal change in y, even though he defined infinitesimals as being zero.) He then wrote

$$p = (dV/dx),$$

where the parentheses about the quotient remind one that p equals the differential of V (with only the x being variable) divided by dx. Similar meanings apply to $q = (dV/dy)$ and $r = (dV/dz)$. This was Euler's notation and understanding of the concept of partial derivatives. The current symbol ∂ dates from the 1840's [14]. Obviously, (6) becomes

$$dV = (dV/dx)\,dx + (dV/dy)\,dy + (dV/dz)\,dz,$$

although Euler did not explicitly write this.

It is worth noting that Euler's exposition of differentials for functions of several variables immediately followed his work with differentials for functions of one variable. Exploring the differential calculus for both single and multivariable functions before passing on to integration is an old idea which I think has merit. It gives the calculus sequence a stronger focus and unity, by concentrating effort on one basic concept (the derivative) in various settings before moving on to its inverse. A recent textbook by Small and Hosack [17] takes this approach. Perhaps we will see more of this, especially since computer algebra systems such as *Derive*, *Maple*, and *Mathematica* have taken the pain out of such tasks as surface sketching.

Differentials in Multiple Integrals

Euler frequently let his readers in on his thought processes, even when the procedures seemed fruitless. This was mathematics being done for all to see, not a slick modern textbook treatment. There was no taking down the scaffolding *à la* Gauss.

Euler, in *De formulis integralibus duplicatis* (1769), gave one of the first clear discussions of double integrals. In the first half of the eighteenth century, $\iint f(x, y)\,dx\,dy$ denoted the solution of $\partial^2 z/\partial x\,\partial y = f(x, y)$ obtained by antidifferentiation. Euler supplemented this by providing a (thoroughly modern) procedure for evaluating definite double integrals over a bounded domain **R** enclosed by arcs in the xy plane. Euler used iterated integrals:

$$\iint_{\mathbf{R}} f(x, y)\, dx\, dy = \int_a^b dx \int_{f_1(x)}^{f_2(x)} z\, dy,$$

where $z = f(x, y)$. For $z > 0$, Euler saw this as a volume, since $\int z\, dy$ gives the area of a "slice" (parallel to the y-axis) of the three-dimensional region above **R** and under $z = f(x, y)$, and the following integration with respect to x "adds up the slices" to yield the volume [8, p. 293]. This is perhaps the first time Leibniz's powerful differential notation was used in tandem with a volume argument employing Cavalieri's method of indivisibles [2, p. 361].

Euler also interpreted $dx\, dy$ as an "area element" of **R**. That is, **R** is made up of an infinite set of infinitesimal area elements $dx\, dy$. This is most clearly seen when Euler attempted to change variables [8, pp. 302–303]. And it was here that Euler ran into difficulties.

He reasoned that if $dx\, dy$ is an area element and we change variables via the transformation

$$x = x(t, v) = a + mt + v\sqrt{1 - m^2}$$
$$y = y(t, v) = b + t\sqrt{1 - m^2} - mv$$

(a translation by the vector (a, b), a clockwise rotation through the angle α, where $\cos \alpha = m$, and a reflection through the x-axis), then $dx\, dy$ should equal $dt\, dv$. But

$$dx = m\, dt + dv\sqrt{1 - m^2}$$
$$dy = dt\sqrt{1 - m^2} - m\, dv$$

and multiplication gives

$$dx\, dy = m\sqrt{1 - m^2}\,(dt)^2 + (1 - 2m^2)\, dt\, dv - m\sqrt{1 - m^2}\,(dv)^2.$$

Euler rejected this as wrong and meaningless. (How many calculus students wonder, explicitly or implicitly, why we cannot just multiply the differential forms for dx and dy?) Euler decided to attack the problem in a formal non-geometric way, not using area elements but rather by changing variables one at a time (for details, see [13]). In this way he arrived at the correct general result:

$$\iint f(x, y)\, dx\, dy = \iint f\big(x(t, v), y(t, v)\big) \left| \frac{\partial(x, y)}{\partial(t, v)} \right| dt\, dv.$$

In 1899, another great mathematician with a computational flair, Élie Cartan, arrived at the straightforward multiplicative result Euler sought, by using Grassmann's exterior product with differential forms. This is a formal product where the usual distributive laws hold but with the conditions that

$$dx\, dx = dy\, dy = 0$$

and

$$dx\, dy = -dy\, dx$$

(see [17, p. 514], and [13]). Thus, for Euler's differentials

$$dx \, dy = \left(m \, dt + dv\sqrt{1-m^2}\right)\left(dt\sqrt{l-m^2} - m \, dv\right)$$
$$= dt \, dt \, m\sqrt{1-m^2} - m^2 \, dt \, dv + (1-m^2) \, dv \, dt - dv \, dv \, m\sqrt{1-m^2}$$
$$= -m^2 \, dt \, dv + dv \, dt(1-m^2)$$
$$= -m^2 \, dt \, dv - dt \, dv(1-m^2)$$
$$= -dt \, dv.$$

The minus sign appears because the transformation (involving a reflection) does not preserve orientation. In general, given any transformation from the tv-plane to the xy-plane, the exterior product yields

$$dx \, dy = \frac{\partial(x, y)}{\partial(t, v)} \, dt \, dv.$$

Conclusion

Even in this rudimentary survey of Euler's work with differentials in calculus, it is fascinating to watch a genius grapple with an ambiguous concept (infinitesimal) and attempt to clarify it (absolute zero)—however flawed the attempt. Reading Euler has enriched my teaching of the calculus by keeping me mindful that my students are tackling a subject whose foundations humbled the greatest minds of the past. Even the seemingly fruitless paths can be instructive, as we have seen. It took mathematicians about 150 years to come up with the exterior product for differential forms that Euler needed for the change of variables formula in multiple integrals. How many other Eulerian dead ends may be worth pursuing? Again, the advice of Harold Edwards [7] points the way for the teacher and the researcher: "Read the masters!"

References

1. C. Boyer, *The History of Calculus*, Dover, New York, 1959, p. 244.

2. ——, *A History of Mathematics*, Wiley, New York, 1968.

3. F. Cajori, *A History of Mathematics*, Macmillan, London, 1931, p. 235.

4. J. Child, *The Early Mathematical Manuscripts of Leibniz*, Open Court, Chicago, 1920.

5. W. Dunham, Euler and the fundamental theorem of algebra, *College Mathematics Journal* 22 (1991) 282–293.

6. C. H. Edwards, Jr., *The Historical Development of the Calculus*, Springer-Verlag, New York, 1979.

7. H. M. Edwards, Read the masters! in L. A. Steen, ed., *Mathematics Tomorrow*, Springer-Verlag, New York, 1981.

8. L. Euler, *De formulis integralibus duplicatis*, in A. Gutzmer, ed., *Opera omnia: L. Euler*, Vol. 17, Series 1, B. G. Teubner, Leipzig, 1915.

9. ——, *Institutiones calculi differentialis*, in C. Kowalewski, ed., *Opera Omnia: L. Euler*, Vol. X, Series 1, B. G. Teubner, Leipzig, 1913.

10. P. Flusser, Euler's amazing way to solve equations, *Mathematics Teacher* 85 (1992) 224–227.

11. J. M. Henle and E. M. Kleinberg, *Infinitesimal Calculus*, MIT Press, Cambridge, 1979.

12. D. Hughes-Hallett, A. M. Gleason, et. al., *Calculus: Preliminary Edition*, Wiley, New York, 1992, p. 239.

13. V. Katz, Change of variables in multiple integrals: Euler to Cartan, *Mathematics Magazine* 55 (1982) 3–11.

14. ——, The history of differential forms from Clairaut to Poincaré, *Historia Mathematica* 8 (1981) 161–188, p. 161.

15. A. Robert, *Nonstandard Analysis*, Wiley, New York, 1988, p. 65.

16. A. Robinson, *Non-Standard Analysis*, North-Holland, Amsterdam, 1970.

17. D. B. Small and J. M. Hosack, *Calculus: An Integrated Approach*, McGraw-Hill, New York, 1990.

18. D. J. Struik, *A Source Book in Mathematics 1200–1800*, Harvard University Press, Cambridge, 1969.

Leonhard Euler

Leonhard Euler's Integral: A Historical Profile of the Gamma Function[1]

Philip J. Davis

Many people think that mathematical ideas are static. They think that the ideas originated at some time in the historical past and remain unchanged for all future times. There are good reasons for such a feeling. After all, the formula for the area of a circle was πr^2 in Euclid's day and at the present time is still πr^2. But to one who knows mathematics from the inside, the subject has rather the feeling of a living thing. It grows daily by the accretion of new information, it changes daily by regarding itself and the world from new vantage points, it maintains a regulatory balance by consigning to the oblivion of irrelevancy a fraction of its past accomplishments.

The purpose of this essay is to illustrate this process of growth. We select one mathematical object, the gamma function, and show how it grew in concept and in content from the time of Euler to the recent mathematical treatise of Bourbaki, and how, in this growth, it partook of the general development of mathematics over the past two and a quarter centuries. Of the so-called "higher mathematical functions," the gamma function is undoubtedly the most fundamental. It is simple enough for juniors in college to meet but deep enough to have called forth contributions from the finest mathematicians. And it is sufficiently compact to allow its profile to be sketched within the space of a brief essay.

The year 1729 saw the birth of the gamma function in a correspondence between a Swiss mathematician in St. Petersburg and a German mathematician in Moscow. The former: Leonhard Euler (1707–1783), then 22 years of age, but to become a prodigious mathematician, the greatest of the 18th century. The latter: Christian Goldbach (1690–1764), a savant, a man of many talents and in correspondence with the leading thinkers of the day. As a mathematician he was something of a dilettante, yet he was a man who bequeathed to the future a problem in the theory of numbers so easy to state and so difficult to prove that even to this day it remains on the mathematical horizon as a challenge.

The birth of the gamma function was due to the merging of several mathematical streams. The first was that of interpolation theory, a very practical subject largely the

[1]Reprinted from the *American Mathematical Monthly*, Vol. 66, December, 1959, pp. 849–869.

product of English mathematicians of the 17th century but which all mathematicians enjoyed dipping into from time to time. The second stream was that of the integral calculus and of the systematic building up of the formulas of indefinite integration, a process which had been going on steadily for many years. A certain ostensibly simple problem of interpolation arose and was bandied about unsuccessfully by Goldbach and by Daniel Bernoulli (1700–1784) and even earlier by James Stirling (1692–1770). The problem was posed to Euler. Euler announced his solution to Goldbach in two letters which were to be the beginning of an extensive correspondence which lasted the duration of Goldbach's life. The first letter dated October 13, 1729, dealt with the interpolation problem, while the second dated January 8, 1730, dealt with integration and tied the two together. Euler wrote Goldbach the merest outline, but within a year he published all the details in an article *De progressionibus transcendentibus seu quarum termini generales algebraice dari nequeunt*. This article can now be found reprinted in Volume I_{14} of Euler's *Opera Omnia*.

Since the interpolation problem is the easier one, let us begin with it. One of the simplest sequences of integers which leads to an interesting theory is $1, 1 + 2, 1 + 2 + 3$, $1 + 2 + 3 + 4, \ldots$. These are the triangular numbers, so called because they represent the number of objects which can be placed in a triangular array of various sizes. Call the nth one T_n. There is a formula for T_n which is learned in school algebra: $T_n = \frac{1}{2}n(n + 1)$.

What, precisely, does this formula accomplish? In the first place, it simplifies computation by reducing a large number of additions to three fixed operations: one of addition, one of multiplication, and one of division. Thus, instead of adding the first hundred integers to obtain T_{100}, we can compute $T_{100} = \frac{1}{2}(100)(100 + 1) = 5050$. Secondly, even though it doesn't make literal sense to ask for, say, the sum of the first $5\frac{1}{2}$ integers, the formula for T_n produces an answer to this. For whatever it is worth the formula yields $T_{5\frac{1}{2}} = \frac{1}{2}(5\frac{1}{2})(5\frac{1}{2} + 1) = 17\frac{7}{8}$. In this way, the formula extends the scope of the original problem to values of the variable other than those for which it was originally defined and solves the problem of interpolating between the known elementary values.

This type of question, one which asks for an extension of meaning, cropped up frequently in the 17th and 18th centuries. Consider, for instance, the algebra of exponents. The quantity a^m is defined initially as the product of m successive a's. This definition has meaning when m is a positive integer, but what would $a^{5\frac{1}{2}}$ be? The product of $5\frac{1}{2}$ successive a's? The mysterious definitions $a^0 = 1$, $a^{m/n} = \sqrt[n]{a^m}$, $a^{-m} = 1/a^m$ which solve this enigma and which are employed so fruitfully in algebra were written down explicitly for the first time by Newton in 1676. They are justified by a utility which derives from the fact that the definition leads to continuous exponential functions and that the law of exponents $a^m \cdot a^n = a^{m+n}$ becomes meaningful for all exponents whether positive integers or not.

Other problems of this type proved harder. Thus, Leibnitz introduced the notation d^n for the nth iterate of the operation of differentiation. Moreover, he identified d^{-1} with $\int$ and d^{-n} with the iterated integral. Then he tried to breathe some sense into the symbol d^n when n is any real value whatever. What, indeed, is the $5\frac{1}{2}$th derivative of a function? This question had to wait almost two centuries for a satisfactory answer.

But to return to our sequence of triangular numbers. If we change the plus signs to multiplication signs we obtain a new sequence: $1, 1 \cdot 2, 1 \cdot 2 \cdot 3, \ldots$. This is the sequence

The Factorials								
n:	1	2	3	4	5	6	7	8	$\cdots$
$n!$:	1	2	6	24	120	720	5040	40,320	$\cdots$

Figure 1.

Intelligence Test

Question: What number should be inserted in the lower line half way between the upper 5 and 6? Euler's Answer: 287.8852.... Hadamard's Answer: 280.3002....

of factorials. The factorials are usually abbreviated 1!, 2!, 3!, ... and the first five are 1, 2, 6, 24, 120. They grow in size very rapidly. The number 100! if written out in full would have 158 digits. By contrast, $T_{100} = 5050$ has a mere four digits. Factorials are omnipresent in mathematics; one can hardly open a page of mathematical analysis without finding it strewn with them. This being the case, is it possible to obtain an easy formula for computing the factorials? And is it possible to interpolate between the factorials? What should $5\frac{1}{2}!$ be? (See Figure 1.) This is the interpolation problem which led to the gamma function, the interpolation problem of Stirling, of Bernoulli, and of Goldbach. As we know, these two problems are related, for when one has a formula there is the possibility of inserting intermediate values into it. And now comes the surprising thing. There is no, in fact there can be, no formula for the factorials which is of the simple type found for T_n. This is implicit in the very title Euler chose for his article. Translate the Latin and we have *On transcendental progressions whose general term cannot be expressed algebraically.* The solution to factorial interpolation lay deeper than "mere algebra." Infinite processes were required.

In order to appreciate a little better the problem confronting Euler it is useful to skip ahead a bit and formulate it in an up-to-date fashion: find a reasonably simple function which at the integers 1, 2, 3, ... takes on the factorial values 1, 2, 6, ... Now today, a function is a relationship between two sets of numbers wherein to a number of one set is assigned a number of the second set. What is stressed is the relationship and not the nature of the rules which serve to determine the relationship. To help students visualize the function concept in its full generality, mathematics instructors are accustomed to draw a curve full of twists and discontinuities. The more of these the more general the function is supposed to be. Given, then, the points $(1, 1)$, $(2, 2)$, $(3, 6)$, $(4, 24)$, ... and adopting the point of view wherein "function" is what we have just said, the problem of interpolation is one of finding a curve which passes through the given points. This is ridiculously easy to solve. It can be done in an unlimited number of ways. Merely take a pencil and draw some curve—any curve will do—which passes through the points. Such a curve automatically defines a function which solves the interpolation problem. In this way, too free an attitude as to what constitutes a function solves the problem trivially and would enrich mathematics but little. Euler's task was different. In the early 18th century, a function was more or less synonymous with a formula, and by a formula was meant an expression which could be derived from elementary manipulations with addition, subtraction, multiplication, division, powers, roots, exponentials, logarithms, differentiation, integration, infinite series, i.e., one which came from the ordinary processes of mathematical analysis. Such a formula was called an *expressio analytica*, an analytical expression. Euler's task was to find, if he could, an analytical expression arising naturally from the corpus of mathematics which would yield factorials when a positive integer was inserted, but which would still be meaningful for other values of the variable.

It is difficult to chronicle the exact course of scientific discovery. This is particularly true in mathematics where one traditionally omits from articles and books all accounts of false starts, of the initial years of bungling, and where one may develop one's topic forward or backward or sideways in order to heighten the dramatic effect. As one distinguished mathematician put it, a mathematical result must appear straight from the heavens as a *deus ex machina* for students to verify and accept but not to comprehend. Apparently, Euler, experimenting with infinite products of numbers, chanced to notice that if n is a positive integer,

$$\left[\left(\frac{2}{1}\right)^n \frac{1}{n+1}\right]\left[\left(\frac{3}{2}\right)^n \frac{2}{n+2}\right]\left[\left(\frac{4}{3}\right)^n \frac{3}{n+3}\right]\cdots = n!. \tag{1}$$

Leaving aside all delicate questions as to the convergence of the infinite product, the reader can verify this equation by cancelling out all the common factors which appear in the top and bottom of the left-hand side. Moreover, the left-hand side is defined (at least formally) for all kinds of n other than negative integers. Euler noticed also that when the value $n = \frac{1}{2}$ is inserted, the left-hand side yields (after a bit of manipulation) the famous infinite product of the Englishman John Wallis (1616–1703):

$$\left(\frac{2\cdot 2}{1\cdot 3}\right)\left(\frac{4\cdot 4}{3\cdot 5}\right)\left(\frac{6\cdot 6}{5\cdot 7}\right)\left(\frac{8\cdot 8}{7\cdot 9}\right)\cdots = \frac{\pi}{2}. \tag{2}$$

With this discovery Euler could have stopped. His problem was solved. Indeed, the whole theory of the gamma function can be based on the infinite product (1) which today is written more conventionally as

$$\lim_{m\to\infty} \frac{m!(m+1)^n}{(n+1)(n+2)\cdots(n+m)}. \tag{3}$$

However, he went on. He observed that his product displayed the following curious phenomenon: for some values of n, namely integers, it yielded integers, whereas for another value, namely $n = \frac{1}{2}$, it yielded an expression involving π. Now π meant circles and their quadrature, and quadratures meant integrals, and he was familiar with integrals which exhibited the same phenomenon. It therefore occurred to him to look for a transformation which would allow him to express his product as an integral.

He took up the integral $\int_0^1 x^e(1-x)^n \, dx$. Special cases of it had already been discussed by Wallis, by Newton, and by Stirling. It was a troublesome integral to handle, for the indefinite integral is not always an elementary function of x. Assuming that n is an integer, but that e is an arbitrary value, Euler expanded $(1-x)^n$ by the binomial theorem, and without difficulty found that

$$\int_0^1 x^e(1-x)^n \, dx = \frac{1\cdot 2\cdots n}{(e+1)(e+2)\cdots(e+n+1)}. \tag{4}$$

Euler's idea was now to isolate the $1\cdot 2\cdots n$ from the denominator so that he would have an expression for $n!$ as an integral. He proceeds in this way. (Here we follow Euler's own formulation and nomenclature, marking with an * those formulas which occur in the original paper. Euler wrote a plain $\int$ for $\int_0^1$.) He substituted f/g for e and found

$$\int_0^1 x^{f/g}(1-x)^n\,dx = \frac{g^{n+1}}{f+(n+1)g}\cdot\frac{1\cdot 2\cdots n}{(f+g)(f+2\cdot g)\cdots(f+n\cdot g)}. \tag{5}$$

And so,

$$\frac{1\cdot 2\cdots n}{(f+g)(f+2\cdot g)\cdots(f+n\cdot g)} = \frac{f+(n+1)g}{g^{n+1}}\int x^{f/g}\,dx(1-x)^n \tag{6}*$$

He observed that he could isolate the $1\cdot 2\cdots n$ if he set $f=1$ and $g=0$ in the left-hand member, but that if he did so, he would obtain on the right an indeterminate form which he writes quaintly as

$$\int \frac{x^{1/0}\,dx(1-x)^n}{0^{n+1}}. \tag{7}*$$

He now proceeded to find the value of the expression $(7)^*$. He first made the substitution $x^{g/(f+g)}$ in place of x. This gave him

$$\frac{g}{f+g}x^{-f/(f+g)}\,dx \tag{8}*$$

in place of dx and hence, the right-hand member of $(6)^*$ becomes

$$\frac{f+(n+1)g}{g^{n+1}}\int\frac{g}{f+g}\,dx\left(1-x^{g/(f+g)}\right)^n. \tag{9}*$$

Once again, Euler made a trial setting of $f=1$, $g=0$ having presumably reduced this integral first to

$$\frac{f+(n+1)g}{(f+g)^{n+1}}\int_0^1\left(\frac{1-x^{g/(f+g)}}{g/(f+g)}\right)^n\,dx, \tag{10}$$

and this yielded the indeterminate

$$\int dx\frac{(1-x^0)^n}{0^n}. \tag{11}*$$

He now considered the related expression $(1-x^x)/z$, for vanishing z. He differentiated the numerator and denominator, as he says, by a known (l'Hôspital's) rule and obtained

$$\frac{-x^2\,dx\,lx}{dz}\qquad (lx=\log x), \tag{12}*$$

which for $z=0$ produced $-lx$. Thus,

$$\frac{1-x^0}{0} = -lx \tag{13}*$$

and

$$\frac{(1-x^0)^n}{0^n} = (-lx)^n. \tag{14}*$$

He therefore concluded that

$$n! = \int_0^1 (-\log x)^n \, dx. \tag{15}$$

This gave him what he wanted, an expression for $n!$ as an integral wherein values other than positive integers may be substituted. The reader is encouraged to formulate his own criticism of Euler's derivation.

Students in advanced calculus generally meet Euler's integral first in the form

$$\Gamma(x) = \int_0^\infty e^{-t} t^{x-1} \, dt, \qquad e = 2.71828\ldots. \tag{16}$$

This modification of the integral (15) as well as the Greek Γ is due to Adrien Marie Legendre (1752–1833). Legendre calls the integral (4) with which Euler started his derivation the first Eulerian integral and (15) the second Eulerian integral. The first Eulerian integral is currently known as the beta function and is now conventionally written

$$B(m,n) = \int_0^1 x^{m-1} (1-x)^{n-1} \, dx. \tag{17}$$

With the tools available in advanced calculus, it is readily established (how easily the great achievements of the past seem to be comprehended and duplicated!) that the integral (16) possesses meaning when $x > 0$ and thus yields a certain function $\Gamma(x)$ defined for these values. Moreover,

$$\Gamma(n+1) = n! \tag{18}$$

whenever n is a positive integer.[2] It is further established that for all $x > 0$

$$x\Gamma(x) = \Gamma(x+1). \tag{19}$$

This is the so-called recurrence relation for the gamma function and in the years following Euler it plays, as we shall see, an increasingly important role in its theory. These facts, plus perhaps the relationship between Euler's two types of integrals

$$B(m,n) = \frac{\Gamma(m)\Gamma(n)}{\Gamma(m+n)} \tag{20}$$

and the all important Stirling formula

$$\Gamma(x) \sim e^{-x} x^{x-1/2} \sqrt{2\pi}, \tag{21}$$

which gives us a relatively simple approximate expression for $\Gamma(x)$ when x is large, are about all that advanced calculus students learn of the gamma function. Chronologically speaking, this puts them at about the year 1750. The play has hardly begun.

Just as the simple desire to extend factorials to values in between the integers led to the discovery of the gamma function, the desire to extend it to negative values and to complex values led to its further development and to a more profound interpretation.

---

[2]Legendre's notation shifts the argument. Gauss introduced a notation $\pi(x)$ free of this defect. Legendre's notation won out, but continues to plague many people. The notations Γ, π, and ! can all be found today.

Naive questioning, uninhibited play with symbols may have been at the very bottom of it. What is the value of $(-5\frac{1}{2})!$? What is the value of $\sqrt{-1}!$? In the early years of the 19th century, the action broadened and moved into the complex plane (the set of all numbers of the form $x + iy$, where $i = \sqrt{-1}$) and there it became part of the general development of the theory of functions of a complex variable that was to form one of the major chapters in mathematics. The move to the complex plane was initiated by Karl Friedrich Gauss (1777–1855), who began with Euler's product as his starting point. Many famous names are now involved and not just one stage of action but many stages. It would take too long to record and describe each forward step taken. We shall have to be content with a broader picture.

Three important facts were now known: Euler's integral, Euler's product, and the functional or recurrence relationship $x\Gamma(x) = \Gamma(x + 1)$, $x > 0$. This last is the generalization of the obvious arithmetic fact that for positive integers, $(n + 1)n! = (n + 1)!$. It is a particularly useful relationship inasmuch as it enables us by applying it over and over again to reduce the problem of evaluating a factorial of an arbitrary real number whole or otherwise to the problem of evaluating the factorial of an appropriate number lying between 0 and 1. Thus, if we write $n = 4\frac{1}{2}$ in the above formula we obtain $(4\frac{1}{2} + 1)! = 5\frac{1}{2}(4\frac{1}{2})!$. If we could only find out what $(4\frac{1}{2})!$ is, then we would know that $(5\frac{1}{2})!$ is. This process of reduction to lower numbers can be kept up and yields

$$\left(5\frac{1}{2}\right)! = \left(\frac{3}{2}\right)\left(\frac{5}{2}\right)\left(\frac{7}{2}\right)\left(\frac{9}{2}\right)\left(\frac{11}{2}\right)\left(\frac{1}{2}\right)! \tag{22}$$

and since we have $(\frac{1}{2})! = \frac{1}{2}\sqrt{\pi}$ from (1) and (2), we can now compute our answer. Such a device is obviously very important for anyone who must do calculations with the gamma function. Other information is forthcoming from the recurrence relationship. Though the formula $(n + 1)n! = (n + 1)!$ as a condensation of the arithmetic identity $(n+1)\cdot 1\cdot 2\cdots n = 1\cdot 2\cdots n\cdot(n+1)$ makes sense only for $n = 1, 2$, etc., blind insertions of other values produce interesting things. Thus, inserting $n = 0$, we obtain $0! = 1$. Inserting successively $n = -5\frac{1}{2}$, $n = -4\frac{1}{2}$, ... and reducing upwards, we discover

$$\left(-5\frac{1}{2}\right)! = \left(\frac{2}{1}\right)\left(-\frac{2}{1}\right)\left(-\frac{2}{3}\right)\left(-\frac{2}{5}\right)\left(-\frac{2}{7}\right)\left(-\frac{2}{9}\right)\left(\frac{1}{2}\right)! \tag{23}$$

Since we already know what $(\frac{1}{2})!$ is, we can compute $(-5\frac{1}{2})!$. In this way the recurrence relationship enables us to compute the values of factorials of negative numbers.

Turning now to Euler's integral, it can be shown that for values of the variable less than 0, the usual theorems of analysis do not suffice to assign a meaning to the integral, for it is divergent. On the other hand, it is meaningful and yields a value if one substitutes for x any complex number of the form $a + bi$ where $a > 0$. With such substitutions the integral therefore yields a complex-valued function which is defined for all complex numbers in the right-half of the complex plane and which coincides with the ordinary gamma function for real values. Euler's product is even stronger. With the exception of $0, -1, -2, \ldots$ any complex number whatever can be inserted for the variable and the infinite product will converge, yielding a value. And so it appears that we have at our disposal a number of methods, conceptually and operationally different for extending the domain of definition of the gamma function. Do these different methods yield the same result? They do. But why?

The answer is to be found in the notion of an analytic function. This is the focal point of the theory of functions of a complex variable and an outgrowth of the older notion of an analytical expression. As we have hinted, earlier mathematics was vague about this notion, meaning by it a function which arose in a natural way in mathematical analysis. When later it was discovered by J. B. J. Fourier (1768–1830) that functions of wide generality and functions with unpleasant characteristics could be produced by the infinite superposition of ordinary sines and cosines, it became clear that the criterion of "arising in a natural way" would have to be dropped. The discovery simultaneously forced a broadening of the idea of a function and a narrowing of what was meant by an analytic function.

Analytic functions are not so arbitrary in their behavior. On the contrary, they possess strong internal ties. Defined very precisely as functions which possess a complex derivative or equivalently as functions which possess power series expansions $a_0 + a_1(z - z_0) + a_2(z - z_0)^2 + \cdots$ they exhibit the remarkable phenomenon of "action at a distance." This means that the behavior of an analytic function over any interval no matter how small is sufficient to determine completely its behavior everywhere else; its potential range of definition and its values are theoretically obtainable from this information. Analytic functions, moreover, obey the principle of the permanence of functional relationships; if an analytic function satisfies in some portions of its region of definition a certain functional relationship, then it must do so wherever it is defined. Conversely, such a relationship may be employed to extend its definition to unknown regions. Our understanding of the process of analytic continuation, as this phenomenon is known, is based upon the work of Bernhard Riemann (1826–1866) and Karl Weierstrass (1815–1897). The complex-valued function which results from the substitution of complex numbers into Euler's integral is an analytic function. The function which emerges from Euler's product is an analytic function. The recurrence relationship for the gamma function if satisfied in some region must be satisfied in any other region to which the function can be "continued" analytically and indeed may be employed to effect such extensions. All portions of the complex plane, with the exception of the values $0, -1, -2, \ldots$ are accessible to the complex gamma function which has become the unique, analytic extension to complex values of Euler's integral (see Figure 3).

To understand why there should be excluded points observe that $\Gamma(x) = \Gamma(x + 1)/x$, and as x approaches 0, we obtain $\Gamma(0) = 1/0$. This is $+\infty$ or $-\infty$ depending whether 0 is approached through positive or negative values. The functional equation (19) then, induces this behavior over and over again at each of the negative integers. The (real) gamma function comprises an infinite number of disconnected portions opening up and down alternately. The portions corresponding to negative values are each squeezed into an infinite strip one unit in width, but the major portion which corresponds to positive x and which contains the factorials is of infinite width (see Figure 2). Thus, there are excluded points for the gamma function at which it exhibits from the ordinary (real variable) point of view a somewhat unpleasant and capricious behavior. But from the complex point of view, these points of singular behavior (singular in the sense of Sherlock Holmes) merit special study and become an important part of the story. In pictures of the complex gamma

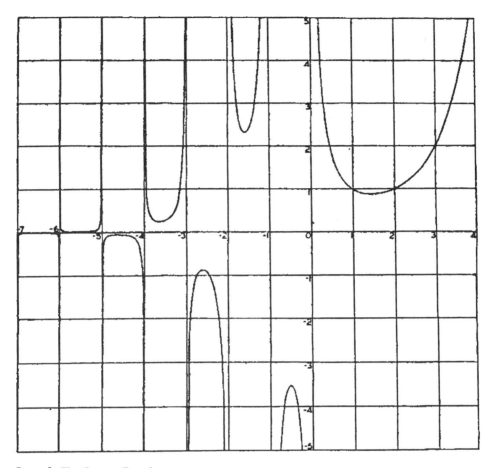

Figure 2. The Gamma Function
From: H. T. Davis, *Tables of the Higher Mathematical Functions*, vol. I, Bloomington, Indiana, 1933.

function they show up as an infinite row of "stalagmites," each of infinite height (the ones in the figure are truncated out of necessity) which become more and more needlelike as they go out to infinity (see Figure 3). They are known as poles. Poles are points where the function has an infinite behavior of especially simple type, a behavior which is akin to that of such simple functions as the hyperbola $y = 1/x$ at $x = 0$ or of $y = \tan x$ at $x = \pi/2$. The theory of analytic functions is especially interested in singular behavior, and devotes much space to the study of the singularities. Analytic functions possess many types of singularity but those with only poles are known as meromorphic. There are also functions which are lucky enough to possess no singularities for finite arguments. Such functions form an elite and are known as entire functions. They are akin to polynomials while the meromorphic functions are akin to the ratio of polynomials. The gamma function is meromorphic. Its reciprocal, $1/\Gamma(x)$, has on the contrary no excluded points. There is no trouble anywhere. At the points $0, -1, -2, \ldots$ it merely becomes zero. And the zero value which occurs an infinity of times, is strongly reminiscent of the sine.

In the wake of the extension to the complex many remarkable identities emerge, and though some of them can and were obtained without reference to complex variables, they

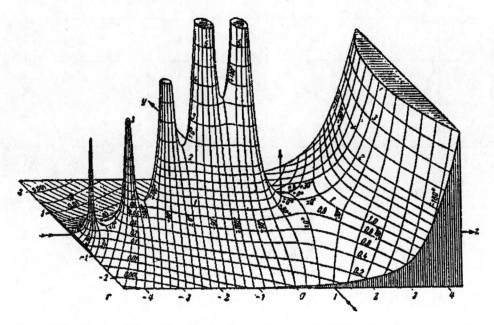

Figure 3. The Absolute Value of the Complex Gamma Function, Exhibiting the Poles at the Negative Integers
From: E. Jahnke and F. Emde, Tafeln höherer Funktionen, 4th ed., Leipzig, 1948.

acquire a far deeper and richer meaning when regarded from the extended point of view. There is the reflection formula of Euler

$$\Gamma(z)\Gamma(1-z) = \frac{\pi}{\sin \pi z}. \tag{24}$$

It is readily shown, using the recurrence relation of the gamma function, that the product $\Gamma(z)\Gamma(1-z)$ is a periodic function of period 2; but despite the fact that $\sin \pi z$ is one of the simplest periodic functions, who could have anticipated the relationship (24)? What, after all, does trigonometry have to do with the sequence 1, 2, 6, 24 which started the whole discussion? Here is a fine example of the delicate patterns which make the mathematics of the period so magical. From the complex point of view, a partial reason for the identity lies in the similarity between zeros of the sine and the poles of the gamma function.

There is the duplication formula

$$\Gamma(2z) = (2\pi)^{-1/2}2^{2z-1/2}\Gamma(z)\Gamma\left(z + \frac{1}{2}\right) \tag{25}$$

discovered by Legendre and extended by Gauss in his researches on the hypergeometric function to the multiplication formula

$$\Gamma(nz) = (2\pi)^{1/2(1-n)}n^{nz-1/2}\Gamma(z)\Gamma\left(z + \frac{1}{n}\right)\Gamma\left(z + \frac{2}{n}\right)\cdots\Gamma\left(\frac{z+n-1}{n}\right). \tag{26}$$

There are pretty formulas for the derivatives of the gamma function such as

$$d^2 \log \Gamma(z)/dz^2 = \frac{1}{z^2} + \frac{1}{(z+1)^2} + \frac{1}{(z+2)^2} + \cdots. \tag{27}$$

This is an example of a type of infinite series out of which G. Mittag-Leffler (1846–1927) later created his theory of partial fraction developments of meromorphic functions. There is the intimate relationship between the gamma function and the zeta function which has been of fundamental importance in studying the distribution of the prime numbers,

$$\zeta(z) = \zeta(1-z)\Gamma(1-z)2^z \pi^{z-1} \sin \frac{1}{2}\pi z, \qquad (28)$$

where

$$\zeta(z) = 1 + \frac{1}{2^z} + \frac{1}{3^z} + \cdots . \qquad (29)$$

This formula has some interesting history related to it. It was first proved by Riemann in 1859 and was conventionally attributed to him. Yet in 1894 it was discovered that a modified version of the identity appears in some work of Euler which had been done in 1749. Euler did not claim to have proved the formula. However, he "verified" it for integers, for $\frac{1}{2}$, and for $\frac{3}{2}$. The verification for $\frac{1}{2}$ is by direct substitution, but for all the other values, Euler works with divergent infinite series. This was more than 100 years in advance of a firm theory of such series, but with unerring intuition, he proceeded to sum them by what is now called the method of Abel summation. The case $\frac{3}{2}$ is even more interesting. There, invoking both divergent series and numerical evaluation, he came out with numerical agreement to 5 decimal places! All this work convinced him of the truth of his identity. Rigorous modern proofs do not require the theory of divergent series, but the notions of analytic continuation are crucial.

In view of the essential unity of the gamma function over the whole complex plane it is theoretically and aesthetically important to have a formula which works for all complex numbers. One such formula was supplied in 1848 by F. W. Newman:

$$1/\Gamma(z) = ze^{\gamma z}\left[(1+z)e^{-z}\right]\left[(1+z/2)e^{-z/2}\right]\ldots, \quad \text{where } \gamma = .5772156649\ldots . \quad (30)$$

This formula is essentially a factorization of $1/\Gamma(z)$ and is much the same as a factorization of polynomials. It exhibits clearly where the function vanishes. Setting each factor equal to zero we find that $1/\Gamma(z)$ is zero for $z = 0$, $z = -1$, $z = -2$, ... In the hands of Weierstrass, it became the starting point of his particular discussion of the gamma function. Weierstrass was interested in how functions other than polynomials may be factored. A number of isolated factorizations were then known. Newman's formula (30) and the older factorization of the sine

$$\sin \pi z = \pi z(1-z^2)\left(1 - \frac{z^2}{4}\right)\left(1 - \frac{z^2}{9}\right)\cdots \qquad (31)$$

are among them. The factorization of polynomials is largely an algebraic matter but the extension to functions such as the sine which have an infinity of roots required the systematic building up of a theory of infinite products. In 1876 Weierstrass succeeded in producing an extensive theory of factorizations which included as special cases these well-known infinite products, as well as certain doubly periodic functions.

In addition to showing the roots of $1/\Gamma(z)$, formula (30) does much more. It shows immediately that the reciprocal of the gamma function is a much less difficult function to deal with than the gamma function itself. It is an entire function, that is, one of those distinguished functions which possesses no singularities whatever for finite arguments.

Weierstrass was so struck by the advantages to be gained by starting with $1/\Gamma(z)$ that he introduced a special notation for it. He called $1/\Gamma(u+1)$ the *factorielle* of u and wrote $Fc(u)$.

The theory of functions of a complex variable unifies a hotch-potch of curves and a patchwork of methods. Within this theory, with its highly developed studies of infinite series of various types, was brought to fruition Stirling's unsuccessful attempts at solving the interpolation problem for the factorials. Stirling had done considerable work with infinite series of the form

$$A + Bz + Cz(z-1) + Dz(z-1)(z-2) + \cdots .$$

This series is particularly useful for fitting polynomials to values given at the integers $z = 0, 1, 2, \ldots$. The method of finding the coefficients $A, B, C, \ldots$ was well known. But when an infinite amount of fitting is required, much more than simple formal work is needed, for we are then dealing with a *bona fide* infinite series whose convergence must be investigated. Starting from the series $1, 2, 6, 24, \ldots$, Stirling found interpolating polynomials via the above series. The resultant infinite series is divergent. The factorials grow too rapidly in size. Stirling realized this and put out the suggestion that if perhaps one started with the logarithms of the factorials instead of the factorials themselves the size might be cut down sufficiently for one to do something. There the matter rested until 1900 when Charles Hermite (1822–1901) wrote down the Stirling series for $\log\Gamma(1+z)$:

$$\log\Gamma(1+z) = \frac{z(z-1)}{1\cdot 2}\log 2 + \frac{z(z-1)(z-2)}{1\cdot 2\cdot 3}(\log 3 - 2\log 2) + \cdots \quad (32)$$

and showed that this identity is valid whenever z is a complex number of the form $a + ib$ with $a > 0$. The identity itself could have been written down by Stirling, but the proof would have been another matter. An even simpler starting point is the function $\psi(z) = (d/dz)\log\Gamma(z)$, now known as the digamma or psi function. This leads to the Stirling series

$$\frac{d}{dz}\log\Gamma(z) = -\gamma + (z-1) - \frac{(z-1)(z-2)}{2\cdot 2!} + \frac{(z-1)(z-2)(z-3)}{3\cdot 3!} - \cdots, \quad (33)$$

which in 1847 was proved convergent for $a > 0$ by M. A. Stern, a teacher of Riemann. All these matters are today special cases of the extensive theory of the convergence of interpolation series.

Functions are the building blocks of mathematical analysis. In the 18th and 19th centuries mathematicians devoted much time and loving care to developing the properties and interrelationships between special functions. Powers, roots, algebraic functions, trigonometric functions, exponential functions, logarithmic functions, the gamma function, the beta function, the hypergeometric function, the elliptic functions, the theta function, the Bessel function, the Mathieu function, the Weber function, Struve function, the Airy function, Lamé functions, literally hundreds of special functions were singled out for scrutiny and their main features were drawn. This is an art which is not much cultivated these days. Times have changed and emphasis has shifted. Mathematicians on the whole prefer more abstract fare. Large classes of functions are studied instead of individual ones. Sociology has replaced biography. The field of special functions, as it is now known, is left largely to a small but ardent group of enthusiasts plus those whose work in physics or engineering confronts them directly with the necessity of dealing with such matters.

The early 1950s saw the publication of some very extensive computations of the gamma function in the complex plane. Led off in 1950 by a six-place table computed in England, it was followed in Russia by the publication of a very extensive six-place table. This in turn was followed in 1954 by the publication by the National Bureau of Standards in Washington of a twelve-place table. Other publications of the complex gamma function and related functions have appeared in this country, in England, and in Japan. In the past, the major computations of the gamma function had been confined to real values. Two fine tables, one by Gauss in 1813 and one by Legendre in 1825, seemed to answer the mathematical needs of a century. Modern technology had also caught up with the gamma function. The tables of the 1800s were computed laboriously by hand, and the recent ones by electronic digital computers.

But what touched off this spate of computational activity? Until the initial labors of H. T. Davis of Indiana University in the early 1930s, the complex values of the gamma function had hardly been touched. It was one of those curious turns of events wherein the complex gamma function appeared in the solution of various theoretical problems of atomic and nuclear theory. For instance, the radial wave functions for positive energy states in a Coulomb field lead to a differential equation whose solution involves the complex gamma function. The complex gamma function enters into formulas for the scattering of charged particles, for the nuclear forces between protons, in Fermi's approximate formula for the probability of β-radiation, and in many other places. The importance of these problems to physicists has had the side effect of computational mathematics finally catching up with two and a quarter centuries of theoretical development.

As analysis grew, both creating special functions and delineating wide classes of functions, various classifications were used in order to organize them for purposes of convenient study. The earlier mathematicians organized functions from without, operationally, asking what operations of arithmetic or calculus had to be performed in order to achieve them. Today, there is a much greater tendency to look at functions from within, organically, considering their construction as achieved and asking what geometrical characteristics they possess. In the earlier classification we have at the lowest and most accessible level, powers, roots, and all that could be concocted from them by ordinary algebraic manipulation. These came to be known as algebraic functions. The calculus, with its characteristic operation of taking limits, introduced logarithms and exponentials, the latter encompassing, as Euler showed, the sines and cosines of trigonometry which had been available from earlier periods of discovery. There is an impassable wall between the algebraic functions and the new limit-derived ones. This wall consists in the fact that try as one might to construct, say, a trigonometric function out of the finite material of algebra, one cannot succeed. In more technical language, the algebraic functions are closed with respect to the processes of algebra, and the trigonometric functions are forever beyond its pale. (By way of a simple analogy: the even integers are closed with respect to the operations of addition, subtraction, and multiplication; you cannot produce an odd integer from the set of even integers using these tools.) This led to the concept of transcendental functions. These are functions which are not algebraic. The transcendental functions count among their members, the trigonometric functions, the logarithms, the exponentials, the elliptic functions, in short, practically all the special functions which had been singled out for special study. But such an indiscriminate dumping produced too large a class to handle. The transcendentals had to be split further for convenience. A major tool of analysis is the differential equation, expressing the relationship between a function and its rate of growth. It was found that some functions, say the trigonometric functions, although they are transcendental and do not therefore satisfy an algebraic equation, nonetheless satisfy a differential equation whose coefficients are algebraic. The solutions of algebraic differential equations are an extensive though not all-encompassing class of transcendental functions. They count among their members a good many of the special functions which arise in mathematical physics.

Where does the gamma function fit into this? It is not an algebraic function. This was recognized early. It is a transcendental function. But for a long while it was an open question whether the gamma function satisfied an algebraic differential equation. The question was settled negatively in 1887 by O. Hölder (1859–1937). It does not. It is of a higher order of transcendency. It is a so-called transcendentally transcendent function, unreachable by solving algebraic equations, and equally unreachable by solving algebraic differential equations. The subject has interested many people through the years and in 1925 Alexander Ostrowski, now Professor Emeritus of the University of Basel, Switzerland, gave an alternate proof of Hölder's theorem.

Problems of classification are extremely difficult to handle. Consider, for instance, the following: Can the equation $x^7 + 8x + 1$ be solved with radicals? Is π transcendental? Can $\int dx / \sqrt{x^3 + 1}$ be found in terms of specified elementary functions? Can the differential equation $dy/dx = (1/x) + (1/y)$ be resolved with quadratures? The general problems of which these are representatives are even today far from solved and this despite famous theories such as Galois theory, Lie theory, theory of Abelian integrals which have derived from such simple questions. Each individual problem may be a one-shot affair to be solved by individual methods involving incredible ingenuity.

We return once again to our interpolation problem. We have shown how, strictly speaking, there are an unlimited number of solutions to this problem. To drive this point home, we might mention a curious solution given in 1894 by Jacques Hadamard (1865–1963). Hadamard found a relatively simple formula involving the gamma function which also produces factorial values at the positive integers. (See Figures 1 and 4.) But Hadamard's function

$$y = \frac{1}{\Gamma(1-x)} \frac{d}{dx} \log \left[\frac{\Gamma\left(\frac{1-x}{2}\right)}{\Gamma\left(1 - \frac{x}{2}\right)} \right], \qquad (34)$$

in strong contrast to the gamma function itself, possesses no singularities anywhere in the finite complex plane. It is an entire analytic solution to the interpolation problem and hence, from the function theoretic point of view, is a simpler solution. In view of all this ambiguity, why then should Euler's solution be considered the solution *par excellence*?

From the point of view of integrals, the answer is clear. Euler's integral appears everywhere and is inextricably bound to a host of special functions. Its frequency and simplicity make it fundamental. When the chips are down, it is the very form of the integral and of its modifications which lend it utility and importance. For the interpolatory point of view, we can make no such claim. We must take a deeper look at the gamma

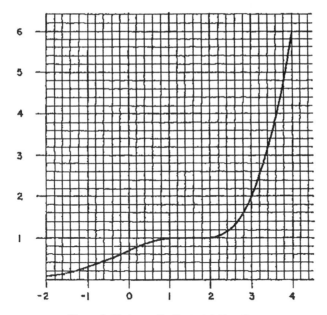

Figure 4. Hadamard's Factorial Function.

There are infinitely many functions which produce factorials. The function

$$F(x) = \frac{1}{\Gamma(1-x)} \frac{d}{dx} \log \left[\frac{\Gamma\left(\frac{1-x}{2}\right)}{\Gamma\left(1 - \frac{x}{2}\right)} \right]$$

is an entire analytic function which coincides with the gamma function at the positive integers. It satisfies the functional equation $F(x+1) = xF(x) + 1/\Gamma(1-x)$.

function and show that of all the solutions of the interpolation problem, it, in some sense, is the simplest. This is partially a matter of mathematical aesthetics.

We have already observed that Euler's integral satisfies the fundamental recurrence equation, $x\Gamma(x) = \Gamma(x+1)$, and that this equation enables us to compute all the real values of the gamma function from knowledge merely of its values in the interval from 0 to 1. Since the solution to the interpolation problem is not determined uniquely, it makes sense to add to the problem more conditions and to inquire whether the augmented problem then possesses a unique solution. If it does, we will hope that the solution coincides with Euler's. The recurrence relationship is a natural condition to add. If we do so, we find that the gamma function is again not the only function which satisfies this recurrence relation and produces factorials. One may easily construct a "pseudo" gamma function $\Gamma_s(x)$ by defining it between, say, 1 and 2 in any way at all (subject only to $\Gamma_s(1) = 1$, $\Gamma_s(2) = 1$), and allowing the recurrence relationship to extend its values everywhere else.

If, for instance, we let $\Gamma_s(x)$ be 1 everywhere between 1 and 2, the recurrence relation leads us to the function (see Figure 5).

$$
\begin{aligned}
\Gamma_s(x) &= 1/x & 0 &< x \le 1; \\
\Gamma_s(x) &= 1, & 1 &\le x \le 2; \\
\Gamma_s(x) &= x - 1, & 2 &\le x \le 3; \\
\Gamma_s(x) &= (x-1)(x-2), & 3 &\le x \le 4; \dots.
\end{aligned}
\tag{35}
$$

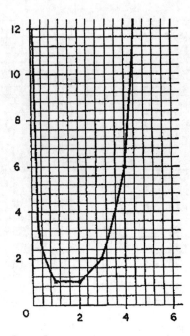

Figure 5. A Pseudogamma Function.
The function illustrated produces factorials, satisfies the functional equation of the gamma function, and is convex.

We might end up with a fairly weird result, depending upon what we start with. Even if we require the final result to be an analytic function, there are ways of doing it. For instance, take any function which is both analytic and periodic with period 1. Call it $p(x)$. Make sure that $p(1) = 1$. The function $1 + \sin 2\pi x$ will do for $p(x)$. Now multiply the ordinary gamma function $\Gamma(x)$ by $p(x)$ and the result $\Gamma(x)p(x)$ will be a "pseudo" gamma function which is analytic, satisfies the recurrence relation, and produces factorials! Thus, we still do not have enough conditions. We must augment the problem again. But what to add?

By the middle of the 19th century it was recognized that Euler's gamma function was the only continuous function which satisfied simultaneously the recurrence relationship, the reflection formula and the multiplication formula. Weierstrass later showed that the gamma function was the only continuous solution of the recurrence relationship for which $\Gamma(x + n)/(n - 1)n^x \to 1$ for all x. These conditions added to the interpolation problem will serve to produce a unique solution and one which coincides with Euler's. But they appear too heavy and too much like Monday morning quarterbacking. That is to say, the added conditions are hardly "natural" for they are tied in with the deeper analytical properties of the gamma function. The search went on.

Aesthetic conditions were not to be found in the older, analytic considerations, but in a newer, inner, organic approach to function theory which was developing at the turn of the century. Backed up by Cantor's set theory and an emerging theory of topology, the new function theory looked not so much at equations and identities as at the fundamental geometrical properties. The desired condition was found in notions of convexity. A curve is convex if the following is true of it: take any two points on the curve and join them by a straight line; then the portion of the curve between the points lies below the line. A convex curve does not wiggle; it cannot look like a camel's back. At the turn of the century, convexity was in the mathematical air. It was found to be intrinsic to many diverse phenomena. Over the period of a generation, it was sought out, it was generalized, it was abstracted, it was investigated for its own sake, it was applied. Called to attention by the work of H. Brunn in 1887 and of H. Minkowski in 1903 on convex bodies and given an independent interest in 1906 by the work of J. L. W. V. Jensen, the idea of convexity spread and established itself in mean value theory, in potential theory, in topology, and most recently in game theory and linear programming. At the turn of the century then, an application of convexity to the gamma function would have been natural and in order.

The individual curves which make up the gamma function are all convex. A glance at Figure 2 shows this to be true. If, as in the previous paragraph, a pseudogamma function satisfying the recurrence formula were produced by introducing the ripple $1 + \sin 2\pi x$ as a factor, it would no longer be true. It must have occurred to many mathematicians to find out whether the gamma function is the only function which yields the factorial values, satisfies the recurrence relation, and is convex downward for $x > 0$. Unfortunately, this is not true. Figure 5 shows a pseudogamma function which possesses just these properties. It remained until 1922 to discover a correct formulation. But it was not at too far a distance. The gamma function is not only convex, it is also logarithmically convex. That is to say, the graph of $\log \Gamma(x)$ is also convex down for $x > 0$. This fact is implicit in formula (27). Logarithmic convexity is a stronger condition than ordinary convexity for logarithmic convexity implies, but is not implied by, ordinary convexity. Now Harald Bohr and J. Mollerup were able to show the surprising fact that the gamma function is the only function which satisfies the recurrence relationship and is logarithmically convex. The original proof was simplified several years later by Emil Artin, now professor at Princeton University, and the theorem together with Artin's method of proof now constitute the Bohr-Mollerup-Artin theorem. Its precise wording is this:

The Euler gamma function is the only function defined for $x > 0$ which is positive, is 1 at $x = 1$, satisfies the functional equation $x\Gamma(x) = \Gamma(x + 1)$, and is logarithmically convex.

This theorem is at once so striking and so satisfying that the contemporary synod of abstractionists who write mathematical canon under the pen name of N. Bourbaki has adopted it as the starting point for its exposition of the gamma function. The proof: one page; the discovery: 193 years.

There is much that we know about the gamma function. Since Euler's day more than 400 major papers relating to it have been written. But a few things remain that we do not know and that we would like to know. Perhaps the hardest of the unsolved problems deal with questions of rationality and transcendentality. Consider, for instance, the number $\gamma = .57721\ldots$ which appears in formula (30). This is the Euler-Mascheroni constant. Many different expressions can be given for it. Thus,

$$\gamma = -d\Gamma(x)/dx|_{x=1}, \tag{36}$$

$$\gamma = \lim_{n \to \infty} \left(1 + \frac{1}{2} + \frac{1}{3} + \cdots + \frac{1}{n}\right) - \log n. \tag{37}$$

Though the numerical value of γ is known to hundreds of decimal places, it is not known at the time of writing whether γ is or is not a rational number. Another problem of this sort deals with the values of the gamma function itself. Though, curiously enough, the product $\Gamma(1/4)/\sqrt[4]{\pi}$ can be proved to be transcendental, it is not known whether $\Gamma(1/4)$ is even rational.

George Gamow, the distinguished physicist, quotes Laplace as saying that when the known areas of a subject expand, so also do its frontiers. Laplace evidently had in mind the picture of a circle expanding in an infinite plane. Gamow disputes this for physics and has in mind the picture of a circle expanding on a spherical surface. As the circle expands, its boundary first expands, but later contracts. This writer agrees with Gamow as far as mathematics is concerned. Yet the record is this: each generation has found something of interest to say about the gamma function. Perhaps the next generation will also.

The writer wishes to thank Professor C. Truesdell for his helpful comments and criticism and Dr. H. E. Salzer for a number of valuable references.

References

1. E. Artin, *Einführung in die Theorie der Gammafunktion*, Leipzig, 1931.

2. N. Bourbaki, *Éléments de Mathématique*, Book IV, Ch. VII, La Fonction Gamma, Paris, 1951.

3. H. T. Davis, *Tables of the Higher Mathematical Functions*, vol. I, Bloomington, Indiana, 1933.

4. L. Euler, *Opera omnia*, vol. I$_{14}$, Leipzig-Berlin, 1924.

5. P. H. Fuss, Ed., *Correspondance Mathématique et Physique de Quelques Célèbres Géomètres du XVIIIéme Siècle*, Tome I, St. Petersbourg, 1843:

6. G. H. Hardy, *Divergent Series*, Oxford, 1949, Ch. II.

7. F. Lösch and F. Schoblik, *Die Fakultät und verwandte Funktionen*, Leipzig, 1951.

8. N. Nielsen, *Handbuch der Theorie der Gammafunktion*, Leipzig, 1906.

9. *Table of the Gamma Function for Complex Arguments*, National Bureau of Standards, Applied Math. Ser. 34, Washington, 1954. (Introduction by Herbert E. Salzer.)

10. E. T. Whittaker and G. N. Watson, *A Course of Modern Analysis*, Cambridge, 1947, Ch. 12.

Change of Variables in Multiple Integrals: Euler to Cartan[1]

Victor J. Katz

Leonhard Euler first developed the notion of a double integral in 1769 [7]. As part of his discussion of the meaning of a double integral and his calculations of such an integral, he posed the obvious question: what happens to a double integral if we change variables? In other words, what happens to $\iint_A f(x, y)\, dx\, dy$ if we let $x = x(t, v)$ and $y = y(t, v)$ and attempt to integrate with respect to t and v? The answer is provided by the change-of-variable theorem, which states that

$$\iint_A f(x, y)\, dx\, dy = \iint_B f\big(x(t, v), y(t, v)\big) \left| \frac{\partial x}{\partial t} \frac{\partial y}{\partial v} - \frac{\partial x}{\partial v} \frac{\partial y}{\partial t} \right| dt\, dv \qquad (1)$$

where the regions **A** and **B** are related by the given functional relationship between (x, y) and (t, v). This result, and its generalization to n variables, are extremely important in allowing one to transform complicated integrals expressed in one set of coordinates to much simpler ones expressed in a different set of coordinates. Every modern text in advanced calculus contains a discussion and proof of the theorem. (For example, see [5], [1], [18].)

Euler interpreted this result formally; namely, he considered $dx\, dy$ as an "area element" of the plane. So his aim was to show that his area element transformed into a new "area element"

$$\left| \frac{\partial x}{\partial t} \frac{\partial y}{\partial v} - \frac{\partial x}{\partial v} \frac{\partial y}{\partial t} \right| dt\, dv$$

under the given change of variables. Obviously, if we merely change coordinates by a translation, rotation, and/or reflection, the area element is transformed into a congruent one. So Euler noted that if t and v are new orthogonal coordinates related to x and y by a translation through constants a and b, a clockwise rotation through the angle θ whose

[1] Reprinted from the *Mathematics Magazine*, Vol. 55, January, 1982, pp. 3–11.

cosine is m, and a reflection through the x-axis, i.e.,

$$x = a + mt + v\sqrt{1 - m^2}$$
$$y = b + t\sqrt{1 - m^2} - mv,$$

then $dx\,dy$ should be equal to $dt\,dv$. Unfortunately, when he performed the obvious formal calculation

$$dx = m\,dt + dv\sqrt{1 - m^2},$$
$$dy = dt\sqrt{1 - m^2} - m\,dv$$

and multiplied the two equations, he arrived at

$$dx\,dy = m\sqrt{1 - m^2}\,dt^2 + (1 - 2m^2)\,dt\,dv - m\sqrt{1 - m^2}\,dv^2,$$

which, he noted, was obviously wrong and even meaningless (see Figure 1). Even more so, then, would a similar calculation be wrong if t and v were related to x and y by more complicated transformations. It was thus necessary for Euler to develop a workable method; i.e., one that in the above situation gives $dx\,dy = dt\,dv$ and, in general, gives $dx\,dy = Z\,dt\,dv$, where Z is a function of t and v.

To see how he arrived at his method, we must first consider his definition and calculation of double integrals. After noting that $\iint Z\,dx\,dy$ means an "indefinite" double integral, i.e., a function of x and y which when differentiated first with respect to x and with respect to y gives $Z\,dx\,dy$, Euler proceeded to calculate "definite" integrals over specified planar regions $\mathbf{A}$ in the way familiar to calculus students. Thus, he wrote the integral as $\int dx \int Z\,dy$ and holding x constant, he integrated with respect to y between

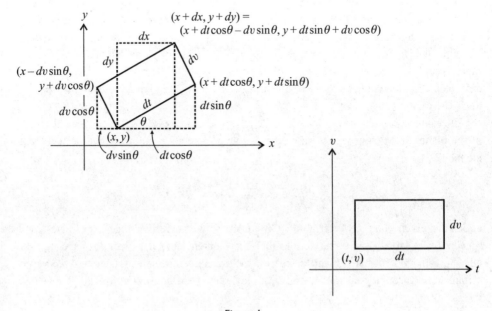

Figure 1.

the functions $y = f_1(x)$ and $y = f_2(x)$ which bounded the region **A**; finally he integrated with respect to x between its minimum and maximum values in **A**. He interpreted this integral in the obvious way as a volume. In particular, he integrated

$$\iint \sqrt{c^2 - x^2 - y^2} \, dx \, dy$$

over various regions to calculate volumes of portions of a sphere. Finally, he noted that $\iint_A dx \, dy$ is precisely the area of **A** and explicitly calculated the area of the circle given by $(x - a)^2 + (y - b)^2 = c^2$ to be πc^2.

Since the method of double integration involves leaving one variable fixed while dealing with the other, Euler proposed a similar method for the change-of-variable problem: change variables one at a time. First he introduced the new variable v and assumed that y could be represented as a function of x and v. So $dy = P \, dx + Q \, dv$ where P and Q are the appropriate partial derivatives. Now by assuming x fixed, he obtained $dy = Q \, dv$ and

$$\iint dx \, dy = \iint Q \, dx \, dv = \int dv \int Q \, dx$$

(Figure 2). Next, he let x be a function of t and v and put $dx = R \, dt + S \, dv$. So by holding v constant, he calculated

$$\int dv \int Q \, dx = \int dv \int QR \, dt = \iint QR \, dt \, dv.$$

This gave Euler the first solution to his problem: $dx \, dy = QR \, dt \, dv$.

Obviously, this was not completely satisfactory, since Q may well depend on x, and, in addition, the method was not symmetric. So Euler continued, now representing y as a function of t and v, hence $dy = T \, dt + V \, dv$. Then, formally,

$$dy = P \, dx + Q \, dv = P(R \, dt + S \, dv) + Q \, dv = PR \, dt + (PS + Q) \, dv.$$

So $PR = T$ and $PS + Q = V$, which gives $QR = VR - ST$. Euler's final answer was that $dx \, dy = (VR - ST) \, dt \, dv$. He noted again that simply multiplying the expressions for dx and dy together and rejecting the terms in dt^2 and dv^2 gives $(RV + ST) \, dt \, dv$, which differs by a sign from the correct answer. After a further note that one must always

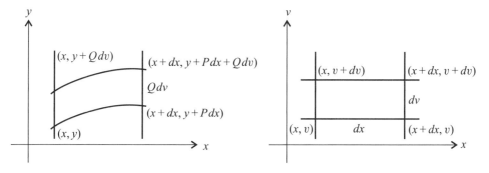

Figure 2.

take the absolute value of the expression $VR - ST$ (since area is positive) he proceeded to confirm the correctness of his result through several increasingly complex examples.

This "proof" was typical of Euler's use of formal methods in many parts of his vast mathematical work. As a developer of algorithms to solve problems of various sorts, Euler has never been surpassed. (We can see that Euler's method, in modern notation, amounts to first factoring the transformation $x = x(t, v)$, $y = y(t, v)$ into two transformations, the first being $x = x(t, v)$, $v = v$ and the second $x = x$, $y = y(x, v)$. This can be done by "solving" $x = x(t, v)$ for t in the form $t = h(x, v)$ and then writing $y = y(h(x, v), v)$. Then

$$P = y_1 \frac{\partial h}{\partial x}, \quad Q = y_1 \frac{\partial h}{\partial v} + y_2, \quad R = x_1, \quad S = x_2, \quad T = y_1, \quad \text{and} \quad V = y_2,$$

where subscripts denote partial derivatives. Since $x(h(x, v), v) = x$ and $h(x(t, v), v) = t$, we calculate that

$$\frac{\partial h}{\partial x} x_1 = 1 \quad \text{and} \quad \frac{\partial h}{\partial x} x_2 + \frac{\partial h}{\partial v} = 0,$$

so $PR = T$ and $PS + Q = V$.)

In 1773 J. L. Lagrange also had need of a change-of-variable formula—this time for triple integrals [12]. He was interested in determining the attraction which an elliptical spheroid exercised on any point placed on its surface or in the interior. Since the general expression for attraction at any point was well known, the difficulty lay in integrating over the entire body. Even though the problem had already been solved geometrically, Lagrange, as part of his general philosophy of treating mathematics analytically, attempted a different solution.

To solve his problem, Lagrange had to calculate a triple integral. Since, following Euler's method, this had to be done by first holding two variables constant, integrating with respect to the third from one surface of the body to another, then evaluating the ensuing double integrals, he was quickly led to very complicated integrands. He realized that new coordinates were needed to replace the rectangular ones in order to make the integration tractable. Thus he proceeded to develop a general formula for changing variables in a triple integral. Lagrange's method was similar to Euler's in that he let vary only one variable at a time, but the details differed.

Given, then, x, y, and z as functions of new variables p, q, r, Lagrange wrote

$$
\begin{aligned}
dx &= A \, dp + B \, dq + C \, dr \\
dy &= D \, dp + E \, dq + F \, dr \\
dz &= G \, dp + H \, dq + I \, dr
\end{aligned}
\tag{2}
$$

where $A, B, \ldots, I$ are, of course, the appropriate partial derivatives. His aim was to calculate the volume of the infinitesimal parallelepiped $dx \, dy \, dz$ (the "volume element") in terms of $dp \, dq \, dr$. To do this, he calculated each "difference" (i.e., edge of the parallelepiped) separately, regarding the other two variables as constant. First x and y are held constant; thus $dx = 0$ and $dy = 0$; the first two equations in (2) become

$$
\begin{aligned}
A \, dp + B \, dq + C \, dr &= 0 \\
D \, dp + E \, dq + F \, dr &= 0.
\end{aligned}
$$

Lagrange solved these two equations for dp and dq in terms of dr and substituted in the expression for dz in (2) to get

$$dz = \frac{G(BF - CE) + H(CD - AF) + I(AE - BD)}{AE - BD}\, dr.$$

Next, x and z are assumed constant and only y varies; so $dx = 0$ and $dz = 0$. It follows immediately that $dr = 0$ and $A\, dp + B\, dq = 0$; therefore, $dp = -(B/A)\, dq$ and

$$dy = \frac{AE - BD}{A}\, dq.$$

Finally, y and z are taken as constant, so $dy = 0$ and $dz = 0$. Thus $dr = 0$ and $dq = 0$, which implies that $dx = A\, dp$. By multiplying together the expressions obtained for dx, dy, and dz, Lagrange calculated his result

$$dx\, dy\, dz = (AEI + BFG + CDH - AFH - BDI - CEG)\, dp\, dq\, dr. \qquad (3)$$

This is, of course, our standard formula. The result for three-dimensional integrals is analogous to (1), and in modern notation, is written as

$$\iiint_A f(x, y, z)\, dx\, dy\, dz = \iiint_B f\big(x(p,q,r), y(p,q,r), z(p,q,r)\big) \left| \frac{\partial(x, y, z)}{\partial(p, q, r)} \right| dp\, dq\, dr$$

where $\frac{\partial(x,y,z)}{\partial(p,q,r)}$ is the functional determinant of x, y, z with respect to p, q, r. (Figure 3 illustrates Lagrange's idea for the case of two variables and polar coordinates.)

We note that Lagrange, like Euler, dealt with the differential forms formally; there is absolutely no infinitesimal approximation that we would require in a similar proof today. But this formalism is typical of some of Lagrange's other work, in particular, his attempt to develop the calculus without limits by the use of algebra and infinite series [11], [13]. Also like Euler, Lagrange noted that the most obvious thing to do to try to obtain the change-of-variable formula would be to multiply together the original expressions (2) for dx, dy, and dz. However, he wrote, this product would contain squares and cubes of dp, dq, and dr and so would not be valid in an expression of a triple integral. Hence he had to use the step-by-step formal approach already outlined.

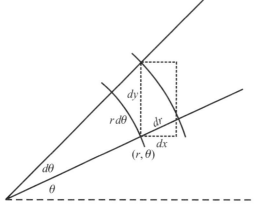

$$x = r \cos \theta$$
$$y = r \sin \theta$$
$$dx = \cos \theta\, dr - r \sin \theta\, d\theta$$
$$dy = \sin \theta\, dr + r \cos \theta\, d\theta$$
$$dx = 0 \Rightarrow dy = r \sec \theta\, d\theta$$
$$dy = 0 \Rightarrow dx = \cos \theta\, dr$$
$$dx\, dy = r\, dr\, d\theta$$

Figure 3.

Stamp issued by the USSR to recognize the 250th anniversary of Euler's birth

Lagrange applied his result to the case of spherical coordinates and was then able to perform the integrations he needed. Similarly, A. Legendre [15] and Pierre S. Laplace [14] soon after used essentially the same method to get similar results. These men were also interested in the change-of-variable formula in order to determine the attraction exercised by solids of various shapes, for which they needed to compute complicated integrals.

In 1813 Carl F. Gauss gave a geometric argument for a special case of the change-of-variable theorem for two variables, although in a somewhat different context [8]. Gauss' method of proof contrasts sharply with that of Euler. Gauss was developing the idea of a surface integral in connection with studying attractions. As part of this he gave a method for finding the element of surface in three-space so that he could integrate over such a surface. He started by parametrizing the surface using three functions x, y, z of the two variables p, q. He then noted that given an infinitesimal rectangle in the p-q plane whose vertices were (p, q), $(p + dp, q)$, $(p, q + dq)$, $(p + dp, q + dq)$, there was a corresponding "parallelogram" element in the surface whose vertices were (x, y, z), $(x + \lambda dp, y + \mu dp, z + \nu dp)$, $(x + \lambda' dq, y + \mu' dq, z + \nu' dq)$, and $(x + \lambda dp + \lambda' dq, y + \mu dp + \mu' dq, z + \nu dp + \nu' dq)$, where

$$
\begin{aligned}
dx &= \lambda \, dp + \lambda' dq \\
dy &= \mu \, dp + \mu' dq \\
dz &= \nu \, dp + \nu' dq.
\end{aligned}
\tag{4}
$$

(One can easily calculate the above result from the definitions and properties of the relevant partial derivatives.) It follows that the projection of the infinitesimal parallelogram onto the x-y plane is the parallelogram whose vertices are (x, y), $(x + \lambda dp, y + \mu dp)$, $(x + \lambda' dq, y + \mu' dq)$, $(x + \lambda dp + \lambda' dq, y + \mu dp + \mu' dq)$ and whose area is clearly $\pm(\lambda\mu' - \mu\lambda')dp \, dq$. (See Figure 4.) Gauss was therefore able to compute the element of surface area as $dp \, dq \big((\mu\nu' - \nu\mu')^2 (\nu\lambda' - \lambda\nu')^2 (\lambda\mu' - \mu\lambda')^2\big)^{1/2}$ and thus to integrate this over the p-q region corresponding to his surface. (In this paper, Gauss used his special cases of the divergence theorem and his parametric method for calculating a

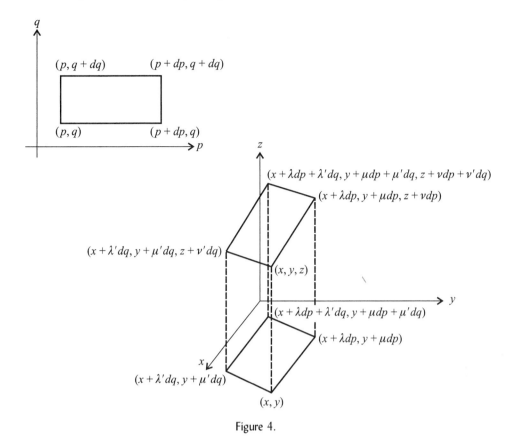

Figure 4.

surface element to evaluate certain "surface integrals" for the case of an ellipsoid given by $x = A\cos(p)$, $y = B\sin(p)\cos(q)$, $z = C\sin(p)\sin(q)$ for $0 \leq p \leq \pi$, $0 \leq q \leq 2\pi$.)

If we let $z = 0$ so that the "surface" is part of the x-y plane, then Gauss' argument shows that the new "area element" is $|\lambda\mu' - \mu\lambda'|\, dp\, dq$, hence that

$$\iint dx\, dy = \iint |\lambda\mu' - \mu\lambda'|\, dp\, dq,$$

a special case of the change-of-variable theorem from which the general case may easily be derived. Gauss' argument differs considerably from those of Euler and Lagrange. He essentially made use of analytic and geometric methods instead of using the formal approach of his predecessors. But as was typical of Gauss, he did not provide all the steps necessary to complete his analytic argument, especially since he was dealing with infinitesimals. The missing parts can, however, be readily supplied.

The next mathematician to break new ground in this field was Mikhail Ostrogradskii, in 1836. A Russian mathematician who studied in France in the 1820s, he later returned to St. Petersburg where he produced many works in applied mathematics. Unfortunately, some of his most important discoveries appear to have been totally ignored, at least in Western Europe. Not only did he give the first generalization of the change-of-variable theorem to n variables, but he also first proved and later generalized the divergence theorem [10], wrote integrals of n-forms over n-dimensional "hypersurfaces," and, as we shall see below, gave the first proof of the change-of-variable theorem for double integrals using infinitesimal concepts. All of these results were eventually repeated by other mathematicians with no credit to Ostrogradskii.

In his 1836 paper [16], Ostrogradskii generalized to n dimensions the change-of-variable theorem and Lagrange's proof of it. Given that $X, Y, Z, \ldots$ are all functions of $x, y, z \ldots$, Ostrogradskii first calculated $dX, dY, dZ, \ldots$ in terms of $dx, dy, dz, \ldots$. Then by holding all variables except X constant, he had $dY = dZ = \cdots = 0$, so he could solve for dX in terms of dx by using determinants; continuing with each variable in turn he calculated expressions for $dY, dZ, \ldots$ in terms of $dy, dz, \ldots$ and by multiplying showed that $dX \, dY \, dZ \cdots = \Delta \, dx \, dy \, dz \ldots$ where Δ is the functional determinant of $X, Y, Z, \ldots$ with respect to $x, y, z, \ldots$. Ostrogradskii did not state this result as a formula for transforming multiple integrals, but he did apply it to convert a hypersurface integral with $n + 1$ terms of the form $dx \, dy \ldots$, to an ordinary n-dimensional integral in n new variables.

Both Carl Jacobi [9] and Eugene Catalan [4] published papers in 1841 giving clearly the general change-of-variable theorem for n-dimensional integrals. Catalan's proof was also similar to Lagrange's in its use of formal manipulations on one variable at a time. Jacobi's paper was the culmination of a series of articles concerning this theorem; it contained additional results such as the multiplication rule for the composition of several changes of variable. Jacobi's work was referred to shortly thereafter by Cauchy and soon his name became tied to the theorem. In fact, the functional determinant Δ is now known as the Jacobian rather than the "Ostrogradskian."

Two years after his 1836 paper, Ostrogradskii published in [17] a proof of the change-of-variable formula in two variables which used the same basic idea as had Gauss. He first criticized the proofs of Euler and Lagrange, and, by implication, his own earlier proof. He claimed that, assuming that x and y were functions of u and v, if one first used $dx = 0$ to solve for dy in terms of du (that is, to evaluate one side of the differential rectangle) one could not then assume that du would be 0 when one tried to evaluate dx by setting $dy = 0$ (to find the other side of the rectangle). In fact, he wrote, you would have to use a new set of differentials, ∂u and ∂v, in evaluating the other side, and, once you did that, you came up with an incorrect result.

So Ostrogradskii returned to the meaning of $\iint V \, dx \, dy$ as a sum of differential elements. Using a method similar to that of Gauss, although staying strictly in two dimensions, he proceeded to recalculate the area of these elements. He carefully chose each element to be bounded by two curves where u was constant and two curves where v was constant. If ω denotes the area of such an element, he noted that by the definition of the definite integral, $\iint V \, dx \, dy = \iint V\omega$. It is easy to calculate ω (see Figure 5) since the four vertices have coordinates

$$(x, y), \quad \left(x + \frac{\partial x}{\partial u} \, du, \, y + \frac{\partial y}{\partial u} \, du\right),$$

$$\left(x + \frac{\partial x}{\partial v} \, dv, \, y + \frac{\partial y}{\partial v} \, dv\right), \quad \text{and} \quad \left(x + \frac{\partial x}{\partial u} \, du + \frac{\partial x}{\partial v} \, dv, \, y + \frac{\partial y}{\partial u} \, du + \frac{\partial y}{\partial v} \, dv\right).$$

By elementary geometry, the area of this parallelogram is $\pm \left(\frac{\partial x}{\partial u} \frac{\partial y}{\partial v} - \frac{\partial x}{\partial v} \frac{\partial y}{\partial u}\right) \, du \, dv$ and

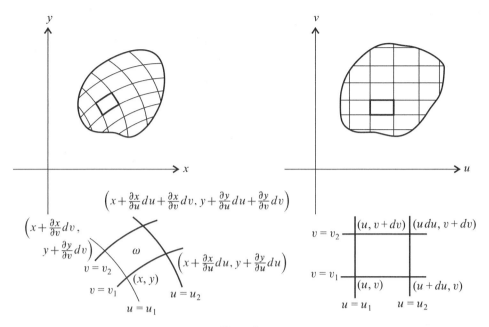

Figure 5.

so the integral formula becomes

$$\iint V \, dx \, dy = \pm \iint V \left(\frac{\partial x}{\partial u} \frac{\partial y}{\partial v} - \frac{\partial x}{\partial v} \frac{\partial y}{\partial u} \right) du \, dv.$$

Ostrogradskii further noted that this method could be easily extended to three dimensions but not more, since there is not a corresponding geometrical result in four dimensions. We must note, of course, that Ostrogradskii had not explicitly justified using the standard formula for the area of a parallelogram when, in fact, the area is actually that of a "curvilinear" parallelogram. However, it was common practice at that time (as we noted also about Gauss' proof), to ignore explicit arguments about infinitesimal approximation.

Only four years later, a proof similar to that of Ostrogradskii appeared in Augustus De Morgan's text *Differential and Integral Calculus* [6], one of the first "analytic" text-books to appear in English. It is doubtful that De Morgan had read Ostrogradskii's work, for his approach is somewhat different; he was considering how to calculate a double integral over a plane region bounded by four curves, where the standard method of integrating, first with respect to one variable between two functions of the other and then with respect to the second between constant limits, will not work. But his method of attack, via the definition of the double integral as a limit, the division of the given region into subregions bounded by curves where u was constant and where v was constant, and the calculation of areas of curvilinear quadrilaterals, is very close to that of Ostrogradskii. De Morgan went even further, however, to provide detailed reasoning as to why the errors of approximation—third order infinitesimals—may be safely ignored.

It is also interesting that De Morgan prefaced his results by stating that Legendre's proof (which was identical to that of Lagrange) was "so obscure in its logic as to be nearly unintelligible, if not dubious."

Ostrogradskii and De Morgan, then, had moved away from the formal symbolic approach of Euler and Lagrange. But we should emphasize that the former had not justified equating the "elements of area" $dx\,dy$ and $\left|\frac{\partial(x,y)}{\partial(u,v)}\right|\,du\,dv$ themselves, as the latter had attempted to do. They had only showed the equality of the integrals over the appropriate regions. A new justification for the formal symbolic approach only came with Elie Cartan and his theory of differential forms.

Beginning in the mid 1890s, Cartan wrote a series of papers in which he formalized the subject of differential forms, namely the expressions which appear under the integral sign in line and surface integrals. As part of this formalization, he used the Grassmann rules of exterior algebra for calculations with such forms. In a paper of 1896 [2], as an example of such a calculation, he was able to do what Euler could not; namely, if $x = x(t, v)$ and $y = y(t, v)$, he could multiply

$$dx = \frac{\partial x}{\partial t}\,dt + \frac{\partial x}{\partial v}\,dv \quad \text{and} \quad dy = \frac{\partial y}{\partial t}\,dt + \frac{\partial y}{\partial v}\,dv$$

using the rules $dt\,dt = dv\,dv = 0$ and $dt\,dv = -dv\,dt$ to show that

$$dx\,dy = \left(\frac{\partial x}{\partial t}\frac{\partial y}{\partial v} - \frac{\partial x}{\partial v}\frac{\partial y}{\partial t}\right)\,dt\,dv.$$

In 1899 [3], Cartan went into much more detail on the rules for operating with these differential forms. And again, one of his first examples was the change-of-variable formula.

As a final point, we note that proofs using the methods of Euler, Lagrange, and Ostrogradskii all appeared in textbooks through the first third of the twentieth century. There were, naturally, attempts to make all three methods more rigorous. A readily available example of this (for the proofs of Euler and Ostrogradskii) occurs in Courant's *Differential and Integral Calculus* [5]. Most current textbooks, on the other hand, use an entirely different proof based on Green's theorem.

References

1. R. C. Buck, *Advanced Calculus*, 2nd ed., McGraw-Hill, New York, 1965, p. 296 ff.

2. E. Cartan, Le principe de dualité et certaines intégrales multiples de l'espace tangentiel et de l'espace réglé, *Bull. Soc. Math. France*, 24 (1896) 140–177; Oeuvres (2) 1, 265–302.

3. ——, Sur certaines expressions différentielles et sur le problème de Pfaff, *Annales École Normale*, 16 (1899) 239–332; *Oeuvres* (2) 1, 303–397.

4. E. Catalan, Sur la transformation des variables dans les intégrales multiples, *Mémoires couronnés par l'Académie Royale des Sciences et Belles-Lettres de Bruxelles*, 14 (1841) 1–48.

5. R. Courant, *Differential and Integral Calculus*, vol. 2, Interscience, New York, 1959, p. 247 ff.

6. A. De Morgan, *The Differential and Integral Calculus*, Robert Baldwin, London, 1842.

7. L. Euler, De formulis integralibus duplicatis, *Novi comm. acad. scient. Petropolitanae*, 14 (1769) 72–103; *Opera* (1) 17, 289–315.

8. C. Gauss, Theoria attractionis corporum sphaeroidicorum ellipticorum homogeneorum methodo novo tractata, *Comm. soc. reg. scient. Gottingensis*, 2 (1813): *Werke* 5, 1–22.

9. C. Jacobi, De determinantibus functionalibus, *Crelle*, 22 (1841) 319–359; *Werke* 3, 393–438.

10. V. Katz, The history of Stokes' theorem, *Math. Mag.*, 52 (1979) 146–156.

11. J. Lagrange, *Théorie des Fonctions Analytiques*, Paris, 1797; *Oeuvres* 9.

12. ———, Sur l'attraction des sphéroides elliptiques, *Nouveaux Mémoires de l'Académie Royale des Sciences et Belles-Lettres de Berlin*, (1773); *Oeuvres* 3, 619–658.

13. ———, *Leçons sur le Calcul des Fonctions*, Paris, 1801.

14. P. Laplace, *Traité de Mécanique Céleste*, Paris, 1799; *Oeuvres* 2.

15. A. Legendre, Mémoire sur les intégrales doubles, *Histoire de l'Académie Royale des Sciences avec les Mémoires de Mathématique et de Physique*, (1788) 454–486.

16. M. Ostrogradskii, Mémoire sur le calcul des variations des intégrales multiples, *Mémoires de l'Académie Impériale des Sciences de St. Petersbourg*, (6) 3 (part 1) (1836), 36–58; also in *Crelle*, 15 (1836) 332–354.

17. ———, Sur la transformation des variables dans les intégrales multiples, *Mémoires de l'Académie Impériale des Sciences de St. Petersbourg*, cl. math., (1838) 401–407.

18. A. Taylor and W. R. Mann, *Advanced Calculus*, 2nd ed., Xerox, Lexington, 1972, p. 490 ff.

Two stamps honoring Euler on his 250th birthday, one from Switzerland (note Euler's formula on the left) and the other from East Germany

Euler's Vision of a General Partial Differential Calculus for a Generalized Kind of Function[1]

Jesper Lützen

The vibrating string controversy involved most of the analysts of the latter half of the 18th century. The dispute concerned the type of functions which could be allowed in analysis, particularly in the new partial differential calculus. Leonhard Euler held the bold opinion that all functions describing any curve, however irregular, ought to be admitted in analysis. He often stressed the importance of such an extended calculus, but did almost nothing to support his point of view mathematically. After having been abandoned during the introduction of rigor in the latter part of the 19th century, Euler's ideas began to take more concrete form during the early part of the 20th century, and they have now been incorporated into L. Schwartz's theory of distributions.

The algebraic function concept

Euler's radical stand in the dispute over the vibrating string is surprising since he had canonized the narrower range of analysis which his main opponent, J. B. R. d'Alembert (1717–1783), adhered to. This was done in the influential book *Introductio in analysin infinitorum* [12], in which Euler chose to determine the relation between the variable quantities by way of functions instead of using curves, as had been universally done earlier (cf. [22] and [7]). He defined a function as follows (see photo on page 201):

> A function of a variable quantity is an analytical expression composed in one way or another of this variable quantity and numbers or constant quantities [12, ch. 1, §4].

In forming the analytical expressions, Euler allowed the use of the standard transcendental operations such as log, exp, sin and cos in addition to algebraic operations. Still, all the

[1]Reprinted from the *Mathematics Magazine*, Vol. 56, November, 1983, pp. 299–306.

LIBER PRIMUS.

CAPUT PRIMUM.

DE FUNCTIONIBUS IN GENERE.

1. Uantitas conſtans eſt quantitas determinata, perpetuo eumdem valorem ſervans.

Ejuſmodi quantitates ſunt numeri cujuſvis generis, quippe qui eumdem, quem ſemel obtinuerunt, valorem conſtanter conſervant: atque ſi hujuſmodi quantitates conſtantes per characteres indicare convenit, adhibentur litteræ Alphabethi initiales a, b, c, &c. In Analyſi quidem communi, ubi tantum quantitates determinatæ conſiderantur, hæ litteræ Alphabethi priores quantitates cognitas denotare ſolent, poſteriores vero quantitates incognitas; at in Analyſi ſublimiori hoc diſcrimen non tantopere ſpectatur, cum hic ad illud quantitatum diſcrimen præcipue reſpiciatur, quo aliæ conſtantes, aliæ vero variabiles ſtatuuntur.

A 2   2. Quan-

First page, Chapter 1 of *Introductio in analysin infinitorum*
with Euler's discussion of "constant quantities"

rules in the theory of functions were taken over from algebra, so that Euler's function concept was in essence entirely algebraic. Thus *Introductio* marked a shift in the setting of analysis from geometry to algebra. Euler even accepted, and treated algebraically, infinite expressions such as infinite series, infinite products and continued fractions. Lebesgue [25] later showed that when such infinite limit procedures are accepted, the class of functions is very extensive, namely, equal to the class of Borel Functions. However, Euler did not realize the immense generality of his function concept and in theoretical considerations he conferred on them all the nice properties he needed—such as differentiability and even analyticity in the modern sense. Still, it would be off the mark to identify Euler's functions with one of the modern classes of functions such as differentiable functions or analytical functions because their definition involves topological (geometrical) ideas which are foreign to Euler's way of thinking.

Most important among the nice properties shared by all of Euler's functions was the possibility of expanding them in a power series:

$$f(x + i) = f(x) + pi + qi^2 + ri^3 + \cdots,$$

for, in all differentiations actually carried out in Euler's second influential textbook *Institutiones calculi differentialis* [16], the differential quotient is found as the coefficient p of the first power term. Later in the century J. L. Lagrange [24] *defined* the derivative of a function in this way and gave a "proof" that the expansion always exists. In the mid-18th century, however, power series were only used as a practical tool whereas the metaphysical basis for the calculus was found elsewhere. For example, d'Alembert defined the derivative using limits, and Euler's definition of the differential rested on a theory of zeros of different order. Yet, these foundational differences were not reflected in the domain they assigned to the ordinary calculus; both agreed that

> ...[calculus] as it has been treated until now can only be applied to curves, whose nature can be contained in one analytical equation" [18, §7].

Euler's generalized functions

The discussion of the vibrating string brought an end to this agreement. D'Alembert, who in 1747 [1] found his famous solution

$$y = f(x, t) = \phi(x + t) + \psi(x - t)$$

of the wave equation

$$\frac{\partial^2 f}{\partial x^2} = \frac{\partial^2 f}{\partial t^2}$$

governing the displacement y of the string, required that the "arbitrary" functions ϕ, and ψ be analytical expressions.

> In all other cases the problem cannot be solved, at least not with my method, and I do not even know whether it will not be beyond the powers of the known analysis. In fact, it seems to me that one cannot express y analytically in a more general way than supposing it to be a function of x and t [2, p. 358].

Euler, on the other hand, pointed out that this requirement restricted the initial displacement $\phi(x) + \psi(x)$ of the string too much; for example, he believed that the plucked string (Figure 1) would be excluded from d'Alembert's solution. (However, the plucked string can be described analytically by a slight modification of Cauchy's example: $\sqrt{x^2} = |x|$ [9].) Therefore he argued that one had to allow the functions ϕ and ψ to represent arbitrarily given curves. In this way physical reality led Euler to generalize the function concept so as to be in one-to-one correspondence with the geometrical concept of curve which he had earlier abandoned as the basic concept in analysis.

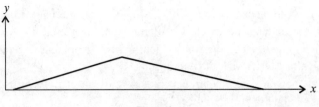

Figure 1.

It is surprising that Euler never provided a proper definition of the more general notion of function. His many papers on the vibrating string (particularly [17]) made clear that a generalized function was something corresponding to a general hand-drawn curve, but he never explicitly stated what this something was supposed to be. To judge from the classification of the new functions he seems to have had an algebraic definition in mind. He divided the general functions into the continuous and the discontinuous. The former were identical with the functions defined in *Introductio*, whereas the latter could not be expressed by one analytical expression. Euler was quite explicit about the continuity of a function having nothing to do with the connectedness of the curve; for example $1/x$ is continuous but its graph is disconnected at $x = 0$. Thus Euler's concept of continuity must be distinguished from the modern concept, due to Cauchy [8], and so we shall term the former **E-continuity**. In [17] Euler further divided the E-discontinuous functions into mixed functions, whose graph can be represented piecewise by finitely many analytical expressions, and the functions corresponding to arbitrary hand-drawn curves, whose analytical expressions may, so to speak, change from point to point.

Thus Euler's division of functions into classes was entirely algebraic and so was his distinction between even and odd functions. For example, in his critique [15] of D. Bernoulli's [6] description of the vibrating string as a trigonometric series, Euler argued that an E-discontinuous function of the form

$$\begin{cases} f(x) & \text{for } x > 0 \\ -f(-x) & \text{for } x < 0 \end{cases}$$

is only odd if f is odd and by that he meant that its power series contains only odd powers of x. To conclude: even when the consequences were absurd, Euler continued to think algebraically about his new functions, which, implicitly, he defined as the collection of the (possibly infinitely many) analytical expressions describing the corresponding curve.

Strangely enough, Euler himself had introduced a way of thinking about functions which he could have used to define his E-discontinuous functions as separate entities. In his *Institutiones calculi differentialis*, he defined functions in the following way:

> If, therefore, x denotes a variable quantity, all quantities which depend in some way on x or are determined by it, are called functions of this variable [16, Preface].

As it stands, this is almost the modern function definition and it clearly encompasses the E-discontinuous functions. However, Euler did not realize its generality. In *Institutiones calculi differentialis* only E-continuous functions occur, and the E-discontinuous functions are not even mentioned. Neither did he refer to his 1755 definition in any of his later papers on E-discontinuous functions. This indicates that Euler thought of his 1755 function definition as being equivalent to the definition given in *Introductio*. In fact,

Page from *Introductio in analysin infinitorum* with Euler's definition of "function" in 4

Euler's statement from 1765 that analysis until then had exclusively been concerned with analytical expressions only makes sense under this assumption. (This point of view is different from the one put forward by Youschkevich [34].)

Euler's vision of a generalized calculus

The lack of a proper definition of the E-discontinuous functions suggests that Euler's main concern was not the foundation of the generalized function concept itself but the analysis it made possible. We saw that initially Euler had introduced his new functions for physical reasons. Later [17] he stressed that the E-discontinuous functions were not forced onto analysis from outside but inevitably emerged as arbitrary functions in the partial integral calculus. For example [20, book 2, sect. 1, §33], the solution of the partial differential equation

$$\frac{\partial u(x, y)}{\partial x} = 0$$

is an arbitrary constant under the variation of x, but the constant can vary as a function f of y. It does not matter whether the constants for different values of y are connected by an analytical expression or not; therefore f must be allowed to be E-discontinuous. Since the functions ϕ and ψ in the solution of the wave equation arise in this way when $x + t$ and $x - t$ are used as independent variables, these functions are by their nature general functions.

Euler only used the E-discontinuous functions in the calculus of functions of several variables, but within that theory he would apparently blaze the trail for their unrestricted application. In contrast to the conservative d'Alembert, Euler argued that the development of a calculus of E-discontinuous functions is particularly desirable because all earlier calculus had been restricted to analytic expressions:

> But if the theory [of the vibrating string] leads us to a solution so general that it extends to all discontinuous as well as continuous figures, one must admit that this research opens to us a new road in analysis by enabling us to apply the calculus to curves which are not subject to any law of continuity, and if that has appeared impossible until now the discovery is so much more important [18, §8].

Euler's insistence that calculus should be applicable within the whole new function domain instead of being restricted to some—possibly varying—subclass(es) (as is the case in modern analysis) was supported not only by the mentioned physical reasons. It was also in agreement with the fundamental belief in the generality of mathematics. For algebraic rules were considered universally valid because they operated on abstract quantities, and since analysis was just infinite algebra, its rules had to be generally applicable as well.

> For, because this calculus applies to variable quantities, that is, quantities considered generally, if it were not generally true... one could never make use of this rule, since the truth of the differential calculus is based on the generality of the rules of which it consists [14, 1. Objection].

This basic belief in the generality of mathematics forced Euler to extend calculus to all E-discontinuous functions as soon as he had allowed them to enter his mathematical universe. Initially it probably also made him believe that this extension would come down to a simple admission of all the well-known rules to the extended domain. However, he soon had to realize that d'Alembert's exclusion of E-discontinuous functions was not only due to plain conservatism but was supported by mathematical arguments.

In many examples d'Alembert showed that the mathematical analysis of the vibrating string broke down at points where ϕ or ψ changed their analytical expression. For example, d'Alembert [3, §7] proved that if ψ is composed of two symmetric parabolas as in Figure 2

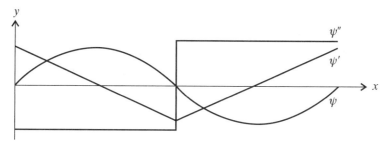

Figure 2.

and $\phi \equiv 0$ then $\psi(x - t)$ does not satisfy the wave equation

$$\frac{\partial^2 f}{\partial x^2} = \frac{\partial^2 f}{\partial t^2}$$

at points where $x - t = 0$. This and other difficulties can be explained in modern terminology by the fact that ϕ or ψ are not twice differentiable. D'Alembert came close to such an insight towards the end of his life [4], but while the controversy was at its highest, he believed that he had proved that ϕ and ψ must be E-continuous.

Euler was not convinced by d'Alembert's arguments and tried to refute them with a few counterarguments [19] of which I shall reproduce the most convincing. He remarked that the trouble was due to the sharp bend in the first derivative of ψ. Therefore, one had only to smooth out ψ' which could be done by changing ψ infinitely little to $\tilde{\psi}$. Since $\tilde{\psi}(x - t)$ would then satisfy the wave equation, one also had to admit $\psi(x - t)$ as a solution since infinitely small changes were always ignored in analysis.

In Eulerian calculus this argument is not completely off the mark, and even in modern analysis it contains the germ of a good idea. Still Euler seems to have realized that he had not overcome all objections to his new general analysis, and so he often encouraged the younger mathematicians to work on these problems.

> This part of analysis [of two or more variables] is essentially different from the former [of one variable], and extends even to functions void of all law of continuity. This part, of which we so far know barely the first elements, certainly deserves the united efforts of all geometers for its investigation and development [19, §32].

The fate of Euler's vision

In order to follow how subsequent geometers cultivated this new branch of analysis it is useful to divide the complex of problems, seen by Euler as a unity, into three separate parts:

(1) The generalization of the concept of function.
(2) The generalization of analysis.
(3) The development of the theory of partial differential equations.

The last and most important point of this research programme (3) was enthusiastically taken up by most of the mathematical community and was probably the most important mathematical discipline during the following half century. However, a discussion of it is far beyond the scope of this paper (see [23, ch. 22, 28]).

The generalization of the function concept (1) was also gradually accepted. In this process Euler's 1755 function definition was influential, regardless of his own interpretation of it. For after 1755 it became normal to reproduce this definition in textbooks on analysis, and slowly mathematicians began to realize its true generality. But this process took almost a century. For example, Lagrange [24] and Cauchy [8] defined functions generally as correspondences between variables, but they both thought of them as analytical expressions. It is natural in Lagrange's case, because he carried Euler's algebraic approach to its extreme, but it is surprising that the father of modern analysis, Cauchy, had a similar way of thinking. Still, this is evident from many remarks in his famous *Cours d'Analyse* [8], for example, the talk about "the constants or variables contained in a given function" [8, ch. 8, §1].

In J. Fourier's works [21, §417], one can find some comprehension of the generality of Euler's 1755 definition but the first mathematician who really took it seriously and understood the implications of the permissible pathologies was J. P. G. Lejeune-Dirichlet [11], after whom our function concept is justly named.

The generalization of analysis (2) suffered the opposite fate. At first it gained widespread acceptance but during the 19th century the idea was entirely abandoned. It happened as follows. In 1787 the St. Petersburg Academy officially terminated the controversy over the vibrating string by awarding L. Arbogast the first prize for a paper on the irregularities of arbitrary functions in the solutions of partial differential equations. Arbogast came out in favor of Euler's point of view, but he added nothing new to the foundational difficulties [5].

However, this official support of a general calculus was brushed aside by Cauchy, whose partial rigorization of analysis was a frontal attack on the principle of the generality of algebraic and analytical rules which had philosophically supported Euler's point of view. Cauchy explicitly pointed out this fundamental shift in the introduction to his famous *Cours d'Analyse* [8]:

> As for the methods, I have tried to give them all the rigour that one demands in geometry, so as never to have recourse to reasoning drawn from the generality of algebra.

Therefore nothing in his philosophy prevented him from confining calculus to a subclass of the class of functions, and in essence he restricted its use to the continuous functions (in the modern sense). In some of his papers he realized the inadequacy of this restriction, but a clear idea of the spaces $C^{(n)}(\mathbb{R})$ as the domain of d^n/dx^n did not crystalize until the 1870s in the Weierstrass school.

As a whole, mathematics benefited from this rigorization of analysis, but the corresponding restriction in the allowable solutions to partial differential equations made life complicated for the applied mathematician. Thus when irregular physical situations occurred (as, for example, a sharp bend in a string), the differential equation could not be used and a new mathematical model of the system had to be found. Such alternative models were set up, for example, by E. Christoffel [10].

However, in the beginning of the 20th century this procedure was felt to be so cumbersome and unnatural that several definitions of generalized solutions to partial differential equations were suggested, beginning in 1899 with H. Petrini's generalization of Poisson's equation [28]. Of the many generalization procedures I shall mention only the "sequence definition" implicitly used by N. Wiener in 1926 [33] and explicitly introduced by Sobolev (1935) [32]. According to this definition, f is a generalized solution to a (partial) differential equation if there exists a sequence of ordinary solutions (f_n) converging, in a suitable topology, to f. This definition is particularly interesting because it leads to a sensible interpretation of Euler's argument against d'Alembert (pp. 303–304); for, if instead of one smooth function $\tilde{\psi}$ infinitely close to ψ, we think of a sequence ψ_n of such functions, then Euler's argument shows that $\psi(x-t)$ is a generalized solution to the wave equation.

All the *ad hoc* definitions of generalized solutions from the first half of this century were incorporated in the theory of distributions created by L. Schwartz during the period 1945–1950 [31] as a result of his work with generalized solutions to the polyharmonic equation [30]. The theory of distributions probably constitutes the closest approximation to Euler's vision of a general calculus one can obtain, for in that theory any generalized function is infinitely often differentiable. However, in many respects the reality has turned out to be different from the dream. In one respect the reality is more satisfactory since it not only generalizes partial differential calculus which Euler had imagined but encompasses ordinary differential calculus as well. In other respects it is less perfect; for example, the general use of the algebraic operations, such as multiplication of two generalized functions, has been sacrificed in the theory of distributions. Moreover, the necessary generalization of the function concept has turned out to be much more extensive than the one Euler suggested.

Concluding remarks

Surely the realization of Euler's vision of a general calculus was different from what he had imagined—and more difficult. This can only increase our admiration for his readiness to overthrow his own framework of analysis when physical reality called for it. His conduct reveals an undogmatic and flexible attitude toward the foundational problems, from which much could be learned by modern mathematicians. On the other hand, it is worth noting that the creation of the theory of distributions made extensive use of the classical theory of differential operators created more in the spirit of d'Alembert; one can even argue that the establishment of a secure foundation for the more restricted classical calculus was a necessary condition for the realization of Euler's vision of a general calculus.

As further reading on the development of the concept of function I can recommend [34], [29] and, for those who want to brush up their Danish, [26]. The book [27] contains more information on the history of generalized solutions to partial differential equations and other aspects of the prehistory of the theory of distributions.

The author and the editor appreciate the assistance of Dr. Dorothy Tyler in translating passages from Euler's works.

References

1. J. d'Alembert, Recherches sur la courbe qui forme une corde tendue mise en vibration, *Mém. Acad. Sci. Berlin*, 3 (1747) 214–219.

2. ——, Addition au mémoire sur la courbe qui forme une corde tendue mise en vibration, *Mém. Acad. Sci. Berlin*, 6 (1750) 355–366.

3. ——, Recherches sur les vibrations des cordes sonores, *Opuscules Mathématiques*, 1 (1761) 1–73.

4. ——, Sur les fonctions discontinues, *Opuscules Mathématiques*, 8 (1780) 302–308.

5. L. F. A. Arbogast, Mémoire sur la nature de fonctions arbitraires qui entrent dans les intégrales des équations aux différences partielles, St. Petersburg, 1791.

6. D. Bernoulli, Réflexions et éclaircissemens sur les nouvelles vibrations des cordes, *Mém. Acad. Sci. Berlin*, 9 (1753 publ. 1755) 147–172 (see also 173–195).

7. H. J. M. Bos, Differentials, higher-order differentials and the derivative in the Leibnizian calculus, *Arch. Hist. Exact Sci.*, 14 (1974) 1–90.

8. A.-L. Cauchy, *Cours d'analyse de l'école roy. Polytechnique*, 1re partie; Analyse algébrique, Paris, 1821 = *Oeuvres* (2) 3.

9. ——, Mémoire sur les fonctions continues ou discontinues, *Comp. Rend. Acad. Roy. Sci. Paris*, 18 (1844) 145–160 = *Oeuvres* (1) 8, 145–160.

10. E. Christoffel, Untersuchungen über die mit Fortbestehen linearer partieller Differentialgleichungen verträglichen Unstetigkeiten, *Ann. Mat. Pura Appl.*, (2) 8 (1876) 81–112 = *Gesammelte Math. Abh.*, 2, 51–80.

11. J. P. G. Lejeune Dirichlet, Sur la convergence des séries trigonométriques qui servent à représenter une fonction arbitraire entre les limites données, *J. Reine Angew. Math.*, 4 (1829) 157–169 = Werke I, 117–132.

12. L. Euler, *Introductio in analysin infinitorum* (2 vols), Lausanne, 1748 = *Opera Omnia* (1) 8–9.

13. ——, Sur la vibration des cordes, *Mém. Acad. Sci. Berlin*, 4 (1748, publ. 1750) 69–85 = *Opera Omnia* (2) 10, 63–77.

14. ——, De la controverse entre Messieurs Leibniz et Bernoulli sur les logarithmes des nombres négatifs et imaginaires, *Mém. Acad. Sci. Berlin*, 5 (1749) 139–179 = *Opera Omnia* (1) 17, 195–232.

15. ——, Remarques sur les mémoires précédens de M. Bernoulli, *Mém. Acad. Sci. Berlin*, 9 (1753, publ. 1755) 196–222 = *Opera Omnia* (2) 10, 233–254.

16. ——, *Institutiones calculi differentialis*, St. Petersburg, 1755 = *Opera Omnia* (1) 10.

17. ——, De usu functionum discontinuarum in analysi, *Novi Comm. Acad. Sci. Petrop.*, 11 (1763, publ. 1768) 67–102 = *Opera Omnia* (1) 23, 74–91.

18. ——, Eclaircissemens sur le mouvement des cordes vibrantes, *Miscellanea Tourinensia*, 3 (1762–1765 publ. 1766) math. cl., 1–26 = *Opera Omnia* (2) 10, 377–396.

19. ——, Sur le mouvement d'une corde qui au commencement n'a été ébranlée que dans une partie, *Mém. Acad. Sci. Berlin*, 21 (1765 publ. 1767) 307–334 = *Opera Omnia* (2) 10, 426–450.

20. ——, *Institutiones calculi integralis* (3 vols), St. Petersburg, 1768–1770.

21. J. B. J. Fourier, *Théorie Analytique de la Chaleur*, Paris, 1822 = *Oeuvres* I.

22. G. F. A. Hospital, *Analyse des Infiniments Petits pour l'intelligence des lignes courbes*, Paris, 1696.

23. M. Kline, *Mathematical Thought from Ancient to Modern Times*, Oxford Univ. Press, 1972.

24. J. L. Lagrange, *Théorie des Fonctions Analytiques*, Paris, 1797, 2nd ed. 1813 = *Oeuvres* 9.

25. H. Lebesgue, Sur les fonctions représentables analytiquement, *J. Math. Pures Appl.*, 1 (1905) 139–216.

26. J. Lützen, Funktionsbegrebets udvikling fra Euler til Dirichlet, *Nordisk Mat. Tidsskr.*, 25–26 (1978) 5–32.

27. ——, *The Prehistory of the Theory of Distributions*, Springer, 1982.

28. H. Petrini, Démonstration générale de l'équation de Poisson $\Delta V = -4\pi\rho$ en ne supposant que ρ soit continu, *K. Vet. Akad. Oeuvres*, Stockholm, 1899.

29. J. R. Ravetz, Vibrating strings and arbitrary functions, Logic of personal knowledge, Essays presented to M. Pólanyi on his 70th birthday, London, 1961, 71–88.

30. L. Schwartz, Sur certaines familles non fondamentales de fonctions continues, *Bull. Soc. Math. France*, 72 (1944), 141–145.

31. ——, *Theorie des Distributions* (2 vols), Hermann, Paris, 1950, 1951.

32. S. L. Sobolev, Obshchaya teoriya difraktsü voln na rimanovykh poverkhnostyakh, *Travaux Inst. Steklov. Tr. Fiz.-Mat. in-ta*, 9 (1935) 433–438.

33. N. Wiener, The operational calculus, *Math. Ann.*, 95 (1926) 557–585.

34. A. P. Youschkevich, The concept of function up to the middle of the 19th century, *Arch. Hist. Exact Sci.*, 16 (1976) 37–85.

LEONH.ᴰ EULER

NAT. BASILEÆ MDCCVII.

MORT. PETROP. MDCCLXXXIII.

Leonhard Euler
Reproduction courtesy of the Library of Congress collection.

On the Calculus of Variations and Its Major Influences on the Mathematics of the First Half of Our Century[1]

Erwin Kreyszig

1. Johann Bernoulli's Brachystochrone. Carathéodory's Method

The calculus of variations evolved from the differential and integral calculus, "the calculus," for short. An initial motivation of the latter was the determination of extrema of functions, as shown by the title of the earliest relevant *published* paper "A new method for the determination of maxima and minima" (Leibniz, 1684). However, the calculus had to be extended to the calculus of variations in order to take care of more general problems involving the determination of stationary values of *functionals*,[2] given in the simplest case by a definite integral involving an unknown function and boundary conditions. The earliest (not very simple) solved problem of this kind was Newton's determination of the shape of a gun shell of least air resistance (letter to Gregory of July 14, 1694).

The earliest problem that received general publicity [due to a rather bombastic advertisement in *Acta Eruditorum* by Johann Bernoulli (1667–1748)] in 1696 was the problem of determining the *brachystochrone*, the curve along which a particle will fall from one given point to another in the shortest time. This problem was solved by Newton, Leibniz, and Johann Bernoulli as well as by his brother Jacob (1654–1705), the solution being a *cycloid*. Thus 1696 can be called the birthyear of the calculus of variations. Johann Bernoulli not only posed that problem but also gave a solution capable of extensive generalization worked out in 1908 by Carathéodory. The resulting general method was later named after Carathéodory.

---

[1] Reprinted from the *American Mathematical Monthly*, Vol. 101, August-September, 1994, pp. 674–678.

[2] A functional is a function defined on a set of functions. A stationary value of a functional is its value at a "point" (= function) that satisfies the necessary conditions for an extremum (see Section 3 below).

2. Simplest General Problems

Although they arose from different geometric and physical applications, many of the early problems led to functionals that depended on real functions defined on an interval and satisfying boundary conditions, and all functionals were of the *same form* (as needed to create a general theory).

$$J[y] = \int_{x_0}^{x_1} L(x, y, y') \, dx, \quad y(x_0) = y_0, \quad y(x_1) = y_1, \quad x_0 < x_1 \tag{1}$$

The task was to determine a function $y(x)$ that satisfied the boundary conditions in (1) and rendered $J[y]$ stationary, possibly yielding a minimum or a maximum of $J[y]$. At this early stage, the existence and uniqueness of solutions was "obvious" for physical reasons because a solution could be verified experimentally if desired. Also, with the concept of function not yet sharply defined, nobody made an attempt to characterize the set of functions in which such a $y(x)$ was to be found. This was accomplished well over one hundred years later in the works of Jacobi (1804–51) around 1835 and, especially, of Weierstrass (1815–97) around 1880.

3. Euler and Lagrange

As the birthyear of the *theory* of the calculus of variations one usually considers 1744, the year in which Euler published his famous book *Methodus inveniendi lineas curvas maximi minimive proprietate gaudentes, sive solutio problematis isoperimetrici latissimo sensu accepti* (A method for discovering curved lines that enjoy a maximum or minimum property, or the solution of the isoperimetric problem taken in its widest sense). Thus Euler replaced "art of invention" (*ars inveniendi*), a very popular term in the works of Tschirnhaus and in other works of Leibniz's time, by "method of invention," a remarkable turn toward systematization. This book, a landmark in the development of the subject, contained the *Euler equation*

$$\frac{\partial L}{\partial y} - \frac{d}{dx} \left(\frac{\partial L}{\partial y'} \right) = 0, \tag{2}$$

(first published by Euler in 1736) as a necessary condition for $y(x)$ satisfying (1) to yield a minimum of $J[y]$. In more explicit form it is the equation

$$L_{y'y'} y'' + L_{y'y} y' + L_{y'x} - L_y = 0. \tag{3}$$

This equation suggests calling (1) a *regular problem* when $L_{y'y'}$ is never zero, and then assuming that $L_{y'y'} > 0$.

Euler's book also contains a fascinating collection of 66 problems. Carathéodory, the editor of the book as a volume of Euler's *Opera Omnia*, said that it

> is one of the most beautiful mathematical works ever written. We cannot emphasize enough the extent to which that *Lehrbuch* over and over again served later generations as a prototype in the endeavor of presenting special mathematical material in its [logical, intrinsic] connection.

METHODUS

INVENIENDI

LINEAS CURVAS

Maximi Minimive proprietate gaudentes,

SIVE

SOLUTIO

PROBLEMATIS ISOPERIMETRICI

LATISSIMO SENSU ACCEPTL

AUCTORE

LEONHARDO EULERO,

Profeſſore Regio, & Academiæ Imperialis Scientia-
rum PETROPOLITANÆ *Socio.*

LAUSANNÆ & GENEVÆ,

Apud MARCUM-MICHAELEM BOUSQUET & Socios.

MDCCXLIV.

Title page of *Methodus inveniendi lineas curvas...*

Euler's inspiration came from geometry and even more from the *principle of least action*, according to which nature realizes all motions in the most economical manner; more precisely, among all possible ways of reaching a given goal, nature chooses the one which minimizes the *action integral* $\int mv\, ds$ over the path (m = mass, v = speed, s = arc length). The beginning of the principle is often dated back to Leibniz because of a (lost) letter he is supposed to have written on the principle in 1707, but the question is still an open one. The principle is usually named after de Maupertuis (1698–1759), president of the Berlin Academy under Frederick the Great. Actually, Euler most likely discovered it earlier, formulated it mathematically more rigorously, and applied it to a nontrivial problem (involving central forces). In contrast to this, Maupertuis published the principle (in 1744 and 1746) in a vague and almost theological form. He defended vigorously his (questionable) priority, but failed to realize that a rigorization of the principle would call for specfication of conditions to be satisfied by the motions with which the actual motion was to be compared. Accordingly his main merit seems to be that he was *searching for* a minimum principle.

The great significance of the calculus of variations in mathematical physics is due to the transparent and coordinate-free form that the laws of nature take in this calculus. That fact became apparent in the work of Euler, and even more impressively in that of Lagrange. In his path-breaking memoir *Essai d'une nouvelle méthode pour déterminer les maxima et les minima des formules intégrales indéfinies* (1760–61) Lagrange substantially overtook Euler (as Euler was well aware). His work was a milestone in the development of the field and in its application to geometry and analytical mechanics. In it he invented the "method of variations" together with the symbol δ. His new idea was to use *"comparison functions,"*

$$\tilde{y} = y + \varepsilon\eta, \quad \eta \in C^2([x_0, x_1]), \quad \eta(x_0) = \eta(x_1) = 0, \tag{4}$$

in (1) and to conclude from the vanishing of the first variation of (1),

$$\delta J = \varepsilon \left.\frac{\partial J[\tilde{y}]}{\partial \varepsilon}\right|_{\varepsilon=0} = \varepsilon \cdot \int_{x_0}^{x_1} (L_y\eta + L_{y'}\eta') \, dx = 0, \tag{5}$$

(and an integration by parts) that Euler's equation (2) gave a necessary condition for $y(x)$ to render $J[y]$ stationary. (For a detailed and transparent explanation of this vital point see pp. 505–508 of G. F. Simmons, *Differential Equations*, second ed., McGraw-Hill, 1991.) In that paper Lagrange also started working on problems with variable endpoints, with application to brachystochrone and other problems. As another important step forward, he explicitly formulated his *multiplier rule* (without proof). This rule became a basic tool in his *Méchanique analitique*, in which he also included his theory of the calculus of variations and derived from the principle of least action his *equations of motion*, equivalent to *Newton's second law* and constituting analogues of Euler's equation. They are:

$$\frac{\partial V}{\partial x_i} + \frac{d}{dt}\left(\frac{\partial T}{\partial x_i}\right) = 0, \quad i = 1, 2, 3, \tag{6}$$

where V and T are the potential and kinetic energy, respectively.

Whereas Euler's interest in the calculus of variations centered around applications, Lagrange's emphasis was on algorithmic aspects of analysis. Lagrange's entire work excels by its wealth of original discoveries as well as by the outstanding assimilation of the historical material.

> By generalizing Euler's method he arrived at his remarkable formulas which in one line contain the solution of all problems of analytical mechanics.

> [In his *Memoir* of 1760–61] he created the whole calculus of variations with one stroke. This is one of the most beautiful articles that have ever been written. The ideas follow one another like lightning with the greatest rapidity...

These enthusiastic lines are taken from lecture notes by C. G. J. Jacobi (1804–51).

4. Minimal Surfaces

Euler's *Methodus inveniendi* of 1744 marked not only the beginning of the *theory* of the calculus of variations but also of one of its most fascinating geometric applications related to the creation of a remarkable class of surfaces called *minimal surfaces*. These were originally obtained from the calculus of variations as (portions of) surfaces of least area among all surfaces bounded by a given space curve. Nowadays we define them as surfaces with vanishing mean curvature H,

$$H = \frac{1}{2}(\kappa_1 + \kappa_2) = \frac{1}{2}\left(\frac{GL - 2FM + EN}{EG - F^2}\right) = 0, \tag{7}$$

using the discovery of Meusnier (1756–93) in 1776 that (7) is a necessary condition for least area. Here κ_1 and κ_2 are the principal curvatures, and E, F, G and L, M, N are the respective coefficients of the first and second fundamental forms of the surface.

What is most important to us here is Euler's discovery of the first non-trivial minimal surface, the *catenoid*. Euler obtained it by minimizing area, as the surface generated by rotating a *catenary* [a cosh curve, the curve of a hanging chain (*catena*) or cable], say,

$$r = A \cosh x, \tag{8}$$

where r is the distance, in 3-dimensional space, from the x-axis.

Euler's discovery of the catenoid was a major accomplishment in his geometric work and marked the beginning of the study of minimal surfaces. It was followed by Lagrange's systematic theory developed in his memoir of 1760–61. In this paper and in a subsequent one he extended his method to *double integrals* for functions of two variables.

$$J[z] = \int_\Omega \int L(x, y, z, p, q)\, dx\, dy \quad (p = z_x, \ q = z_y) \tag{9}$$

over a domain Ω in the xy-plane subject to given boundary conditions; the corresponding *Euler-Lagrange equation* [taking the place of (2)] is

$$L_z - \frac{\partial}{\partial x} L_p - \frac{\partial}{\partial y} L_q = 0. \tag{10}$$

5. Legendre, Jacobi, Weierstrass

In the calculus, $y' = 0$ is only a necessary condition for a minimum of a function $y(x)$, and for a decision one must also consider y''. Similarly, in the calculus of variations Euler's equation is only a necessary condition for a minimum, and for a decision one must also consider the *second variation* of (1),

$$\delta^2 J = \frac{\varepsilon^2}{2} \left.\frac{\partial^2 J[\tilde{y}]}{\partial \varepsilon^2}\right|_{\varepsilon=0} = \frac{\varepsilon^2}{2} \int_{x_0}^{x_1} \left(L_{yy}\eta^2 + 2L_{yy'}\eta\eta' + L_{y'y'}\eta'^2\right) dx, \tag{11}$$

introduced by Legendre (1752–1833) in 1786. It was formally motivated by Taylor's theorem

$$J[y + \varepsilon\eta] = J[y] + \delta J + \delta^2 \tilde{J}, \tag{12}$$

where the tilde means that the arguments are $y + \tilde{\varepsilon}\eta$, $y' + \tilde{\varepsilon}\eta'$ with $\tilde{\varepsilon} \in (0, \varepsilon]$. Legendre obtained the condition $L_{y'y'} \geq 0$ along a minimizing curve and $L_{y'y'} \leq 0$ along a maximizing curve (very similar to the calculus!) but he did not justify his analysis completely.

In fact, it took another fifty years before Jacobi succeeded in rigorously demonstrating that $L_{y'y'} > 0$ and the so-called *Jacobi condition*, which asserts that x_1 should be closer to x_0 than the so-called conjugate point[3] of x_0, suffice for a local minimum, that is a minimizing $\tilde{y}$ among $y \in C^1[x_0, x_1]$ satisfying the boundary conditions in (1) and lying close to $\tilde{y}$ in the C^1 sense, that is, satisfying

$$\text{(a)} \quad |y - \tilde{y}| < \rho, \qquad \text{(b)} \quad |y' - \tilde{y}'| < \rho \quad \text{for small positive } \rho. \tag{13}$$

Jacobi's discovery of the conjugate point and of its significance closed a substantial gap. However, condition (13b) seemed to be too retrictive and in no way suggested by the nature of the problem. Weierstrass emphasized that one should extend the domain of (1) and consider *strong minima*, that is, one should drop (13b).

For this program of obtaining a sufficient condition for a strong minimum, Weierstrass set up entirely new machinery centering around two ingenious concepts. The first was a *field of extremals* of (1), which he defined as a domain Ω in the xy-plane such that through each of its points there passes precisely one extremal of a one-parameter family of extremals of (1) (solution curves of Euler's equation) depending continuously on the parameter and the second was the so-called E-function, a turning point in the history of the calculus of variations.

To define the E-function, Weierstrass started from the *slope function* $p = p(x, y)$, the slope at (x, y) of the extremal of a field of extremes $y = h(x, \alpha)$; thus

$$p(x, y) = h'(x, \alpha)\big|_{\alpha = \alpha(x,y)}, \tag{14}$$

where $'$ refers to differentiation with respect to x. Then he defined the E-function by

$$E(x, y, p, y') = L(x, y, y') - L(x, y, p) - (y' - p)L_{y'}(x, y, p), \tag{15}$$

where $y = y(x)$ is any C_1-curve in the region covered by the field of extremals. Now he could prove that if for an extremal $y = \tilde{y}(x)$ of the field, the above sufficient conditions for a local minimum are satisfied, and if $E \geq 0$ at every point in the field and for every y', then $\tilde{y}(x)$ gives a strong minimum of (1).

In addition to his path-breaking new method, Weierstrass also revolutionized the calculus of variations by stressing—practically for the first time—the importance of a precise definition of the domain $D(J)$ of the function $J[y]$ and of *admissible functions*, the functions $y \in D(J)$ satisfying the side conditions.

---

[3] The conjugate point of x_0 is the first value $x > x_0$ where a nonzero solution of

$$\frac{d}{dx}\left(L_{y'y'}\frac{dw}{dx}\right) - \left(L_{yy} - \frac{d}{dx}L_{yy'}\right)w = 0, \quad w(x_0) = 0 \quad (x \geq x_0)$$

vanishes. Here $w(x) = \partial y / \partial \alpha|_{\alpha=0}$ and $\alpha = 0$ corresponds to $\tilde{y}$ in the family of extremals $y = y(x, \alpha)$.

Some Remarks and Problems in Number Theory Related to the Work of Euler[1]

Paul Erdős and Underwood Dudley

Motto. *One is mathematics and Euler is its prophet.* This phrase was coined half as a joke at a mathematical party in Budapest about 50 years ago by Tibor Gallai. In these remarks we mention some of the things the prophet Euler has handed down to us and sometimes give some later developments. Many of the recollections and conjectures in these remarks are those of the first author, and first person references are used to keep the exposition informal.

In 1737 Euler proved that the number of primes was infinite by showing that the sum of their reciprocals diverges, i.e.,

$$\sum_{p \text{ prime}} \frac{1}{p} = \infty. \tag{1}$$

He did this by using the (invalid) identity

$$\sum_{n=1}^{\infty} \frac{1}{n} = \prod_{p} \left(1 - \frac{1}{p}\right)^{-1}$$

Though invalid—Euler rarely worried about convergence—it can be fixed by looking at

$$\sum_{n=1}^{\infty} \frac{1}{n^s} = \prod_{p} \left(1 - \frac{1}{p^s}\right)^{-1}$$

as $s \to 1$. For this, see Ayoub [1], who said elsewhere [2] that Euler "laid the foundations of analytic number theory."

Denote by $\pi(x)$ the **number of primes** $p \le x$. It is curious that Euler after having proved (1) never asked himself: how does $\pi(x)$ behave for large x? For (1) immediately implies that for infinitely many $x, \pi(x) > x^{1-\varepsilon}$. In fact, for infinitely many x,

[1] Reprinted from the *Mathematics Magazine*, Vol. 56, November, 1983, pp. 292–298.

$\pi(x) > x/(\log x)^{1+\varepsilon}$. It seems to me that with a little experimentation Euler could have discovered the prime number theorem

$$\lim_{x \to \infty} \frac{\pi(x)}{x/\log x} = 1.$$

After all, he did discover the quadratic reciprocity theorem by observation, and that seems to be at least as hard to see. But as we will see again later, such questions did not seem to occur to Euler. The prime number theorem was first conjectured shortly before Euler's death by Legendre in 1780 in the form

$$\pi(x) \approx \frac{x}{\log x - c},$$

with $c \approx 1.08$. In 1792 Gauss, who was only 15 at the time, even noticed that

$$\int_2^x \frac{dy}{\log y} = \sum_{k=2}^x \frac{1}{\log k} + O(1)$$

gives a much better approximation to $\pi(x)$ than $x/\log x$, a most remarkable achievement! Again, it is strange that Gauss and others did not prove that

$$\frac{c_1 x}{\log x} < \pi(x) < \frac{c_2 x}{\log x}$$

and that if $\lim_{x \to \infty} \pi(x)/(x/\log x)$ exists, then it must be 1. All these results were proved by Tchebychef around 1850. Both Euler and Gauss could easily have proved all this. The prime number theorem was first proved by Hadamard and de la Vallée Poussin in 1896 using analytic functions, which were not available to Euler and Gauss.

More than 40 years ago, I conjectured that if $1 \le a_1 < a_2 < \cdots$ is a sequence of integers for which

$$\sum_{n=1}^{\infty} \frac{1}{a_n} = \infty, \tag{2}$$

then the sequence $\{a\}$ contains arbitrarily long arithmetic progressions. This conjecture is still not settled; I offer \$3,000 for its proof or disproof. If my conjecture is true, then Euler's result that $\sum 1/p$ diverges immediately implies that the primes contain arbitrarily long arithmetic progressions. Until this year, the longest such progression known was due to Weintraub [33] and has 17 terms: $3430751869 + 87297210t$, $t = 0, 1, \ldots, 16$. With patience and a good computer one could probably find more primes in arithmetic progression. In fact, 18 such primes were found by P. Pritchard, who reported this in January 1983 at the AMS meeting in Denver. The discovery was also described in *The Chronicle of Higher Education*, 2/9/83, p. 27.

It often happens that a problem on primes can be solved by generalizing it, and proving it for some more general sequences which share some property with the primes, such as being equally numerous. Even using this idea, my \$3,000 problem really seems to be very deep. Schur conjectured, and van der Waerden proved [30], that if we divide the integers into two classes, then at least one of the classes contains arbitrarily long arithmetic progressions. Fifty years ago, Turan and I conjectured [13] that if $r_k(n)$ is the smallest integer such that every sequence of integers of the form

$$1 \le a_1 < a_2 < \cdots < a_{r_k(n)} \le n$$

contains an arithmetic progression of k terms, then for every k,

$$\lim_{n \to \infty} \frac{r_k(n)}{n} = 0.$$

This conjecture is clearly stronger than van der Waerden's theorem, but weaker than (2). About 30 years ago K. F. Roth [28] proved the conjecture for $k = 3$. The general conjecture was finally proved by Szémeredi in 1972 [29]. For further information see [14].

A much stronger conjecture on primes states that for every k there are k *consecutive* primes which form an arithmetic progression. The longest known has only six terms: $121174811 + 30t, t = 0, 1, \ldots, 5$ [19]. This conjecture is undoubtedly true but is completely unattackable by the methods at our disposal.

Denote by $p(n)$ the **number of unrestricted partitions** of n, that is, the number of ways of writing n as a sum of positive integers. For example, $p(5) = 7$ because $1 + 1 + 1 + 1 + 1 = 2 + 1 + 1 + 1 = 2 + 2 + 1 = 3 + 1 + 1 = 3 + 2 = 4 + 1 = 5$. Leibniz asked Bernoulli about $p(n)$ in 1669, but it was not until Euler saw that

$$1 + \sum_{n=1}^{\infty} p(n)x^n = \prod_{n=1}^{\infty} (1 - x^n)^{-1}$$

and ingeniously proved that

$$\prod_{n=1}^{\infty} (1 - x^n) = \sum_{n=-\infty}^{\infty} (-1)^n x^{n(3n+1)/2}$$

that any progress was made. Combining the last two equations gives a recursion relation

$$p(n) = p(n - 1) + p(n - 2) - p(n - 5) - p(n - 7) + p(n - 12) + \cdots$$

that lets values like $p(200) = 397299029388$ [15] be calculated. This was the start of generating functions.

As far as I know, Euler never tried to estimate $p(n)$ as a function of n. Hardy-Ramanujan [15] and Uspensky were the first to obtain the asymptotic formula for $p(n)$,

$$p(n) \sim \frac{1}{4\sqrt{3}n} e^{\pi \sqrt{2n/3}}. \tag{3}$$

In 1937 Rademacher [26] found a convergent series for $p(n)$ and later I [11] and Newman [24] gave an elementary proof of (3). These estimates are complicated, but the inequality

$$e^{c_1 \sqrt{n}} < p(n) < e^{c_2 \sqrt{n}}$$

could very easily have been obtained by Euler. These questions which seem so natural to us now must not have occurred to Euler. It could have been that the idea of function was not yet a natural one. Euler was more concerned with representing integers in various forms. He spent 40 years, off and on, trying to prove that every positive integer is a sum of four squares, only to have Legendre give the first proof in 1770. And think of how much time he must have spent on doing things like the following, finding integers x, y, z, w such that

486   Zweyter Abschnitt

Tabelle

Für die Zahlen welche in der Form $m^4 - n^4$
enthalten sind

mn	n·n	mm − nn	mm + nn	$m^4 - n^4$
4	1	3	5	3·5
9	1	8	10	16·5
9	4	5	13	5·13
16	1	15	17	3·5·17
16	9	7	25	25·7
25	1	24	26	16·3·13
25	9	16	34	16·2·17
49	1	48	50	25·16·2·3
49	16	33	65	3·5·11·13
64	1	63	65	9·5·7·13
81	49	32	130	64·5·13
121	4	117	125	25·9·5·13
121	9	112	130	16·2·5·7·13
121	49	72	170	144·5·17
144	25	119	169	169·7·17
169	1	168	170	16·3·5·7·17
169	81	88	250	25·16·5·11
225	64	161	289	289·7·23

Hier-

Euler's *Algebra* (v. 2, chap. 15, SS235, p. 486) contains this table of squares m^2, n^2, their differences and sum, and $m^4 - n^4$ (the left column heading has a printer's error).

$x \pm y$, $x \pm z$, $y \pm z$ are all squares (see facing page),

$xy \pm x$, $xy \pm y$ are all squares,

$x^2 + y^2$, $x^2 + z^2$, $y^2 + z^2$ are all squares,

$x^2 + y^2 + z^2$, $x^2 + y^2 + w^2$, $x^2 + z^2 + w^2$, $y^2 + z^2 + w^2$ are all squares,

$x + y$ is a square, $x^2 + y^2$ is a cube,

$x + y + z$, $xy + yz + zx$, xyz are all squares,

and so on ([8], ii, XV–XXI). Perhaps not many today are very interested in this.

Von der unbeſtimmten Analytic. 487

Hieraus können wir ſchon einige Auflöſungen geben: man nehme nemlich ff = 9 und kk = 4, ſo wird f⁴−k⁴=13. 5: ferner nehme man gg=81, und hh=49, ſo wird g⁴ − h⁴ = 64. 5. 13, woraus tt = 64. 25. 169; folglich t = 520. Da nun tt=270400; f=3 ; g=9; k=2 ; h = 7, ſo bekommen wir a = 21; b = 18; hieraus p = 117 , q = 765 und r=756; daraus findet man 2 x = tt + pp + qq = 869314 und alſo x=434657; dahero ferner y=x−pp=420968; und endlich z=x−qq= −150568; welche Zahl auch poſitiv genommen werden kann, weil alsdann die Summe in der Differenz und umgekehrt die Differenz in der Summe verwandelt werden; folglich ſind unſere drey geſuchten Zahlen.

$$x = 434657$$
$$y = 420968$$
$$z = 150568$$

dahero wird x + y = 855625 = (925)²
x + z = 585225 = (765)²
y + z = 571536 = (756)²

und weiter x − y = 13689 = (117)²
x − z = 284089 = (533)²
y − z = 270400 = (520)²

♄ 4   Noch

Euler then demonstrates (p. 487), using his table values, how to obtain integers x, y, z such that sums and differences of any pair of these is a square. In his example, he obtains $x = 434,657$, $y = 420,968$, $z = 150,568$.

Euler was the first to consider the function $\phi(n)$, the **number of integers** $1 \le m < n$ **relatively prime to** n, and this function bears his name (see Glossary, p.). Euler derived a formula equivalent to the well-known

$$\phi(n) = n \prod_{p \mid n} \left(1 - \frac{1}{p} \right)$$

but he never investigated the function any further, though a great deal of work has been done on it since. It is one more example of Euler's lack of curiosity about functions. There is still a surprising number of unsolved problems about $\phi(n)$. Carmichael conjectured [4] that the number of solutions of $\phi(n) = m$ can never be 1 (i.e., if $\phi(n_1) = m$ then there is an $n_2 \neq n_1$ with $\phi(n_2) = m$). Though the conjecture is known to be true for $m < 10^{400}$ [17], it is probably unattackable by the methods at our disposal. I proved [10] that if there is an integer m for which $\phi(n) = m$ has k solutions, then there are infinitely many integers with this property. If n is prime, $\phi(n) = n-1$ of course; Lehmer conjectured [21] that $\phi(n)$ divides $(n - 1)$ only if n is prime. This conjecture also seems unattackable. On the other hand, it is an easy exercise to show that $\phi(n)$ divides n if and only if $n = 2^{\alpha}3^{\beta}$. R. L. Graham has conjectured that for every a there are infinitely many n for which $\phi(n)$ divides $n + a$.

The well-known conjecture of Fermat states that $x_1^k + x_2^k = x_3^k$ has no positive integer solutions for $k > 2$. Euler proved the statement for $k = 4$ and almost proved it for $k = 3$ (see [8], ii, XXI, XXII). It has recently been proved by Wagstaff for all $k \leq 125,000$ [31]. The general conjecture seems to be out of reach at present. [Ed. Note: In 1995 Andrew Wiles, with an assist from Richard Taylor, proved Fermat's last theorem.] Euler conjectured the following generalization:

$$x_k^k = x_1^k + x_2^k + \cdots + x_{k-1}^k$$

has no nontrivial solution in integers for $k \geq 3$. This conjecture was disproved by Lander and Parkin [18] who found the equation

$$144^5 = 133^5 + 110^5 + 84^5 + 27^5.$$

This is so far the only known counterexample. The case $k = 4$ seems to be of special interest; in 1948 M. Ward [32] showed that there are no nontrivial solutions for $x_4 \leq 10,000$, and in 1967 Lander, Parkin and Selfridge [20] extended the result to $x_4 < 220,000$. [Ed. Note: In 1988 Noam Elkies found the following counterexample to the $k = 4$ case: $2682440^4 + 15365639^4 + 18796760^4 = 20615673^4$] Euler was not even able to find four fourth powers whose sum is a fourth power and it was only in 1911 that the example

$$353^4 = 315^4 + 272^4 + 120^4 + 30^4$$

was found by R. Norrie ([8]; ii, XXII).

In the same direction, Euler gave a complete parametric solution of the equation

$$x^3 + y^3 = u^3 + v^3,$$

namely,

$$x = 1 - (a - 3b)(a^2 + 3b^2) \qquad u = (a + 3b) - (a^2 + 3b^2)^2$$
$$y = (a + 3b)(a^2 + 3b^2) - 1 \qquad v = (a^2 + 3b^2)^2 - (a - 3b)$$

and proved that for infinitely many integers n, $n = x^4 + y^4 = u^4 + v^4$ by giving a complicated parametric solution [16] which includes the smallest solution

$$133^4 + 134^4 = 158^4 + 59^4 = 635,318,657.$$

After Ramanujan surprised Hardy by knowing that

$$1729 = 10^3 + 9^3 = 12^3 + 1^3$$

was the smallest integer which is the sum of two cubes in more than one way, Hardy asked him if he knew any integer which was the sum of two fourth powers in more than one way. Ramanujan answered that he did not know any such numbers, and if they existed, they must be very large. Thus, both were unaware of the old work by Euler. It is not yet known if there are any integers which are the sum of two fourth powers in *more* than two ways, i.e., if the number of solutions of $n = x^4 + y^4$ is at most 2.

Denote by $f_3^{(2)}(n)$ the number of solutions of $n = x^3 + y^3$. Mordell proved that $\limsup_{n\to\infty} f_3^{(2)}(n) = \infty$ and Mahler [23] proved that $f_3^{(2)}(n) > (\log n)^{1/4}$ for infinitely many n. As far as I know there is no nontrivial upper bound known for $f_3^{(2)}(n)$. Very likely $f_3^{(2)}(n) < c_1(\log n)^{c_2}$ for all n, if c_1 and c_2 are sufficiently large absolute constants.

Euler was the first to evaluate $\sum_{n=1}^{\infty} 1/n^2$. In 1731 he obtained the sum accurate to 6 decimal places, in 1733 to 20, and in 1734 to infinitely many $(= \pi^2/6)$. Ayoub [2] said about his proof that "it opened up the theory of infinite products and partial fraction decomposition of transcendental functions and its importance goes far beyond the immediate application." Euler studied further what we now call the **Riemann ζ-function** $(= \sum_{n=1}^{\infty} n^{-s}$ when $\text{Re}(s) > 1)$ and in 1749 he proved the functional equation

$$\zeta(1 - s) = \pi^{-s} 2^{1-s} \Gamma(s) \cos \frac{\pi s}{2} \zeta(s)$$

for $s = 1, 2, \ldots$ and said that he was certain it was true for all real s. It was not until 1859 that Riemann proved this.

As far as we know, Euler was the first to define **transcendental numbers** as numbers which are not the roots of algebraic equations. It is perhaps curious that he never proved their existence. The proof of Liouville was well within his reach. Maybe Euler considered the existence of transcendental numbers as self-evident, which by our standards, is certainly not the case.

Of course, not even Euler was perfect. His proofs of Fermat's last theorem for exponent 3, as well as his proof that every prime has a primitive root, are considered incomplete by our present standards. He regularly used infinite series without paying any attention to convergence (nevertheless his proofs are almost always correct except for rigor, which is easy to supply).

However, in at least one instance, Euler's intuition completely misled him and he produced a false "proof" which could not be corrected by methods at his disposal. Euler wanted to prove that $\sum_{n=1}^{\infty} \mu(n)/n = 0$, where $\mu(1) = 1$, $\mu(n) = 0$ if n is not square-free and $\mu(n) = (-1)^k$ if n is the product of k distinct primes ($\mu(n)$ is known as the Möbius function). He simply argued as follows:

$$\sum_{n=1}^{\infty} \frac{\mu(n)}{n} = \prod_p \left(1 - \frac{1}{p}\right) = 0. \tag{4}$$

This argument is, of course, inaccurate, since $\sum_{n=1}^{\infty} \mu(n)/n$ is not absolutely convergent and (4) was first proved correctly by von Mangoldt at the end of the nineteenth century. Equation (4) is known to be equivalent to the prime number theorem.

Another error of a different kind was pointed out to me by Mordell. Euler proved that if p divides $x^2 + ny^2$ without dividing both x and y, then p is $u^2 + nv^2$ for $n = 1, 2, 3$. He then used the same arguments for $n = 5$, though Fermat knew long before, and Euler knew too, that the conclusion was not true. (We know now that the reason is that unique factorization fails.) Edwards [9] thinks it was Euler's age, his blindness, or his secretary that may have caused the mistake.

We close with some of the less important things Euler did, to give an idea of his immense range and power. Before Euler, only three pairs of **amicable numbers** were known. These are pairs like 220 and 284, where the sum of the proper divisors of one of the numbers is equal to the other:

$$110 + 55 + 44 + 22 + 20 + 11 + 10 + 5 + 4 + 2 + 1 = 284 \quad \text{and} \quad 142 + 71 + 4 + 2 + 1 = 220.$$

The pair (220, 284) was known to Pythagoras; another pair, (17296, 18416), was found by Fermat in 1636; and the pair (9363584, 9437056) was found by Descartes in 1638. In 1750, Euler gave 62 new pairs ([8], i, I). Amicable pairs are still studied. There were 1095 pairs known in 1972 [22] and a 152-digit pair was found in 1974 [27]. In 1955 I showed [12] that if $A(x)$ is the number of amicable pairs (m, n) with $m \le n \le x$, then $\lim_{x \to \infty} A(x)/x = 0$; Pomerance showed in 1981 [25] that $A(x) < x e^{-(\log x)^{1/3}}$. In the other direction, I conjecture that there are infinitely many pairs; in fact, it is likely that $A(x) > c x^{1-\varepsilon}$.

In a letter, Goldbach called Euler's attention to **multigrades**: sets of integers with equal sums of different powers, as in

$$1^k + 5^k + 9^k + 17^k + 18^k = 2^k + 3^k + 11^k + 15^k + 19^k$$

for $k = 0, 1, 2, 3, 4$, and Euler proved the first theorems about them. They have been studied a great deal since then. It was also in a letter to Euler that Goldbach made his famous conjecture that every even integer greater than 4 is a sum of two primes, and that has been studied even more than multigrades. There has not been much progress since Chen showed in 1966 [5] that every sufficiently large even integer is a sum of a prime and a product of at most two primes.

Euler discovered that if $p = 4k + 1$ is a prime, then p can be written $p = x^2 + y^2$ in exactly one way; this led him to look for numbers d such that if $n = x^2 + dy^2$ with $(x, y) = 1$ in exactly one way, then n is prime. He found 65 of them, with 1848 the largest ([8], ii, XIV). It seems likely that he found them all, since it is known that their number is finite [6] and there is at most one greater than 10^{65} [7]. So in a way, Euler said the first and last words on this subject.

Euler proved that every **even perfect number** (i.e., equal to the sum of its proper divisors, as $28 = 14 + 7 + 4 + 2 + 1$) is of the form $2^{p-1}(2^p - 1)$ for p and $2^p - 1$ prime and gave the first of a long list of necessary conditions that an odd perfect number will have to satisfy ([8], i, I).

Fermat thought all the Fermat numbers $2^{2^n} + 1$ were prime. Euler factored $2^{2^5} + 1$ in 1732; $2^{2^7} + 1$ was not factored until 1971 [3].

Euler was the first to look at that equation that keeps coming up in popular journals, $x^y = y^x$ ([8], ii, XXIII).

And Euler discovered, no one knows how, that the polynomial $n^2 - n + 41$ is a prime for $n = 1, 2, \ldots, 40$.

If Euler had never done anything *except* number theory, he would still be remembered as one of the great mathematicians.

References

1. R. Ayoub, Euler and the Zeta function, *Amer. Math. Monthly*, 81 (1974) 1067–1085. Reprinted here on pp. 113–132.

2. ———, *An Introduction to the Analytic Theory of Numbers*, Amer. Math. Soc., Providence, RI, 1963.

3. J. Brillhart and M. A. Morrison, The factorization of F_7, *Bull. Amer. Math. Soc.*, 77 (1971) 264.

4. R. D. Carmichael, Note on Euler's ϕ-function, *Bull. Amer. Math. Soc.*, 28 (1922) 109–110.

5. J. R. Chen, On the representation of a large even integer as the sum of a prime and the product of at most two primes, *Kexue Tongbao*, 17 (1966) 385–386.

6. S. Chowla, An Extension of Heilbronn's class-number theorem, *Quart. J. Math. Oxford Ser.*, 5 (1934) 304–307.

7. S. Chowla and W. E. Briggs, On discriminants of binary quadratic forms with a single class in each genus, *Canad. J. Math.*, 6 (1954) 463–470.

8. L. E. Dickson, *History of the Theory of Numbers*, Chelsea, New York, 1952 reprint of the 1919 edition.

9. H. M. Edwards, The genesis of ideal theory, *Arch. Hist. Exact Sci.*, 23 (1980) 321–378.

10. P. Erdős, On the normal number of prime factors of $p - 1$ and some related problems concerning Euler's ϕ-function, *Quart. J. Math. Oxford Ser.*, 6 (1935) 205–213.

11. ———, On an elementary proof of some asymptotic formulas in the theory of partitions, *Ann. of Math.*, (2) 43 (1942) 437–450.

12. ———, On amicable numbers, *Publ. Math. Debrecen*, 4 (1955) 108–111.

13. P. Erdős and P. Turan, On some sequences of integers, *J. London Math. Soc.*, 11 (1936) 261–264.

14. R. L. Graham, B. Rothschild, and J. Spencer, *Ramsey Theory*, Wiley-Interscience, New York, 1980.

15. G. H. Hardy and S. Ramanujan, Asymptotic formulae in combinatory analysis, *Proc. London Math. Soc.*, (2) 17 (1918) 75–115.

16. G. H. Hardy and E. M. Wright, *The Theory of Numbers*, 4th ed., Oxford U. Press, 1960.

17. V. J. Klee, Jr., On a conjecture of Carmichael, *Bull. Amer. Math. Soc.*, 53 (1947) 1183–1186.

18. L. J. Lander and T. R. Parkin, A counterexample to Euler's sum of powers conjecture, *Math. Comp.*, 21 (1967) 101–103.

19. ———, Consecutive primes in arithmetic progression, *Math. Comp.*, 21 (1967) 489.

20. L. J. Lander, T. R. Parkin and J. L. Selfridge, A survey of equal sums of like powers, *Math. Comp.*, 21 (1967) 446–453.

21. D. H. Lehmer, On Euler's totient function, *Bull. Amer. Math. Soc.*, 38 (1932) 745–757.

22. E. J. Lee and J. S. Madachy, The history and discovery of amicable numbers I, II, III, *J. of Recreational Math.*, 5 (1972) 77–93, 153–173, 231–249.

23. K. Mahler, On the lattice points of curves of genus 1, *Proc. London Math. Soc.*, (2) 39 (1935) 431–466.

24. D. J. Newman, The evaluation of the constant in the formula for the number of partitions of n, *Amer. J. Math.*, 73 (1951) 599–601.

25. C. Pomerance, On the distribution of amicable numbers II, *J. Reine Angew. Math.*, 325 (1981) 183–188.

26. H. Rademacher, On the partition function $p(n)$, *Proc. London Math. Soc.*, (2) 43 (1937) 241–254.

27. H. J. J. te Riele, Four large amicable pairs, *Math. Comp.*, 28 (1974) 309–312.

28. K. F. Roth, On certain sets of integers II, *J. London Math. Soc.*, 29 (1954) 20–26.

29. E. Szémeredi, On sets of integers containing no k elements in arithmetic progression, *Acta Arith.*, 27 (1975) 199–245.

30. B. L. van der Waerden, Beweis einer baudetschen Vermuting, *Nieuw Arch. Wisk.*, 15 (1928) 212–216.

31. S. S. Wagstaff, Jr., The irregular primes to 125,000, *Math. Comp.*, 32 (1978) 583–591.

32. M. Ward, Euler's problem on fourth powers, *Duke Math. J.*, 15 (1948) 827–837.

33. S. Weintraub, Seventeen primes in arithmetic progression, *Math. Comp.*, 31 (1977) 1030.

$\mathcal{E}uler's\ \mathcal{P}entagonal\ \mathcal{N}umber\ \mathcal{T}heorem$ [1]

George E. Andrews

One of Euler's most profound discoveries, the *Pentagonal Number Theorem* [7], has been beautifully described by André Weil:

> Playing with series and products, he discovered a number of facts which to him looked quite isolated and very surprising. He looked at this infinite product
>
> $$(1 - x)(1 - x^2)(1 - x^3) \cdots$$
>
> and just formally started expanding it. He had many products and series of that kind; in some cases he got something which showed a definite law, and in other cases things seemed to be rather random. But with this one, he was very successful. He calculated at least fifteen or twenty terms; the formula begins like this:
>
> $$\prod(1 - x^n) = 1 - x - x^2 + x^5 + x^7 - x^{12} - x^{15} \cdots$$
>
> where the law, to your untrained eyes, may not be immediately apparent at first sight. In modern notation, it is as follows:
>
> $$\prod_{1}^{\infty}(1 - q^n) = \sum_{-\infty}^{+\infty}(-1)^n q^{n(3n+1)/2} \tag{1}$$
>
> where I've changed x into q since q has become the standard notation in elliptic function theory since Jacobi. The exponents make up a progression of a simple nature. This became immediately apparent to Euler after writing down some 20 terms; quite possibly he calculated about a hundred. He very reasonably says, "this is quite certain, although I cannot prove it;" ten years later he does prove it. He could not possibly guess that both series and product are part of the theory of elliptic modular functions. It is another tie-up between number-theory and elliptic functions [22, pp. 97–98].

---

[1] Reprinted from the *Mathematics Magazine*, Vol. 81, November, 1983, pp. 279–284.

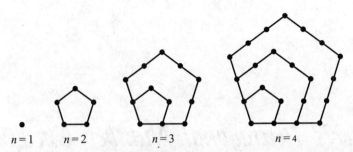

$n=1$ $n=2$ $n=3$ $n=4$

Figure 1. The first four pentagonal numbers are 1, 5, 12, 22.

G. Pólya [16, pp. 91–98] provides a more extensive account of Euler's wonderful discovery together with a translation of Euler's own description [6].

The numbers $n(3n-1)/2$ are called "pentagonal numbers" because of their relationship to pentagonal arrays of points. Figure 1 illustrates this. Legendre [14, pp. 131–133] observed that purely formal multiplication of the terms on the left side of (1) produces the term $\pm q^N$ precisely as often as N is representable as a sum of distinct positive integers; the "+" is taken when there is an even number of summands in the representation and the "−" when the number of summands is odd. For example,

$$(1-q)(1-q^2)(1-q^3)(1-q^4)\cdots$$
$$= 1 - q - q^2 - q^3 - q^4 + q^{1+2} + q^{1+3} + q^{1+4} + q^{2+3}$$
$$+ q^{2+4} + q^{3+4} - q^{1+2+4} - q^{1+3+4} - q^{2+3+4} + q^{1+2+3+4} + \cdots.$$

The term **partition** is usually used to describe the representation of a positive integer as the sum of positive integers. In this article, we are concerned with *unordered* partitions; two such partitions are considered the same if the terms in the sum are the same, e.g., $1 + 2$ and $2 + 1$ are considered as the same partition of 3. Thus Legendre's observation may be matched up with the actual infinite series expansion (1) as follows.

Theorem. *Let $p_e(n)$ denote the number of partitions of n into an even number of distinct summands. Let $p_o(n)$ denote the number of partitions of n into an odd number of distinct summands. Then*

$$p_e^n - p_o(n) = \begin{cases} (-1)^j & \text{if } n = j(3j \pm 1)/2 \\ 0 & \text{otherwise.} \end{cases} \tag{2}$$

The impact of Euler's Pentagonal Number Theorem and Legendre's observations on subsequent developments in number theory is enormous. Both (1) and (2) are justly famous. Indeed, F. Franklin's purely arithmetic proof of (2) [10] (see also [21, pp. 261–263]) has been described by H. Rademacher as the first significant achievement of American mathematics. Franklin's proof is so elementary and lovely that it has been presented many times over in elementary algebra and number theory texts [5, pp. 563–564], [11], [12, pp. 206–207], [13, pp. 286–287], [15, pp. 221–222].

It is, however, interesting to note that Euler's proof of (1) alluded to by Weil remains almost unknown. In recent years only Rademacher's book has contained an exposition of it [17, pp. 224–226]. This book and earlier books [4, pp. 23–24] have presented it more or less as Euler did. While the idea behind Euler's proof is ingenious (as one would expect), the mathematical notation of Euler's day hides the fact that other results of significance are either transparent corollaries of Euler's proof or lie just below the surface. The remainder of this article is devoted to a long overdue modern exposition of Euler's proof and an examination of its consequences.

To begin, we define a function of two variables:

$$f(x,q) = 1 - \sum_{n=1}^{\infty}(1-xq)(1-xq^2)\cdots(1-xq^{n-1})x^{n+1}q^n. \tag{3}$$

(Absolute convergence of all series and products considered is ensured by $|q| < 1$, $|x| < |q|^{-1}$). We first note that

$$f(1,q) = \prod_{n=1}^{\infty}(1-q^n). \tag{4}$$

This is because we may easily establish the identity

$$1 - \sum_{n=1}^{N}(1-q)(1-q^2)\cdots(1-q^{n-1})q^n = \prod_{n=1}^{N}(1-q^n) \tag{5}$$

by mathematical induction on N (a nice exercise for the reader). Thus (4) is the limiting case of (5) as $N \to \infty$.

The main step in Euler's proof is essentially the verification of the following functional equation:

$$f(x,q) = 1 - x^2q - x^3q^2 f(xq,q). \tag{6}$$

Actually Euler does equation (6) over and over, first with $x = 1$, then $x = q$, then $x = q^2$, and so on [7, pp. 473–475]; the title page from his original 1783 paper is shown on the next page. This repetition of special cases of (6) tends to hide what exactly is happening. To prove (6), we take the defining equation (3) through a sequence of algebraic manipulations:

$$f(x,q) = 1 - x^2q - \sum_{n=2}^{\infty}(1-xq)(1-xq^2)\cdots(1-xq^{n-1})x^{n+1}q^n$$

$$= 1 - x^2q - \sum_{n=1}^{\infty}(1-xq)(1-xq^2)\cdots(1-xq^n)x^{n+2}q^{n+1}$$

$$= 1 - x^2q - \sum_{n=1}^{\infty}(1-xq^2)\cdots(1-xq^n)x^{n+2}q^{n+1}(1-xq)$$

$$= 1 - x^2q - \sum_{n=1}^{\infty}(1-xq^2)\cdots(1-xq^n)x^{n+2}q^{n+1}$$

$$+ \sum_{n=1}^{\infty}(1-xq^2)\cdots(1-xq^n)x^{n+3}q^{n+2}$$

$$= 1 - x^2q - x^3q^2 - \sum_{n=2}^{\infty}(1-xq^2)\cdots(1-xq^n)x^{n+2}q^{n+1}$$

$$+ \sum_{n=1}^{\infty}(1-xq^2)\cdots(1-xq^n)x^{n+3}q^{n+2}$$

⁕⁘). 47 (⁘⁕

EVOLVTIO
PRODVCTI INFINITI

$$(1-x)(1-\dot{x}x)(1-x^3)(1-x^4)(1-x^5)(1-x^6)$$

IN SERIEM SIMPLICEM.

Auctore
L. EVLERO.

§. 1.

P osito $s=(1-x)(1-xx)(1-x^3)(1-x^4)$. etc. facile patet fore:

$s=1-x-xx(1-x)-x^3(1-x)(1-xx)-x^4(1-x)1-xx)(1-x^3)-$ etc. quae feries cum iam fit infinita, quaeritur, fi finguli eius termini euoluantur, qualis feries fecundum fimplices pote-ftates ipfius x fit prodicura. Cum igitur duo primi ter-mini $1-x$ iam fint euoluti, loco reliquorum omnium fcri-batur littera A, ita vt fit $s=1-x-A$, ideoque

$A=xx(1-x)+x^3(1-x)(1-xx)+x^4(1-x)(1-xx)(1-x^3)$etc.

§. 2. Quoniam hi termini omnes factorem habent communem $1-x$, eo euoluto finguli termini discerpentur in binas partes quas ita repraefentemus:

$A=xx+x^3(1-xx)+x^4(1-xx)(1-x^3)+x^5(1-xx)(1-x^3)(1-x^4)$
$\qquad -x^3-x^4(1-xx)-x^5(1-xx)(1-x^3)-x^6(1-xx)(1-x^3)(1-x^4)$

Hinc

First page of Euler's paper on the pentagonal number theorem

$$= 1 - x^2 q - x^3 q^2 - \sum_{n=1}^{\infty} (1-xq^2)\cdots(1-xq^{n+1})x^{n+3}q^{n+2}$$

$$+ \sum_{n=1}^{\infty} (1-xq^2)\cdots(1-xq^n)x^{n+3}q^{n+2}$$

$$= 1 - x^2 q - x^3 q^2 - \sum_{n=1}^{\infty} (1-xq^2)\cdots(1-xq^n)x^{n+3}q^{n+2}\{(1-xq^{n+1})-1\}$$

$$= 1 - x^2 q - x^3 q^2 \left\{ 1 - \sum_{n=1}^{\infty} (1-xq^2)\cdots(1-xq^n)x^{n+1}q^{2n+1} \right\}$$

$$= 1 - x^2 q - x^3 q^2 f(xq, q).$$

If you followed the above sequence of steps carefully, you see how at each stage things seem to fit together magically at just the right moment. Also you can appreciate the complication of repeated presentation of the same steps first with $x = 1$, then $x = q$, then $x = q^2$, etc. However, the empirical discovery of (6) by Euler must have come precisely in this repetitive manner.

The rest of Euler's proof is now almost mechanical. Equation (6) is repeatedly iterated; thus

$$
\begin{aligned}
f(x, q) &= 1 - x^2 q - x^3 q^2 \left(1 - x^2 q^3 - x^3 q^5 f(xq^2, q)\right) \\
&= 1 - x^2 q - x^3 q^2 + x^5 q^5 + x^6 q^7 \left(1 - x^2 q^5 - x^3 q^8 f(xq^3, q)\right) \\
&\vdots \\
&= 1 + \sum_{n=1}^{N-1} (-1)^n \left(x^{3n-1} q^{n(3n-1)/2} + x^{3n} q^{n(3n+1)/2}\right) \\
&\quad + (-1)^N x^{3N-1} q^{N(3N-1)/2} + (-1)^N x^{3N} q^{N(3N+1)/2} f(xq^N, q),
\end{aligned}
$$

(7)

another fine exercise in mathematical induction on N. In the limit as $N \to \infty$ we find

$$
f(x, q) = 1 + \sum_{n=1}^{\infty} (-1)^n \left(x^{3n-1} q^{n(3n-1)/2} + x^{3n} q^{n(3n+1)/2}\right).
$$

(8)

Therefore by (4) and (8),

$$
\begin{aligned}
\prod_{n=1}^{\infty} (1 - q^n) = f(1, q) &= 1 + \sum_{n=1}^{\infty} (-1)^n \left(q^{n(3n-1)/2} + q^{n(3n+1)/2}\right) \\
&= \sum_{n=-\infty}^{\infty} (-1)^n q^{n(3n+1)/2},
\end{aligned}
$$

which completes the proof of (1).

It should be noted that several authors (L. J. Rogers [18, pp. 334–335], G. W. Starcher [19], and N. J. Fine [9]) also found formula (8) essentially in the way we have; however, none has noted that he was, in fact, rediscovering Euler's proof in simpler clothing. M. V. Subbarao [20] has also shown the connection between (8) and Franklin's arithmetic proof [10].

Now the reader may naturally ask: Was anything gained in this general formulation of Euler's proof besides simplicity of presentation? Is any new information available that was hidden before? We can answer a resounding YES just by setting $x = -1$ in (3) and (8). This substitution gives the equation

$$
\begin{aligned}
&1 + \sum_{n=1}^{\infty} (1 + q)(1 + q^2) \cdots (1 + q^{n-1})(-1)^n q^n \\
&\quad = f(-1, q) = 1 + \sum_{n=1}^{\infty} \left(-q^{n(3n-1)/2} + q^{n(3n+1)/2}\right).
\end{aligned}
$$

(9)

Equation (9) yields a corollary as appealing as Legendre's Theorem. It implies that *the product*

$$
(1 + q)(1 + q^2) \cdots (1 + q^{n-1}) q^n
$$

when multiplied out produces the term q^N exactly as often as N can be partitioned into distinct summands with largest part equal to n. For example, when $n = 4$,

$$(1 + q)(1 + q^2)(1 + q^3)q^4$$
$$= q^4 + q^{4+1} + q^{4+2} + q^{4+3} + q^{4+2+1} + q^{4+3+1} + q^{4+3+2} + q^{4+3+2+1}$$

Hence the series on the left side of (9) when expanded out yields the term $\pm q^N$ for each partition of N into distinct summands; the "+" occurs if the largest summand is even, and the "−" occurs if the largest summand is odd. In the same manner that (1) yielded Legendre's Theorem, we see that (9) yields the following equally elegant but little publicized result found by N. J. Fine [8] more than 118 years after Legendre's observation.

Theorem. *Let $\pi_e(n)$ denote the number of partitions of n with distinct summands the largest of which is even. Let $\pi_o(n)$ denote the number of partitions of n with distinct summands the largest of which is odd. Then*

$$\pi_e(n) - \pi_o(n) = \begin{cases} 1 & \text{if } n = j(3j + 1)/2, \\ -1 & \text{if } n = j(3j - 1)/2, \\ 0 & \text{otherwise.} \end{cases}$$

Let us check out the theorems with an example. The partitions of $n = 12$ into distinct parts are: $12, 11 + 1, 10 + 2, 9 + 3, 9 + 2 + 1, 8 + 4, 8 + 3 + 1, 7 + 5, 7 + 4 + 1, 7 + 3 + 2, 6 + 5 + 1, 6 + 4 + 2, 6 + 3 + 2 + 1, 5 + 4 + 3, 5 + 4 + 2 + 1$. The partitions enumerated by each of $p_e(12)$, $p_o(12)$, $\pi_e(12)$ and $\pi_o(12)$ are listed in the following table:

$p_e(12)$	$p_o(12)$	$\pi_e(12)$	$\pi_o(12)$
$11 + 1$	12	12	$11 + 1$
$10 + 2$	$9 + 2 + 1$	$10 + 2$	$9 + 3$
$9 + 3$	$8 + 3 + 1$	$8 + 4$	$9 + 2 + 1$
$8 + 4$	$7 + 4 + 1$	$8 + 3 + 1$	$7 + 5$
$7 + 5$	$7 + 3 + 2$	$6 + 5 + 1$	$7 + 4 + 1$
$6 + 3 + 2 + 1$	$6 + 5 + 1$	$6 + 4 + 2$	$7 + 3 + 2$
$5 + 4 + 2 + 1$	$6 + 4 + 2$	$6 + 3 + 2 + 1$	$5 + 4 + 3$
	$5 + 4 + 3$		$5 + 4 + 2 + 1.$

Thus $p_e(12) - p_o(12) = 7 - 8 = -1$ and $\pi_e(12) - \pi_o(12) = 7 - 8 = -1$, as predicted by our theorems.

Beyond this immediate pleasant discovery that Euler's approach, properly modernized, yields Fine's Theorem, we may ask: Are there interesting extensions of Euler's method that yield more than equation (8)? Here again the answer is positive. L. J. Rogers [18, p. 334], apparently unaware of how Euler's proof worked, showed in almost precisely the way we proved (8) that

$$1 + \sum_{n=1}^{\infty} \frac{(1 - a)(1 - aq) \cdots (1 - aq^{n-1})t^n}{1 - b)(1 - bq) \cdots (1 - bq^{n-1})} = \frac{1 - at}{1 - t} + \tag{10}$$

$$\sum_{n=1}^{\infty} \frac{(1-a)(1-aq) \cdots (1-aq^{n-1})\left(1-\frac{atq}{b}\right)\left(1-\frac{atq^2}{b}\right) \cdots \left(1-\frac{atq^n}{b}\right)b^n t^n q^{n^2-2}(1-atq^2 n)}{(1 - b)(1 - bq) \cdots (1 - bq^n)(1 - t)(1 - tq) \cdots (1 - tq^{n+1})}$$

Rogers in fact showed that if $f(a, b, t)$ denotes the left side of (10), then

$$f(a, b, t) = \frac{1 - at}{1 - t} + t \frac{(1 - a)(b - atq)}{(1 - b)(1 - t)} f(aq, bq, tq). \tag{11}$$

This result and deeper extensions of it that require much more than Euler's method have had a major impact in the theory of partitions [1, Secs. 3 and 4], [2, Ch. 7], [3, Ch. 3]. N. J. Fine [9] also independently rediscovered (10).

Surely the story unfolded here emphasizes how valuable it is to study and understand the central ideas behind major pieces of mathematics produced by giants like Euler. The discoveries of theorems as appealing as the two we have described would not be separated by 118 years if students of additive number theory had followed this advice.

This paper was partially supported by National Science Foundation Grant MCS-8201733.

References

1. G. E. Andrews, Two theorems of Gauss and allied identities proved arithmetically, *Pacific J. Math.*, 41 (1972) 563–578.

2. ——, The Theory of Partitions, vol. 2, *Encyclopedia of Mathematics and Its Applications*, Addison-Wesley, Reading, 1976.

3. ——, *Partitions: Yesterday and Today*, New Zealand Math. Soc., Wellington, 1979.

4. P. Bachmann, *Zahlentheorie: Zweiter Teil, Die Analytische Zahlentheorie*, Teubner, Berlin, 1921.

5. G. Chrystal, *Algebra*, Part II, 2nd ed., Black, London, 1931.

6. L. Euler, *Opera Omnia*, (1) 2, 241–253.

7. ——, Evolutio producti infiniti $(1 - x)(1 - xx)(1 - x^3)(1 - x^4)(1 - x^5)(1 - x^6)$ etc., *Opera Omnia*, (1) 3, 472–479.

8. N. J. Fine, Some new results on partitions, *Proc. Nat. Acad. Sci. U.S.A.*, 34 (1948) 616–618.

9. ——, Some Basic Hypergeometric Series and Applications, unpublished monograph.

10. F. Franklin, Sur le développement du produit infini $(1-x)(1-x^2)(1-x^3) \cdots$, *Comptes Rendus*, 82 (1881) 448–450.

11. E. Grosswald, *Topics in the Theory of Numbers*, Macmillan, New York, 1966; Birkhauser, Boston, 1983.

12. H. Gupta, *Selected Topics in Number Theory*, Abacus Press, Tunbridge Wells, 1980.

13. G. H. Hardy and E. M. Wright, *An Introduction to the Theory of Numbers*, 4th ed., Oxford University Press, London, 1960.

14. A. M. Legendre, *Théorie des Nombres*, vol. II, 3rd. ed., 1830 (Reprinted: Blanchard, Paris, 1955).

15. I. Niven and H. Zuckerman, *An Introduction to the Theory of Numbers*, 3rd ed., Wiley, New York, 1972.

16. G. Pólya, *Mathematics and Plausible Reasoning, Induction and Analogy in Mathematics*, vol. 1, Princeton, 1954.

17. H. Rademacher, *Topics in Analytic Number Theory*, Grundlehren series, vol. 169, Springer, New York, 1973.

18. L. J. Rogers, On two theorems of combinatory analysis and some allied identities, *Proc. London Math. Soc.*, (2) 16 (1916) 315–336.

19. G. W. Starcher, On identities arising from solutions of q-difference equations and some interpretations in number theory, *Amer. J. Math.*, 53 (1930) 801–816.

20. M. V. Subbarao, Combinatorial proofs of some identities, *Proc. Washington State University Conf. on Number Theory*, Pullman, 1971, 80–91.

21. J. J. Sylvester, A constructive theory of partitions, arranged in three acts, an interact and an exodion, *Amer. J. Math.*, 5 (1882) 251–330.

22. A. Weil, Two lectures on number theory, past and present, *L'Enseignement Mathématique*, 20 (1974) 87–110.

Euler and Quadratic Reciprocity[1]

Harold M. Edwards

In a letter to Goldbach bearing the date 28 August 1742, Euler described a property of positive whole numbers that was to play a central role in the history of the theory of numbers. (The original is a mixture of Latin and German, which I have translated into English as best I can. The letter can be found in [1] or [3].)

Whether there are series of numbers which either have no divisors of the form $4n + 1$, or which even are prime, I very much doubt. If such series could be found, however, one could use them to great advantage in finding prime numbers.

By the way, the prime divisors of all series of numbers which are given by the formula $\alpha xx \pm \beta yy$ show a very orderly pattern which, although I have no demonstration of it as yet, seems to be completely correct. For this reason I take the liberty of communicating to Your Excellency a few such theorems; from these, infinitely many others can be derived.

I. If x and y are relatively prime, the formula $xx + yy$ has no prime divisors other than those contained in the form $4n + 1$, and these prime numbers are themselves all contained in the form $xx + yy$. I put this known theorem at the beginning in order to make the connection of the others more apparent.

II. The formula $2xx + yy$ has no prime divisors other than those contained in the form $8n + 1$ or $8n + 3$. And whenever $8n + 1$ or $8n + 3$ is prime, it is the sum of a square and twice a square, that is, it is of the form $2xx + yy$.

III. The formula $3xx + yy$ has no prime divisors other than those contained in the forms $12n + 1$ and $12n + 7$ (or the single form $6n + 1$). And whenever $6n + 1$ is a prime number, it is contained in the form $3xx + yy$.

IV. The formula $5xx + yy$ has no prime divisors other than those contained in the forms $20n + 1$, $20n + 3$, $20n + 9$, $20n + 7$, and every prime number contained in one of these four forms is itself a number of the form $5xx + yy$.

V. The formula $6xx + yy$ has no prime divisors other than those contained in one of the four forms $24n + 1$, $24n + 5$, $24n + 7$, $24n + 11$, and every prime number contained in one of these forms is itself a number of the form $6xx + yy$.

[1] Reprinted from the *Mathematics Magazine*, Vol. 56, November, 1983, pp. 285–291.

VI. The formula $7xx + yy$ has no prime divisors other than those contained in one of the 6 forms $28n + 1$, $28n + 9$, $28n + 11$, $28n + 15$, $28n + 23$, $28n + 25$ (or in one of the three $14n + 1$, $14n + 9$, $14n + 11$), and every prime number contained in one of these forms is itself a number of the form $7xx + yy$.

From this it is thus clear that the expression $pxx + yy$ can have no prime divisors other than those contained in a certain number of forms of the type $4pn + s$, where s represents some numbers which, although they appear to have no particular order, actually proceed according to a very beautiful rule, which is clarified by these theorems:

VII. If a prime number of the form $4pn + s$ is a divisor of the formula $pxx + yy$, then likewise every prime number contained in the general form $4pn + s^k$ will be a divisor of the formula $pxx + yy$ and indeed will itself be a number of the form $pxx + yy$. For example, because a prime number $28n + 9$ is a number of the form $7xx + yy$ [$37 = $ prime $= 28 \cdot 1 + 9 = 7 \cdot 4 + 9$] prime numbers $28n + 81$ (i.e., $28n + 25$) and $28n + 729$ (i.e., $28n + 1$) are indeed numbers of the form $7xx + yy$ [53 and 29].

VIII. If two prime numbers $4pn + s$ and $4pn + t$ are divisors of the formula $pxx + yy$, then every prime number of the form $4pn + s^k t^i$ is also a number of the form $pxx + yy$.

Thus when one has found a few prime divisors of such an expression $pxx + yy$, one can easily find all possible divisors using these theorems. For example, let $13xx + yy$ be the given formula, which includes the numbers 14, 17, 22, 29, 38, 49, 62, etc. Thus 1, 7, 11, 17, 19, 29, 31 are prime numbers which divide the formula $13xx + yy$. Therefore all prime numbers of the forms $52n + 1$, $52n + 7$, $52n + 11$ etc. can be divisors of $13xx + yy$. But the formula $52n + 7$ gives, by Theorem VII, also these $52n + 49$, $52n + 343$ (or $52n + 31$), $52n + 7 \cdot 31$, or $52n + 9$, further $52n + 7 \cdot 9$, or $52n + 11$, further $52n + 7 \cdot 11$, or $52n + 25$, further $52n + 7 \cdot 25$, or $52n + 19$, further $52n + 7 \cdot 19$, or $52n + 29$, further $52n + 7 \cdot 29$, or $52n + 47$, further $52n + 7 \cdot 47$, or $52n + 17$, further $52n + 7 \cdot 17$, or $52n + 15$, further $52n + 7 \cdot 15$, or $52n + 1$ and at this point the numbers cease to be different which when added to $52n$ give prime numbers of the form $13xx + yy$. Thus from the single fact that 7 can be a divisor of the form $13xx + yy$ the last two theorems imply that all prime numbers of any of the forms

$52n + 1$;	$52n + 31$;	$52n + 25$;	$52n + 47$
$52n + 7$;	$52n + 9$;	$52n + 19$;	$52n + 17$
$52n + 49$;	$52n + 11$;	$52n + 29$;	$52n + 15$

have the form $13xx + yy$ and also can be divisors of such numbers $13xx + yy$, and also more formulas can not be derived using the theorems. From this it is known that no prime number can be a divisor of the form $13xx + yy$ other than those contained in the 12 formulas that have been found. Now every prime number of the form $4pn + 1$ can be a divisor of $pxx + yy$. From this, beautiful properties can be derived, as, for example, because 17 is prime and also of the form $2xx + yy$ it follows that whenever $17^m \pm 8n$ is prime it must also be of the form $2xx + yy$.

And when $17^m \pm 8n$ is a number of the form $2xx + yy$ which admits no divisors of this form, it is certainly a prime number.

The same situation occurs with the divisors of the forms $pxx - yy$ or $xx - pyy$, which, when they are prime, must be contained in the form $4np \pm s$, where s represents certain determined numbers. Namely, in a few cases, one will have

1. All prime divisors of the form $xx - yy$ contained in the form $4n \pm 1$, which is clear.

2. All prime divisors of the form $2xx - yy$ contained in the form $8n \pm 1$.

Coroll. Therefore a prime number of the form $8n \pm 3$ is not a number of the form $2xx - yy$.

3. All prime divisors of the form $3xx - yy$ contained in the form $12n \pm 1$.

4. All prime divisors of the form $5xx - yy$ contained in either the form $20n \pm 1$ or the form $20n \pm 9$ (or in the single one $10n \pm 1$).

etc.

And if a prime number $4pn + s$ divides the form $pxx - yy$ or $xx - pyy$, then $\pm 4np \pm s^k$ will itself be of the form $pxx - yy$ or $xx - pyy$, whenever it is prime. If two prime numbers s and t are numbers of the form $pxx - yy$, then whenever $4np \pm s^\mu t^\nu$ is prime it will also be a number of the form $pxx - yy$. Thus, because 7 and 17 are prime numbers and of the form $2xx - yy$, $\pm 8n \pm 7^\mu \cdot 17^\nu$ will also be of this form whenever it is prime. Let $\mu = 1$, $\nu = 1$, so $7 \cdot 17 = 119$ and $119 + 8 = 127 = \text{prime}$, and consequently $127 = 2xx - yy = 2 \cdot 64 - 1$. From this it is now clear that it is not possible to find sequences of numbers of the type $pxx + qyy$ which do not admit divisors of the form $4n + 1$.

But I am convinced that I have not exhausted this material, rather, that there are countless wonderful properties of numbers to be discovered here, by means of which the theory of divisors could be brought to much greater perfection; and I am convinced that if Your Excellency were to consider this subject worthy of some attention He would make very important discoveries in it. The greatest advantage would show itself, however, when one could find proofs for these theorems.

This passage is vintage Euler in that the basic idea is an insight so profound that it is crucial to much of algebraic number theory, yet at the same time many of the individual statements are patently false. The last statement of Theorem IV, for example, is clearly wrong. Not only is it not true that *all* prime numbers of the form $20n + 3$ are of the form $5x^2 + y^2$, but *no* prime numbers $20n + 3$ are $5x^2 + y^2$. To prove this it suffices to note that, since p is to be odd, x and y must have opposite parity, that is, either $x = 2j + 1$, $y = 2k$ or $x = 2c$, $y = 2d + 1$. In the first case

$$5x^2 + y^2 = (4 + 1)(4j^2 + 4j + 1) + 4k^2 = 4(4j^2 + 4j + 1 + j^2 + k^2) + 1$$

and in the second case

$$5x^2 + y^2 = 4(5c^2 + d^2 + d) + 1,$$

so in either case p is 1 more than a multiple of 4 and cannot have the form $4n + 3$, much less the form $20n + 3$, or the form $20n + 4 + 3$.

Fortunately, the letter to Goldbach is only the first of many passages in his known writings where Euler deals with this subject, and in later versions the obvious mistakes are corrected. For example, in his main exposition [2] of these ideas he corrects the second part of Theorem IV to say that if p is a prime of the form $p = 20n + 1$ or $20n + 9$ then $p = 5x^2 + y^2$, and if it is a prime of the form $20n + 3$ or $20n + 7$ then $2p = 5x^2 + y^2$. (Examples: $2 \cdot 3 = 5 \cdot 1^2 + 1^2$, $2 \cdot 7 = 5 \cdot 1^2 + 3^2$, $2 \cdot 23 = 5 \cdot 3^2 + 1^2$, $2 \cdot 43 = 5 \cdot 1^2 + 9^2$, $2 \cdot 47 = 5 \cdot 3^2 + 7^2$.) As restated, the theorem is correct and definitely not easy to prove.

The style of the corrected exposition [2] is similar to the letter above in that Euler first states a number of special theorems—covering the prime divisors of $a^2 + Nb^2$ (a, b relatively prime) for $N = 1, 2, 3, 5, 7, 11, 13, 17, 19, 6, 10, 14, 15, 21, 35, 30$—before he states general theorems. This style has the advantage that the reader, far from having to struggle with the meaning of the general theorem, has probably become impatient with the special cases and has already made considerable progress toward guessing what the general theorem will be. Such a style is not appropriate to the sort of short note I am writing, however, and I will skip to the general case. Moreover, I will state it much more succinctly than Euler does.

Theorem. *Let N be a given positive integer. Then there is a list $s_1, s_2, \ldots, s_m$ of positive integers less than $4N$ and relatively prime to $4N$ with the following properties:*

(1) *Any odd prime number p which divides a number of the form $a^2 + Nb^2$ without dividing either a or Nb is of the form $p = 4Nn + s_i$ for some s_i in the list.*

(2) *Every prime number of the form $p = 4Nn + s_i$ for some s_i in the list divides a number of the form $a^2 + Nb^2$ without dividing either a or Nb.*

(3) *If s_i and s_j are in the list and if $s_i s_j = 4Nn + s$, $0 < s < 4N$, then s is in the list.*

(4) *If x is any integer less than $4N$ and relatively prime to $4N$ then either x or $4N - x$, but not both, are in the list.*

For example, when $N = 13$, the list contains the 12 numbers 1, 7, 49, 31, 9, 11, 25, 19, 29, 47, 17, 15, that Euler gave in his letter to Goldbach. Property (4) becomes clearer if one writes $-x$ in place of $4N - x$ when $2N < 4N - x < 4N$ and reorders the list in order of the size of the absolute values of the entries. In the case $N = 13$ this gives 1, $-3, -5, 7, 9, 11, 15, 17, 19, -21, -23, 25$, and in the general case it gives (by (4)) a list of the positive integers x less than $2N$ and relatively prime to $2N$ with a sign assigned to each. To see that property (3) holds in the case $N = 13$ it suffices to note that Euler, in the letter, derived his list $1, 7, 49, 31, \ldots$ by repeatedly multiplying by 7 and removing multiples of 52. Thus, in the case $N = 13$, the numbers s_i in the list are determined by $7^i = 52n_i + s_i$ for $i = 0, 1, \ldots, 11$, and $7^{12} = 52n_{12} + 1$, from which (3) follows. Here are the lists described in the theorem for a few values of N (see Table 1).

I have included $N = 4, 8, 9, 12$ just to show that the theorem applies in these cases, but Euler omits them for the simple reason that if you have the list for any N then you can trivially derive from it the list for Nk^2 for any k. For if p divides $a^2 + Nk^2b^2$ without dividing either a or Nk^2b then it divides $a^2 + N(kb)^2$ without dividing either a or Nkb, and on the other hand, if it divides $a^2 + Nb^2$ *and* if it does not divide k then it divides $(ka)^2 + Nk^2b^2$ without dividing either ka or Nk^2b.

A modern reader, after he sees the word Theorem, expects to find the word *Proof* soon thereafter. However, customs were different in Euler's day and his paper contains 59 theorems without a single proof. He told Goldbach in his letter that "I have no demon-

Table 1.

N	list										
1	1										
2	1,	3									
3	1,	−5									
4	1,	−3,	5,	−7							
5	1,	3,	7,	9							
6	1,	5,	7,	11							
7	1,	−3,	−5,	9,	11,	−13					
8	1,	3,	−5,	−7,	9,	11,	−13,	−15			
9	1,	5,	−7,	−11,	13,	17					
10	1,	−3,	7,	9,	11,	13,	−17,	19			
11	1,	3,	5,	−7,	9,	−13,	15,	−17,	−19,	−21	
12	1,	−5,	7,	−11,	13,	−17,	19,	−23			
13	1,	−3,	−5,	7,	9,	11,	15,	17,	19,	−21,	−23, 25

stration of it as yet," and the fact is that he never found a demonstration of it or even of a substantial portion of it. His "theorems" were based on nothing but empirical evidence.

In order to test the theorem empirically one needs to be able to test, given a prime number p and a positive integer N not divisible by p, whether there exist integers a and b not divisible by p such that p divides $a^2 + Nb^2$. This at first looks impossible to test because it looks as if one must test an infinite number of values of a and b. However, a moment's reflection shows that one need only test values of a and b that are positive and less than p, because p divides $a^2 + Nb^2$ if and only if it divides $(a + p)^2 + Nb^2$ and the same holds for $a^2 + N(b + p)^2$, so multiples of p can be removed from a and b.

Using this observation, we can illustrate how one can test the theorem, for example, for $N = 30$. Some numbers of the form $a^2 + 30b^2$ are

$31, \ 34 = 2 \cdot 17, \ 39 = 3 \cdot 13, \ 46 = 2 \cdot 23, \ 55 = 5 \cdot 11, 66 = 2 \cdot 3 \cdot 11, \ 79, \ \text{and} \ 94 = 2 \cdot 47.$

Thus the list must contain 31, 17, 13, 23, 11, $79 \equiv -41$, 47, which indicates that 79 appears in the list as -41 when multiples of $4N = 120$ are removed to put the number between -60 and 60. More entries in the list can be found by using products of these. For example, $31 \cdot 17 = 527 \equiv 47$ is already in the list,

$$31 \cdot 13 = 403 \equiv 43, \quad 31 \cdot 23 = 713 \equiv -7, \quad 31 \cdot 11 = 341 \equiv -19,$$
$$31 \cdot (-41) = -1271 \equiv 49, \quad \text{and} \quad 31 \cdot 47 = 1457 \equiv 17.$$

A check shows that this assigns a sign to each positive integer less than 60 and relatively prime to 60 other than 1, 29, 37, 53, and 59. These are resolved by

$$17 \cdot 11 = 187 \equiv -53, \quad 13 \cdot 23 = 299 \equiv 59, \quad 23 \cdot 47 = 1081 \equiv 1,$$
$$11 \cdot (-41) = -451 \equiv -29, \quad \text{and} \quad 31 \cdot (-53) = -1643 \equiv 37.$$

Thus the list for $N = 30$ is

$$1, \ -7, \ 11, \ 13, \ 17, \ -19, \ 23, \ 29, \ 31, \ 37, \ -41, \ 43, \ 47, \ 49, \ -53, \ 59.$$

For any prime p, the theorem now gives a prediction as to whether p does or does not divide a number of the form $a^2 + 30b^2$ without dividing a or $30b$, and this prediction can be checked in a finite number of steps. For example, it predicts that 37 *does* divide a number of this form, and, indeed, $9^2 + 30 = 111 = 3 \cdot 37$. It predicts that 7 *does not* divide a number of this form, and, indeed, a check of the 36 numbers $a^2 + 30b^2$, $0 < a < 7$, $0 < b < 7$, shows that none of them is divisible by 7. It is a long test to determine in this straightforward way whether a given p divides $a^2 + Nb^2$. The work can be greatly reduced by showing that if p divides any number of this type without dividing b then it divides a number of this type in which $b = 1$ and $0 < a < p/2$.[2] Thus in the case $p = 7$, $N = 30$, one need only check that 7 does not divide 31, 34, or 39 in order to conclude that the prediction of the theorem is correct. Similarly, since 19 does not divide 31, 34, 39, 46, 55, 66, 79, 94, 111, the prediction for 19 is correct.

In a few hours one could verify in this way the prediction of the theorem in thousands of cases for dozens of values of N. Because the theorem is so simple and general and withstands these tests so easily, one readily becomes convinced that it is true. Certainly Euler was convinced, so much so that at times he seems to have forgotten that the theorem was completely unproved.

For simplicity, the case of negative N, that is, of prime divisors of $x^2 - Dy^2$ where $D > 0$, was omitted from the statement of the theorem. It is easy to see that if D is a square then every prime p divides a number of this form. (For if $D = k^2$ then $x = k + p$ gives $x^2 - k^2 = p(2k + p)$, and p divides x only if it divides D.) However, if D is not a square then, as Euler already observed in his letter to Goldbach, a similar theorem holds, except that instead of *never* containing both x and $-x$ the list in these cases always contains both whenever it contains either.

Theorem (continued). *If N is negative and not of the form $-k^2$ then there is a list of integers s in the range $0 < s < |4N|$ and relatively prime to $4N$ such that (1), (2), and (3) hold (with $s < 4N$ changed to $s < |4N|$ in (3)). In this case (4) is replaced by*

(4') Exactly half the positive integers less than $|2N|$ and relatively prime to $2N$ are in the list, and x is in the list if and only if $|4N| - x$ is in the list.

For example, here are the lists for a few negative values of N written, as before, with $-x$ in place of $|4N| - x$. The first three are from Euler's letter (see Table 2).

Actually, there is a simple relation between the lists for N and $-N$ which can be summarized by saying that a number x of the form $4n + 1$ is either in *both* lists or it is in *neither*. For example, for $N = 7$, the numbers 1, 9, -3 are in both lists and 5, 13, -11 are in neither. It is possible in this way to find either list once the other is known. The relation is simple to prove[3] and it was well known to Euler.

---

[2] Here is the argument. Since p does not divide b and p is prime, 1 is the greatest common divisor of p and b. The Euclidean algorithm can therefore be used to write $1 = Ap + Bb$ for integers A and B. If p divides $a^2 + Nb^2$ then it also divides $B^2a^2 + NB^2b^2 = c^2 + N(1 - Ap)^2$ and therefore divides $c^2 + N$. Now $c = qp + r$ where the remainder r can be taken in the range $-p/2 < r < p/2$ and p divides $r^2 + N$, as was to be shown.

[3] If $p = 4n + 1$ then, by the case $N = 1$ of the theorem (which is one of the few cases that Euler later succeeded in proving) p divides $y^2 + 1$ for some y. If p also divides $x^2 + N$ for some x not divisible by p—i.e., if p is in the list for N—then p divides $x^2y^2 + Ny^2 = (xy)^2 - N + N(y^2 + 1)$, which shows that p divides $(xy)^2 - N$ and therefore that p is in the list for $-N$. Since N is not assumed to be positive in this argument, the same argument shows that if p is in the list for $-N$ it is also in the list for N.

Table 2.

N	list					
-2	±1					
-3	±1					
-5	$\pm1,$	±9				
-6	$\pm1,$	±5				
-7	$\pm1,$	$\pm3,$	±9			
-10	$\pm1,$	$\pm3,$	$\pm9,$	±13		
-11	$\pm1,$	$\pm5,$	$\pm7,$	$\pm9,$	±19	
-13	$\pm1,$	$\pm3,$	$\pm9,$	$\pm17,$	$\pm23,$	±25

It would be difficult to exaggerate the importance of this theorem in the history of number theory. The effort to prove it surely spurred much of Euler's own later work, and the other two great number theorists of the 18th century, Lagrange and Legendre, also worked on topics around and about the theorem without penetrating the theorem itself. Finally, the young Gauss found a proof in 1796, and published two proofs in his great work, the *Disquisitiones Arithmeticae*, in 1801. Gauss claimed to have discovered the theorem on his own, but he would have needed to be in a cocoon in order not to have had *some* contact with work in this direction by Euler, Lagrange, and Legendre in the preceding half-century. I believe that Gauss was not being dishonest, but that he may have forgotten many subtle influences.

Gauss's formulation of the theorem was very different from Euler's. For Euler, the basic question was whether, given N and p, the prime p divides a number of the form $x^2 + N$. It was noted above that if one can answer this question for N then one can easily deduce the answer for $-N$. A similar argument shows that if N is a product of two numbers $N = mn$ and if the question can be answered for each factor m, n then it can be answered for N. (This becomes clear when the question "Is p in the list for N?" is restated "Is $-N$ a square mod p?" as below. If the answer is known for m and $-n$ then it is known for $N = mn$ because a product is a square if and only if *both* factors are squares or *neither* factor is a square.) Thus it suffices to be able to answer the question for $N = 1$ and N a prime. The cases $N = 1$ and $N = 2$ were resolved by Euler and Lagrange, so the question was reduced to the case where N is an odd prime. Thus the problem is in essence to find the list in Euler's theorem when $\pm N$ is an odd prime. One can find this list without testing a single prime divisor of $x^2 \pm N$ if one observes that *the numbers common to the lists for N and $-N$, when N is prime, are precisely those numbers s, where $-2N < s < 2N$, that can be written in the form $s = t^2 - 4Nk$ where t is a positive odd integer less than N.* This is a simple consequence of the fact that squares are necessarily in the list.[4]

---

[4]To see this, note that if N is an odd prime then each list has $N - 1$ entries and half that many are common to the two lists. Therefore one need only show that all squares (reduced by subtracting multiples of $4N$ to put them between $-2N$ and $2N$) are in both lists, because this would account for all $(N - 1)/2$ common entries. For any of the $2N - 2$ nonzero odd integers x between $-2N$ and $2N$, multiplication by x and reduction by removing multiples of $4N$ is a one-to-one map of this set with $2N - 2$ elements to itself. For either of the two lists, if x is in the list then, by (3), multiplication by x carries elements of the list to elements of the list. Therefore, by counting, it carries elements not in the list to elements not in the list. In other words, if x is in the

Figure 1. Swiss banknote shows engraving of Euler (after Handmann's portrait) on a background of Euler diagrams and interlocking gears.

For example, when $N = 11$, the numbers common to the lists are $1^2 = 1$, $3^2 = 9$, $5^2 \equiv -19$, $7^2 \equiv 5$, $9^2 \equiv -7$; thus the list for -11 is ± 1, ± 9, ± 19, ± 5, ± 7, and the list for 11 is 1, 3, 5, -7, 9, -13, 15, -17, -19, -21. When $N = 13$ the numbers in common are $1^2 = 1$, $3^2 = 9$, $5^2 = 25$, $7^2 \equiv -3$, $9^2 \equiv -23$, $11^2 \equiv 17$ so the lists are ± 1, ± 9, ± 25, ± 3, ± 23, ± 17 and 1, -3, -5, 7, 9, 11, 15, 17, 19, -21, -23, 25.

Gauss approached the subject from a different point of view, asking, for distinct odd primes p and q, whether q is a square mod p, that is, whether there is an integer x such that $x^2 - q$ is divisible by p. His "fundamental theorem," now known as the law of quadratic reciprocity because it describes a reciprocal relationship between the questions "Is q a square mod p?" and "Is p a square mod q?" states:

---

list and y is not then the reduction of xy is not in the list. Therefore multiplication by y carries elements of the list to elements not in the list. Since the list and its complement both have $N - 1$ elements, multiplication by y and reduction carries elements in the list one-to-one onto elements not in the list. By counting, then, it carries elements not in the list to elements of the list. Therefore if y is not in the list, the reduction of y^2 is. Thus the reduction of y^2 is in the list whether or not y is.

If p is of the form $p = 4n + 1$ then q is a square mod p if and only if p is a square mod q.

If p is of the form $p = 4n - 1$ then q is a square mod p if and only if $-p$ is a square mod q.

This is easy to deduce from the theorem above,[5] easy enough that it is not stretching matters very far to say that the law of quadratic reciprocity is a consequence of Euler's theorems. However, for reasons to be explained in a moment, it is not in Euler's interest to stretch matters at all.

The law of quadratic reciprocity is the crowning theorem of elementary number theory. One might almost say that it is the theorem with which elementary number theory ceases to be elementary. Gauss, who did not waste time with trivialities, was fascinated by this theorem, so simple to state and so difficult to prove, and he returned to it many times in his career, giving six different proofs of it.

Gauss also studied higher reciprocity laws, which deal, roughly speaking, with the prime divisors of $x^3 - N$ (cubic reciprocity), $x^4 - N$ (biquadratic reciprocity), etc. The study of higher reciprocity laws was unquestionably the central question of 19th century number theory, engaging the best efforts of Jacobi, Eisenstein, Kummer, Hilbert, and many others, and leading to the creation of algebraic number theory. Two developments in the subsequent history of the subject give further testimony to Euler's genius and the importance of the theorems that he first announced to Goldbach.

First, a manuscript of Euler published in 1849 (he had died in 1783) showed that Gauss was not in fact the first to study higher reciprocity laws, but that Euler had already made some substantial progress on cubic reciprocity as early as 1749, and had not published his "theorems" in this field. For example, he stated the following conjecture:

> Let p be a prime of the form $3n + 1$. Then 5 is a cube mod p if and only if the representation of p in the form $p = x^2 + 3y^2$ satisfies one of the 4 conditions (1) $y = 15m$, (2) $x = 5k$, $y = 3m$, (3) $x \pm y = 15m$, or (4) $2x \pm y = 15m$.

(Theorem III of the letter to Goldbach may or may not assert the existence of such a representation $p = x^2 + 3y^2$ whenever $p = 3n + 1$, depending on one's interpretation of the phrase "contained in the form $3xx + yy$." In any case, Euler later not only asserted the existence of such a representation, he proved it rigorously.) Euler gave no indication of how he arrived at this astounding set of conditions, and the fact that they are correct struck the editor of the relevant volume of his collected works (Vol. 5 of the first series) as "bordering on the incomprehensible." However, the conjecture can be derived by applying the ideas described above to "imaginary primes" of the form $x + y\sqrt{-3}$ and finding the classes of imaginary primes mod $3 \cdot 5$ for which 5 is a cube.

The second testimony to Euler's genius in the history of the subject is that later research showed that the "reciprocity law" approach to the subject was something of a blind alley. Hilbert in the 1890s formulated the quadratic and higher laws in terms of a simple

---

[5]Here is the argument. If $p = 4n + 1$ and p is a square mod q, say $p - z^2$ is divisible by q, then $y = z$ or $z + q$ is odd and $p - y^2$ is divisible by both 4 and q. Therefore p is in the list for $N = -q$ (and also for $N = q$), which means that $x^2 - q$ is divisible by p for some x, that is, q is a square mod p. Conversely, if q is a square mod p then p is in the list for $N = -q$. Therefore, since $p = 4n + 1$, p is in both lists and $p = t^2 - 4qk$, which shows that p is a square mod q. The proof in the case $p = 4n - 1$ is the same with p replaced by $-p$.

product formula which was generally regarded as a more natural way of describing the basic phenomenon, and in which there is no "reciprocity" but, rather, an explicit formula for determining (in the quadratic cases) which classes mod $4N$ contain prime divisors of $x^2 + Ny^2$. Later, in the 1920s, the subject reached what is generally regarded as its culmination in the form of the Artin Reciprocity Law, which, again, has no element of "reciprocity" in it. Moreover, in the quadratic case, Artin's Law is almost exactly the theorem we have stated, which was discovered by Euler nearly 200 years earlier.

Work on this paper was supported in part by a grant from the Vaughn Foundation.

References

1. P.-H. Fuss, ed., *Correspondance Mathématique et Physique*, Imp. Acad. Sci., St. Petersburg, 1843, vol. 1, pp. 144–153, reprint by Johnson Reprint Corp., New York and London, 1968.

2. L. Euler, Theoremata circa divisores numerorum in hac forma $paa \pm qbb$ contentorum, Eneström 164, *Comm. Acad. Sci. Petrop.* 14 (1744/6), 1751, pp. 151–181; also *Opera Omnia*, (1)2, 194–222.

3. A. P. Juikevic and E. Winter, eds., *Leonhard Euler und Christian Goldbach, Briefwechsel 1729–1764*, Akademie-Verlag, Berlin, 1965.

Euler and the Fundamental Theorem of Algebra[1]

William Dunham

A watershed event for all students of mathematics is the first course in basic high school algebra. In my case, this provided an initial look at graphs, inequalities, the quadratic formula, and many other critical ideas. Somewhere near the term's end, as I remember, our teacher mentioned what sounded like the most important result of them all—the fundamental theorem of algebra. Anything with a name like that, I figured, must be (for want of a better term) *fundamental*. Unfortunately, the teacher informed us that this theorem was much too advanced to state, let alone to investigate, at our current level of mathematical development.

Fine. I was willing to wait. However, second year algebra came and went, yet the fundamental theorem occupied only an obscure footnote from which I learned that it had something to do with factoring polynomials and solving polynomial equations. My semester in college algebra/precalculus the following year went a bit further, and I emerged vaguely aware that the fundamental theorem of algebra said that nth-degree polynomials could be factored into n (possibly complex) linear factors, and thus nth-degree polynomial equations must have n (possibly complex and possibly repeated) solutions. Of course, to that point we had done little with complex numbers and less with complex solutions of polynomial equations, so the whole business remained obscure and mysterious. Even in those pre-Watergate days, I began to sense that the mathematical establishment was engaged in some kind of cover-up to keep us ignorant of the true state of algebraic affairs.

"Oh well," I thought, "I'm off to college, where surely I'll get the whole story." Four years later I was still waiting. My undergraduate mathematics training—particularly courses in linear and abstract algebra—examined such concepts as groupoids, eigenvalues, and integral domains, but none of my algebra professors so much as mentioned the fundamental theorem. This was very unsatisfactory—a bit like reading Moby Dick and

[1] Reprinted from the *College Mathematics Journal*, Vol. 22, September, 1991, pp. 282–293. This article received the George Pólya Award in 1992.

never encountering the whale. The cover-up had continued through college, and algebra's superstar theorem was as obscure as ever.

It was finally in a graduate school course on complex analysis that I saw a proof of this key result, and I immediately realized the trouble: the theorem really is a monster to prove in full generality, for it requires some sophisticated preliminary results about complex functions. Clearly a complete proof *is* beyond the reach of elementary mathematics.

So what does a faculty member do if an inquiring student seeks information about the fundamental theorem of algebra? It is hopeless to try to prove the thing for any precalculus student whose I.Q. lies on this side of Newton's; on the other hand, it would more or less continue the cover-up to avoid answering the question—to treat an inquiry about the fundamental theorem of algebra as though the student had asked something truly improper, delicate, or controversial—like a question about one's religion, or one's sex life, or even one's choice of personal computer.

Let me, then, suggest an intermediate option—something less rigorous than a grad school proof, yet something more satisfying than simply telling our inquisitive student to get lost. My suggestion is that we look back to the history of mathematics and to the work of that most remarkable of eighteenth century mathematicians, Leonhard Euler (1707–1783). With Euler's attempted proof of the fundamental theorem of algebra from 1749, we find yet another example of the history of mathematics serving as a helpful ingredient in the successful teaching of the subject. The reasoning is not impossibly difficult; it raises some interesting questions for further discussion; and while his is not a complete proof by any means, it does establish the result for low degree polynomials and suggests to students that this sweeping theorem is indeed reasonable.

Before addressing the subject further, we state the theorem in its modern form:

Any nth-degree polynomial with complex coefficients can be factored into n complex linear factors.

That is,

If $P(z) = c_n z^n + c_{n-1} z^{n-1} + \cdots + c_2 z^2 + c_1 z + c_0$, where $c_n, c_{n-1}, \ldots, c_2, c_1, c_0$ are complex numbers, then there exist complex numbers $\alpha_1, \alpha_2, \ldots, \alpha_n$ such that

$$P(z) = c_n(z - \alpha_1)(z - \alpha_2) \cdots (z - \alpha_n).$$

It may come as a surprise that, to mathematicians of the mid-eighteenth century, the fundamental theorem appeared in the following guise:

Any polynomial with real coefficients can be factored into the product of real linear and/or real quadratic factors.

Note that there is no mention here of complex numbers, either as the polynomial's coefficients nor as parts of its factors. For mathematicians of the day, the theorem described a phenomenon about real polynomials and their real factors.

As an example, consider the factorization

$$3x^4 + 5x^3 + 10x^2 + 20x - 8 = (3x - 1)(x + 2)(x^2 + 4).$$

Here the quartic has been shattered into the product of two linear fragments and one irreducible quadratic one, and all polynomials in sight are real. The theorem stated that such a factorization was possible for any real polynomial, no matter its degree.

Anticipating a bit, we see that we can further factor the quadratic expression—provided we allow ourselves the luxury of complex numbers. That is,

$$ax^2 + bx + c = a\left(x^2 + \frac{b}{a}x + \frac{c}{a}\right)$$

$$= a\left(x - \frac{-b + \sqrt{b^2 - 4ac}}{2a}\right)\left(x - \frac{-b - \sqrt{b^2 - 4ac}}{2a}\right)$$

factors the real quadratic $ax^2 + bx + c$ into two, albeit rather unsightly, linear pieces. Of course, there is no guarantee these linear factors are composed of real numbers, for if $b^2 - 4ac < 0$, we venture into the realm of imaginaries. In the specific example cited above, for instance, we get the complete factorization:

$$3x^4 + 5x^3 + 10x^2 + 20x - 8 = (3x - 1)(x + 2)(x - 2i)(x + 2i).$$

This is "complete" in the sense that the real fourth-degree polynomial with which we began has been factored into the product of four *linear* complex factors, certainly as far as any factorization can hope to proceed.

It was the Frenchman Jean d'Alembert (1717–1783) who gave this theorem its first serious treatment in 1746 [5, p. 99]. Interestingly, for d'Alembert and his contemporaries the result had importance beyond the realm of algebra: its implications extended to the relatively new subject of calculus and in particular to the integration technique we now know as "partial fractions." As an illustration, suppose we sought the indefinite integral

$$\int \frac{28x^3 - 4x^2 + 69x - 14}{3x^4 + 5x^3 + 10x^2 + 20x - 8} \, dx.$$

To be sure, this looks like absolute agony, as all calculus teachers will readily agree. (One would have trouble finding it in the Table of Integrals of a calculus book's inside cover, unless the book is very thorough or its cover is very large.) This problem even gives a good workout to symbolic manipulators such as *Mathematica* (which required 50 seconds to find the antiderivative on my Mac II) and which were not available to eighteenth century mathematicians in any case.

But if, as d'Alembert claimed, the denominator could be decomposed into real linear and/or real quadratic factors, then the difficulties drop away. Here, the integrand becomes

$$\int \frac{28x^3 - 4x^2 + 69x - 14}{(3x - 1)(x + 2)(x^2 + 4)} \, dx.$$

We then determine its partial fraction decomposition, getting

$$\int \frac{28x^3 - 4x^2 + 69x - 14}{3x^4 + 5x^3 + 10x^2 + 20x - 8} \, dx$$

$$= \int \frac{28x^3 - 4x^2 + 69x - 14}{(3x - 1)(x + 2)(x^2 + 4)} \, dx = \int \frac{1}{3x - 1} \, dx + \int \frac{7}{x + 2} \, dx + \int \frac{2x - 3}{x^2 + 4} \, dx$$

$$= \frac{1}{3} \ln|3x - 1| + 7 \ln|x + 2| + \ln(x^2 + 4) - \frac{3}{2} \tan^{-1}\left(\frac{x}{2}\right) + C,$$

and the antiderivative is found.

Thus, if the fundamental theorem were proved in general, we could conclude that for any $P(x)/Q(x)$ where P and Q are real polynomials, the indefinite integral $\int (P(x)/Q(x))dx$ would exist as a combination of fairly simple functions (at least theoretically). That is, we could first perform long division to reduce this rational expression to one where the degree of the numerator was less than the degree of $Q(x)$; next we consider $Q(x)$ as the product of real linear and/or real quadratic factors; then apply the partial fraction technique to break the integral into pieces of the form

$$\int \frac{A}{(ax+b)^n}\, dx \quad \text{and/or} \quad \int \frac{Bx+C}{(ax^2+bx+c)^n}\, dx;$$

and finally determine these indefinite integrals using nothing worse than natural logarithms, inverse tangents, or trigonometric substitutions. Admittedly, the fundamental theorem gives no process for finding the denominator's explicit factors; but, just as the theorem guarantees the *existence* of such a factorization, so too will the *existence* of simple antiderivatives for any rational function be established.

Unfortunately, d'Alembert's 1746 attempt to prove his theorem was unsuccessful, for the difficulties it presented were simply too great for him to overcome (see [4, pp. 196–198]). In spite of this failure, the fundamental theorem of algebra has come to be known as "d'Alembert's Theorem" (especially in France). Attaching his name to this result may seem a bit generous, given that he failed to prove it. This is a bit like designating the Battle of Waterloo as "Napoleon's Victory."

So matters stood when Euler turned his awesome mathematical powers to the problem. At the time he picked up the scent, there was not even universal agreement that the theorem was true. In 1742, for instance, Nicholas Bernoulli had expressed to Euler his conviction that the real quartic polynomial

$$x^4 - 4x^3 + 2x^2 + 4x + 4$$

cannot be factored into the product of real linear and/or real quadratic factors in any fashion whatever [1, pp. 82–83]. If Bernoulli were correct, the game was over; the fundamental theorem of algebra would have been instantly disproved.

However, Bernoulli's skepticism was unfounded, for Euler factored the quartic into the product of the quadratics

$$x^2 - \left(2 + \sqrt{4 + 2\sqrt{7}}\right)x + \left(1 + \sqrt{4 + 2\sqrt{7}} + \sqrt{7}\right) \quad \text{and}$$

$$x^2 - \left(2 - \sqrt{4 + 2\sqrt{7}}\right)x + \left(1 - \sqrt{4 + 2\sqrt{7}} + \sqrt{7}\right).$$

Those with a taste for multiplying polynomials can check that these complicated factors yield the fairly innocent quartic above; far more challenging, of course, is to figure out how Euler derived this factorization in the first place. (Hint: it was not by guessing.)

By 1742, Euler claimed he had proved the fundamental theorem of algebra for real polynomials up through the sixth-degree [3, p. 598], and in a landmark 1749 article titled "Recherches sur les racines imaginaires des équations" [1, pp. 78–169], he presented his proof of the general result which we shall now examine (see also [5, pp. 100–102]). We stress again that his argument failed in its ultimate mission. That is, Euler furnished only a partial proof which, in its full generality, suffered logical shortcomings. Nonetheless, even with these shortcomings, one cannot fail to recognize the deftness of a master at work.

He began with an attack on the quartic:

Theorem. *Any quartic polynomial $x^4 + Ax^3 + Bx^2 + Cx + D$, where A, B, C, and D are real, can be decomposed into two real factors of the second degree.*

Proof. Euler first observed that the substitution $x = y - (A/4)$ reduces the original quartic into one lacking a cubic term—a so-called "depressed quartic." Depressing an nth-degree polynomial by a clever substitution that eliminates its $(n - 1)$st-degree term is a technique whose origin can be traced to the sixteenth century Italian mathematician Gerolamo Cardano in his successful attack on the cubic equation [3, p. 265].

With this substitution, the quartic becomes

$$\left(y - \frac{A}{4}\right)^4 + A\left(y - \frac{A}{4}\right)^3 + B\left(y - \frac{A}{4}\right)^2 + C\left(y - \frac{A}{4}\right) + D,$$

and the only two sources of a y^3 term are

$$\left(y - \frac{A}{4}\right)^4 = y^4 - Ay^3 + \cdots \quad \text{and} \quad A\left(y - \frac{A}{4}\right)^3 = A(y^3 - \cdots) = Ay^3 - \cdots.$$

Upon simplifying, we find that the "y^3" terms cancel and there remains the promised depressed quartic in y.

Not surprisingly, there are advantages to factoring a depressed quartic rather than a full-blown one; yet it is crucial to recognize that any factorization of the depressed quartic yields a corresponding factorization of the original. For instance, suppose we were trying to factor $x^4 + 4x^3 - 9x^2 - 16x + 20$ into a product of two quadratics. The substitution $x = y - \frac{4}{4} = y - 1$ depresses this to $y^4 - 15y^2 + 10y + 24$, and a quick check confirms the factorization:

$$y^4 - 15y^2 + 10y + 24 = (y^2 - y - 2)(y^2 + y - 12).$$

Then, making the reverse substitution $y = x + 1$ yields

$$x^4 + 4x^3 - 9x^2 - 16x + 20 = (x^2 + x - 2)(x^2 + 3x - 10),$$

and the original quartic is factored as claimed.

Having reduced the problem to that of factoring depressed quartics, Euler noted that we need only consider $x^4 + Bx^2 + Cx + D$, where B, C, and D are real.

At this point, two cases present themselves:

Case 1. $C = 0$.

This amounts to having a depressed quartic $x^4 + Bx^2 + D$, which is just a quadratic in x^2. (Euler omitted discussion of this possibility, perhaps because it could be handled in two fairly easy subcases by purely algebraic means.)

First of all, suppose $B^2 - 4D \geq 0$ and apply the quadratic formula to get the decomposition into two second-degree *real* factors as follows:

$$x^4 + Bx^2 + D = \left[x^2 + \frac{B - \sqrt{B^2 - 4D}}{2} \right] \left[x^2 + \frac{B + \sqrt{B^2 - 4D}}{2} \right].$$

For instance, $x^4 + x^2 - 12 = (x^2 - 3)(x^2 + 4)$.

Less direct is the case where we try to factor $x^4 + Bx^2 + D$ under the condition that $B^2 - 4D < 0$. The previous decomposition no longer works, since the factors containing $\sqrt{B^2 - 4D}$ are not real. Fortunately, a bit of algebra shows that the quartic can be written as the difference of squares and thus factored into quadratics as follows:

$$\begin{aligned}
x^4 + Bx^2 + D &= \left[x^2 + \sqrt{D} \right]^2 - \left[x\sqrt{2\sqrt{D} - B} \right]^2 \\
&= \left[x^2 + \sqrt{D} - x\sqrt{2\sqrt{D} - B} \right] \left[x^2 + \sqrt{D} + x\sqrt{2\sqrt{D} - B} \right].
\end{aligned}$$

A few points must be made about this factorization. First, $B^2 - 4D < 0$ implies that $4D > B^2 \geq 0$, and so the expression $\sqrt{D}$ in the preceding factorization is indeed real. Likewise, $4D > B^2$ guarantees that $\sqrt{4D} > \sqrt{B^2}$, or simply $2\sqrt{D} > |B| \geq B$, and so the expression $\sqrt{2\sqrt{D} - B}$ is likewise real. In short, the factors above are two real quadratics, as we hoped.

For example, when factoring $x^4 + x^2 + 4$, we find $B^2 - 4D = -15 < 0$ and the formula yields $x^4 + x^2 + 4 = [x^2 - x\sqrt{3} + 2][x^2 + x\sqrt{3} + 2]$.

Case 2. $C \neq 0$.

Here Euler observed that a factorization of his depressed quartic into real quadratics—if it exists—*must* take the form

$$x^4 + Bx^2 + Cx + D = (x^2 + ux + \alpha)(x^2 - ux + \beta) \tag{1}$$

for some real numbers u, α, and β yet to be determined. Of course, this form is necessary since the "ux" in one factor must have a compensating "$-ux$" in the other.

Euler multiplied out the right-hand side of (1) to get:

$$x^4 + Bx^2 + Cx + D = x^4 + (\alpha + \beta - u^2)x^2 + (\beta u - \alpha u)x + \alpha\beta,$$

and then equated coefficients to generate three equations:

$$B = \alpha + \beta - u^2, \quad C = \beta u - \alpha u = (\beta - \alpha)u, \quad \text{and} \quad D = \alpha\beta.$$

Note that B, C, and D are just the coefficients of the original polynomial, whereas u, α, and β are unknown real numbers whose *existence* Euler had to establish.

From the first two of these we conclude that

$$\alpha + \beta = B + u^2 \quad \text{and} \quad \beta - \alpha = \frac{C}{u}.$$

It may be worth noting that since $0 \neq C = (\beta - \alpha)u$, then u itself is nonzero, so its presence in the denominator above is no cause for alarm.

If we both add and subtract these two equations, we arrive at

$$2\beta = B + u^2 + \frac{C}{u} \quad \text{and} \quad 2\alpha = B + u^2 - \frac{C}{u}. \tag{2}$$

Euler recalled that $D = \alpha\beta$ and consequently:

$$4D = 4\alpha\beta = (2\beta)(2\alpha) = \left(B + u^2 + \frac{C}{u}\right)\left(B + u^2 - \frac{C}{u}\right).$$

In other words, $4D = u^4 + 2Bu^2 + B^2 - (C^2/u^2)$, and multiplying through by u^2 gives us

$$u^6 + 2Bu^4 + (B^2 - 4D)u^2 - C^2 = 0. \tag{3}$$

It may appear that things have gotten worse, not better, for we have traded a fourth-degree equation in x for a sixth-degree equation in u. Admittedly, (3) is also a cubic in u^2, so we can properly conclude that there is a real solution for u^2; this, unfortunately, does not guarantee the existence of a real value for u, which was Euler's objective.

Undeterred, he noticed four critical properties of (3):

(a) B, C, and D are known, so the only unknown here is u.

(b) B, C, and D are real.

(c) the polynomial is even and thus its graph is symmetric about the y-axis.

(d) the constant term of this sixth-degree polynomial is $-C^2$.

Here Euler's mathematical agility becomes especially evident. He was considering a sixth-degree real polynomial whose graph looks something like that shown in Figure 1. This has a negative y-intercept at $(0, -C^2)$ since C is a nonzero real number. Additionally, since the polynomial is monic of even degree, its graph climbs toward $+\infty$ as u becomes unbounded in either the positive or negative direction. By a result from analysis we now call the intermediate value theorem—but which Euler took as intuitively clear—we are guaranteed the existence of real numbers $u_0 > 0$ and $-u_0 < 0$ satisfying this sixth-degree equation.

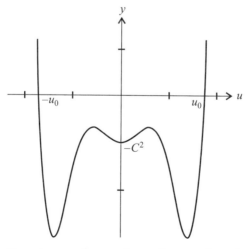

Figure 1. $y = u^6 + 2Bu^4 + (B^2 - 4D)u^2 - C^2$

Using the positive solution u_0 and returning to the equations in (2), Euler solved for β and α, getting real solutions

$$\beta_0 = \frac{1}{2}\left(B + u_0^2 + \frac{C}{u_0}\right) \quad \text{and} \quad \alpha_0 = \frac{1}{2}\left(B + u_0^2 - \frac{C}{u_0}\right)$$

and, since $u_0 > 0$, these fractions are well-defined.

In summary, under the case that $C \neq 0$, Euler had established the existence of real numbers u_0, α_0, and β_0 such that

$$x^4 + Bx^2 + Cx + D = (x^2 + u_0x + \alpha_0)(x^2 - u_0x + \beta_0).$$

We thus see that any depressed quartic with real coefficients—and by extension any real quartic at all—does have a factorization into two real quadratics, whether or not $C = 0$. Q.E.D.

At this point, Euler immediately observed, "...it is also evident that any equation of the fifth degree is also resolvable into three real factors of which one is linear and two are quadratic" [1, p. 95]. His reasoning was simple (see Figure 2). Any *odd-degree* polynomial—and thus any fifth-degree polynomial $P(x)$—is guaranteed by the intermediate value theorem to have at least one real x-intercept, say at $x = a$. We then write $P(x) = (x-a)Q(x)$, where $Q(x)$ is a polynomial of the fourth-degree, and the previous result allows us to decompose $Q(x)$, in turn, into two real quadratic factors.

By now, a general strategy was brewing in his mind. He realized that if he could prove his decomposition for real polynomials of degree 4, 8, 16, 32, and in general of degree 2^n, then he could prove it for any real polynomials whatever.

Why is this? Suppose, for instance, we were trying to establish that the polynomial

$$x^{12} - 3x^9 + 52x^8 + 3x^3 - 2x + 17$$

could be factored into real linear and/or real quadratic factors. We would simply multiply it by x^4 to get

$$x^{16} - 3x^{13} + 52x^{12} + 3x^7 - 2x^5 + 17x^4.$$

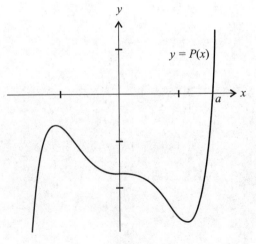

Figure 2.

Assuming that Euler had proved the 16th-degree case, he would know that this latter polynomial would have such a factorization, obviously containing the four linear factors x, x, x, and x. If we merely cancelled them out, we would of necessity be left with the real linear and/or real quadratic factors for the original 12th-degree polynomial.

And so, with typical Eulerian cleverness, he reduced the entire issue to a few simpler cases. Having disposed of the fourth-degree case, he next claimed, "Any equation of the eighth degree is always resolvable into two real factors of the fourth degree" [1, p. 99]. Since each of the fourth-degree factors was itself decomposable into a pair of real quadratics, which themselves can be broken into (possibly complex) linear factors, he would have succeeded in shattering the eighth-degree polynomial into eight linear pieces. From there he went to the 16th-degree before finally tackling the general situation, namely showing that any real polynomial of degree 2^n can be factored into two real polynomials each of degree 2^{n-1} [1, p. 105].

It was a brilliant strategy. Unfortunately, the proofs he furnished left something to be desired. As we shall see, for the higher-degree cases the arguments became hopelessly complicated, and his assertions as to the existence of real numbers satisfying certain equations were unconvincing. Consider, for instance, the eighth-degree case. It began in a fashion quite similar to its fourth-degree counterpart, namely by first depressing the octic and imagining that it has been factored into the two quartics:

$$
\begin{aligned}
& x^8 + Bx^6 + Cx^5 + Dx^4 + Ex^3 + Fx^2 + Gx + H \\
& = (x^4 + ux^3 + \alpha x^2 + \beta x + \gamma)(x^4 - ux^3 + \delta x^2 + \varepsilon x + \phi).
\end{aligned}
\tag{4}
$$

One multiplies the quartics, equates the resulting coefficients with the known quantities $B, C, D, \ldots$ to get seven equations in seven unknowns, and asserts that there exist real values of $u, \alpha, \beta, \gamma, \ldots$ satisfying this system.

The parallels with what he had previously done are evident. But what made this case so much less successful was Euler's admission that for equations of higher degree, "...it will be very difficult and even impossible to find the equation by which the unknown u is determined" [1, p. 97]. In short, he was unwilling or unable to solve this system explicitly for u.

Ever resourceful, Euler decided to look again at the depressed quartic in (1) for inspiration. As it turned out, an entirely different line of reasoning suggested itself, a line that he thought could be extended naturally to the eighth and higher-degree cases:

Assuming that the quartic in (1) has four roots p, q, r, and s, Euler wrote:

$$
\begin{aligned}
(x^2 + ux + \alpha)(x^2 - ux + \beta) & = x^4 + Bx^2 + Cx + D \\
& = (x - p)(x - q)(x - r)(x - s),
\end{aligned}
\tag{5}
$$

and from this factorization he drew three key conclusions.

First, upon multiplying the four linear factors on the right of (5), we see immediately that the coefficient of x^3 is $-(p + q + r + s)$; hence $p + q + r + s = 0$ since the quartic is depressed.

Second, the quadratic factor $(x^2 - ux + \beta)$ must arise as the product of two of the four linear factors. Thus, $(x^2 - ax + \beta)$ could be $(x - p)(x - r) = x^2 - (p + r)x + pr$; it could just as well be $(x - q)(x - r) = x^2 - (q + r)x + qr$; and so on. This implies that, in the first case, $u = p + r$, whereas in the second $u = q + r$. In fact, it is clear that u can take any of the $\binom{4}{2} = 6$ values

$$R_1 = p + q \qquad R_4 = r + s$$
$$R_2 = p + r \qquad R_5 = q + s$$
$$R_3 = p + s \qquad R_6 = q + r.$$

Since u is an unknown having these six possible values, it must be determined by the sixth-degree polynomial $(u - R_1)(u - R_2)(u - R_3)(u - R_4)(u - R_5)(u - R_6)$. This conclusion, of course, is entirely consistent with the explicit sixth-degree polynomial for u that Euler had found in (3).

But Euler made one additional observation. Because $p + q + r + s = 0$, it follows that $R_4 = -R_1$, $R_5 = -R_2$, and $R_6 = -R_3$. Hence the sixth-degree polynomial becomes

$$(u - R_1)(u + R_1)(u - R_2)(u + R_2)(u - R_3)(u + R_3)$$
$$= (u^2 - R_1^2)(u^2 - R_2^2)(u^2 - R_3^2).$$

The constant term here—which is to say, this polynomial's y-intercept—is simply

$$-R_1^2 R_2^2 R_3^2 = -(R_1 R_2 R_3)^2.$$

This constant, Euler stated, was a negative real number, again in complete agreement with his conclusions from equation (3).

To summarize, Euler had provided an entirely different argument to establish that, in the quartic case, u is determined by a $\binom{4}{2} = $ 6th-degree polynomial with a negative y-intercept. This was the critical conclusion he had already drawn, but here he drew it without *explicitly* finding the equation determining u.

The advantage of this alternate proof for the quartic case was that it could be used to analyze the depressed octic in (4). *Assuming* that the octic was decomposed into eight linear factors, Euler mimicked his reasoning above to deduce that for each different combination of four of these eight factors, we would get a different value of u. Thus, u would be determined by a polynomial of degree $\binom{8}{4} = 70$ having a negative y-intercept. He then confidently applied the intermediate value theorem to get his desired *real* root u_0, and from this he claimed that the other real numbers α_0, β_0, γ_0, δ_0, ε_0, and ϕ_0 exist as well.

Euler reasoned similarly in the 16th-degree case, claiming that "... the equation which determines the values of the unknown u will necessarily be of the 12870th degree" [1, p. 103]. The degree of this (obviously unspecified) equation is simply $\binom{16}{8} = 12870$, as his pattern suggested. By this time, Euler's comment, that it was "... very difficult and even impossible..." to specify these polynomials, had become something of an understatement.

From there it was a short and entirely analogous step to the general case: that any real polynomial of degree 2^n could be factored into two real polynomials of degree 2^{n-1}. With that, his proof was finished.

Or was it? Unfortunately, his analyses of the 8th-degree, 16th-degree, and general cases were flawed and left significant questions unanswered. For instance, if we look back at the quartic in (5), how could Euler assert that it has four roots? How could he assert that the octic in (4) has eight?

More significantly, what is the nature of these supposed roots? Are they real? Are they complex? Or are they an unspecified—and perhaps entirely unimagined—new kind of number? If so, can they be added and multiplied in the usual fashion?

These are not trivial questions. In the quartic case above, for example, if we are uncertain about the nature of the roots p, q, r, and s, then we are equally uncertain about the nature of their sums R_1, R_2, R_3. Consequently, there is no guarantee whatever that mysterious expressions such as $-(R_1 R_2 R_3)^2$ are negative real numbers. But if these y-intercepts are not negative reals, then the intermediate value arguments that Euler applied to the 8th-degree, 16th-degree, and general cases fall apart completely.

It appears, then, that Euler had started down a very promising path in his quest of the fundamental theorem. His first proof worked nicely in dealing with fourth- and fifth-degree real polynomials. But as he pursued this elusive theorem deeper into the thicket, complications involving the existence of his desired real factors became overwhelming. In a certain sense, he lost his way among the enormously high degree polynomials that beckoned him on, and his general proof vanished in the wilderness.

So even Euler suffered setbacks, a fact from which comfort may be drawn by lesser mathematicians (a category that includes virtually everybody else in history). Yet, before the dust settles and his attempted proof is consigned to the scrap heap, I think it deserves at least a modest round of applause, for it certainly bears signs of his characteristic cleverness, boldness, and mental agility as he leaps between the polynomial's analytic and algebraic properties. More to the point, the fourth- and fifth-degree arguments are understandable by good precalculus students and can give them not only a deeper look at this remarkable theorem but also a glimpse of a mathematical giant at work. For even when he stumbled, Leonhard Euler left behind signs of great insight. Such, perhaps, is the mark of genius.

Epilogue

The fundamental theorem of algebra—the result that established the complex numbers as the optimum realm for factoring polynomials or solving polynomial equations—thus remained in a very precarious state. D'Alembert had not proved it; Euler had given an unsatisfactory proof. It was obviously in need of major attention to resolve its validity once and for all.

Such a resolution awaited the last year of the eighteenth century and came at the hands of one of history's most talented and revered mathematicians. It was the 22-year old German Carl Friedrich Gauss (1777–1855) who first presented a reasonably complete proof of the fundamental theorem (see [4, p. 196] for an interesting twist on this oft-repeated statement). Gauss' argument appeared in his 1799 doctoral dissertation with the long and descriptive title, "A New Proof of the Theorem That Every Integral Rational Algebraic Function [i.e., every polynomial with real coefficients] Can Be Decomposed into Real Factors of the First or Second Degree" (see [5, pp. 115–122]). He began by reviewing past attempts at proof and giving criticisms of each. When addressing Euler's "proof," Gauss raised the issues cited above, designating Euler's mysterious, hypothesized roots as "shadowy." To Gauss, Euler's attempt lacked "... the clarity which is required in

Figure 3. Ten mark German banknote featuring Carl Friedrich Gauss

mathematics" [2, p. 491]. This clarity he attempted to provide, not only in the dissertation but in two additional proofs from 1816 and another from 1848. As indicated by his return to this result throughout his illustrious career, Gauss viewed the fundamental theorem of algebra as a great and worthy project indeed.

We noted previously that this crucial proposition is seen today in somewhat greater generality than in the early nineteenth century, for we now transfer the theorem entirely into the realm of complex numbers in this sense: the polynomial with which we begin no longer is required to have real coefficients. In general, we consider nth-degree polynomials having complex coefficients, such as

$$z^7 + 6iz^6 - (2 + i)z^2 + 19.$$

In spite of this apparent increase in difficulty, the fundamental theorem nonetheless proves that it can be factored into the product of (in this case seven) linear terms having, of course, complex coefficients. Interestingly, modern proofs of this result almost never appear in algebra courses. Rather, today's proofs rest upon a study of the *calculus* of complex numbers and thus move quickly into the realm of genuinely advanced mathematics (just as my high school algebra teacher had so truthfully said).

And so, we reach the end of our story, a story that can be a valuable tale for us and our students. It addresses an oft-neglected theorem of much importance; it allows the likes of Jean d'Alembert, Leonhard Euler, and Carl Friedrich Gauss to cross the stage; and it gives an intimate sense of the historical development of great mathematics in the hands of great mathematicians.

References

1. Leonhard Euler, Recherches sur les racines imaginaires des équations, *Mémoires de l'académie des sciences de Berlin* (5) (1749), 1751, 222–288 (*Opera Omnia*, (1) 6, 78–147).

2. J. Fauvel and J. Gray, *The History of Mathematics: A Reader*, The Open University, London, 1987.

3. Morris Kline, *Mathematical Thought from Ancient to Modern Times*, Oxford University Press, New York, 1972.

4. John Stillwell, *Mathematics and its History*, Springer-Verlag, New York, 1989.

5. Dirk Struik (Ed.), *A Source Book in Mathematics: 1200–1800*, Princeton University Press, 1986.

An East German stamp honoring Euler

An East German stamp showing Euler, his formula for polyhedra,
and an icosahedron (which is, of course, the logo of the MAA!)

Guessing and Proving[1]

George Pólya

In the "commentatio" (Note presented to the Russian Academy) in which his theorem on polyhedra (on the number of faces, edges and vertices) was first published, Euler gives no proof.[2] In place of proof, he offers an inductive argument: He verifies the relation in a variety of special cases. There is little doubt that he also discovered the theorem, as many of his other results, inductively. Yet he does not give a direct indication of how he was led to his theorem, of how he "guessed" it, whereas in some other cases he offers suggestive hints about the ways and motives of his inductive considerations.

How was Euler led to his theorem on polyhedra? I think that it is not futile to speculate on this question although, of course, we cannot expect a conclusive answer. The question is relevant pedagogically: The theorem is of high interest in itself, and it is so simple; its understanding requires so little preliminary knowledge that its *rediscovery* can be proposed as a stimulating project to an intelligent teenager. Projects of this kind could give young people a first idea of scientific work, a first insight into the interplay of guessing and proving in the mathematician's mind.

One can imagine various approaches to the discovery (rediscovery) of Euler's theorem. I have presented two different approaches on former occasions.[3] I offer here a third one which, I like to think, could have been Euler's own approach. At any rate, I shall stress in the following some points of contact with Euler's text.

1. Analogy suggests a problem

There is a certain analogy between plane geometry and solid geometry which may appear plausible even to a beginner. A circle in the plane is analogous to a sphere in space; the area enclosed by a curve in the plane is analogous to the volume enclosed by a surface in space; polygons enclosed by straight sides in the plane are analogous to polyhedra enclosed by plane faces in space.

---

[1]Reprinted from the *Two-Year College Mathematics Journal*, Vol. 9, January, 1978, pp. 21–27.

[2]See [1], pp. 72–93. There is a following second Note, pp. 94–108 and a preceding summary, pp. 71–72, which mentions the second Note. See also the remarks of the editor, pp. XIV–XVI.

[3]See [2], vol. 1, pp. 35–43 and [3], vol. 2, pp. 149–156, and also the annexed problems and solutions in both books.

Yet there is a difference. If we look closer, the geometry of the plane appears as simpler and easier whereas that of space appears as more intricate and more difficult. Take the polygons and the polyhedra. We have a simple classification of polygons according to the number of their sides. The triangles form the simplest class; they have three sides. Next comes the class of quadrilaterals which have four sides, and so on. The n-sided polygons form a class; two polygons belonging to this class may differ quantitatively, in the lengths of their sides and the openings of their angles, yet they agree in an important respect—should we say "qualitatively" or "morphologically"?[4] We could try to classify the polyhedra according to the number of their faces. Now look at the three polyhedra of Figure 1: the regular octahedron, a prism with a hexagonal base, and a pyramid with a heptagonal base. All three have the same number of faces, namely eight, but they are too different in their whole aspect (morphology) to be classified together.[5]

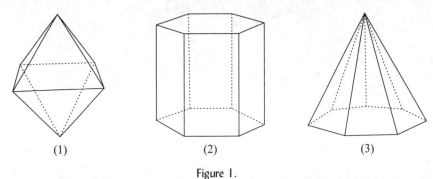

Figure 1.

Here emerges a problem: Let us devise a classification of polyhedra that accomplishes something analogous to the simple classification of polygons according to the number of their sides. Yet, in the case of polyhedra, taking into account just the number of faces is not enough as the example of Figure 1 shows.[6]

2. A first trial

Here is a remark that could be relevant. Two polygons that have the same number of sides also have the same number of vertices, equal to the number of sides. Yet polyhedra are more complicated. All three polyhedra in Figure 1 have eight faces, yet they have different numbers of vertices, namely six, twelve and eight vertices, respectively. Would it be enough for a good classification to take into account the number of faces F and the number of vertices V?

What should we do to answer this question? Survey as many different forms of polyhedra as we can and count their faces and vertices. Figure 2 offers a short survey. Most polyhedra in Figure 2 are named. The abbreviation "n Pyd" means "Pyramid with an n-sided base," and "n Psm" stands for "Prism with an n-sided base." Thus in polyhedron

---

[4] I am intentionally avoiding the standard term which, by the way, did not exist in Euler's time. One of the ugliest outgrowths of the "new math" is the premature introduction of technical terms.

[5] See [1], p. 71.

[6] The classification of polyhedra takes up the major part of the text of Euler's Note. Was this problem of classification his starting point?

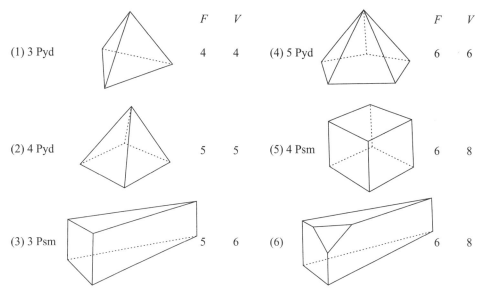

Figure 2.

(5) the 4 Psm is represented by a cube, as quantitative specialization does not matter. Just the last polyhedron is not named; it is a "truncated 3 Psm," and we obtain it from a 3 Psm by cutting off one vertex (in fact, a small tetrahedron topped by that vertex). Polyhedron (6) answers our question, and it answers it negatively: (6) agrees with (5) (the cube) in both numbers considered (of faces and vertices), yet (5) and (6) are essentially (morphologically) different; they should not be put into the same class. The F and V, the numbers of faces and vertices, are *not enough*.[7]

3. Another trial. Disappointment and triumph

What else is here, besides the faces and the vertices, to provide a basis for the classification of polyhedra? The edges.[8] Let us look again at the polyhedra we have surveyed, and let us examine their edges.

Figure 3 omits the names shown in Figure 2, but repeats the rest and adds the number of edges E. Yet it does not help; it provides no distinction between the polyhedra (5) and (6). They agree also in the number of edges, as they have agreed in the number of faces and vertices. And exploring further cases we find invariably: If two polyhedra have the same F and V, they also have the same E. Thus the number of edges contributes nothing to the classification of polyhedra over and above what the faces and vertices have done already. What a disappointment!

Yet there is something else. If the number E of edges is determined by the numbers F and V, of faces and vertices, then E is a function of F and V. Which function? Is

---

[7]A high school student, working on our project, will probably have to survey many more polyhedra before he encounters such a critical pair, providing the negative answer.

[8]Euler was the first to introduce the concept of the "edge of a polyhedron" and to give a name to it (*acies*). He emphasizes this fact, mentions it twice; see [1], p. 71 and p. 73. Perhaps Euler introduced edges in the hope of a better classification, and we follow his example here.

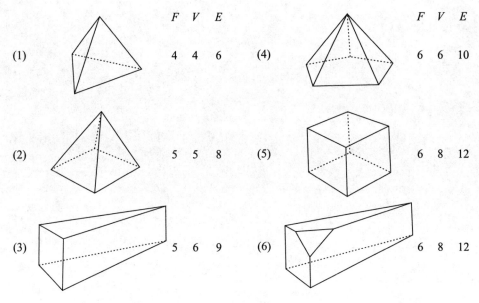

	F	V	E			F	V	E
(1)	4	4	6	(4)		6	6	10
(2)	5	5	8	(5)		6	8	12
(3)	5	6	9	(6)		6	8	12

Figure 3.

it an increasing function? Does E increase whenever F increases? Does E necessarily increase with V? When examples show that neither is the case, the question arises, "Does E increase somehow with F and V jointly?" Such or similar questions (cf. the pages quoted in footnote [2]) may lead to more examples (more than displayed in Figures 2 and 3) and eventually to the guess,

$$E = F + V - 2.$$

An unsuspected, extremely simple relation, unique of its kind. What a triumph!

One is tempted to compare Euler with Columbus. Columbus set out to reach India by a western route across the ocean. He did not reach India, but he discovered a new

PROPOSITIO IV.

§. 33. *In omni folido bedris planis inclufo aggregatum ex numero angulorum folidorum et ex numero bedrarum binario excedit numerum acierum.*

DEMONSTRATIO.

Scilicet fi ponatur vt hactenus :
numerus angulorum folidorum $=$ S
numerus acierum - - - - $=$ A
numerus hedrarum - - - $=$ H
demonſtrandum eſt , eſſe S $+$ H $=$ A $+$ 2.

Euler's polyhedral formula in a 1758 paper. Here S is the number of solid angles (vertices), A is the number of edges, and H is the number of faces.

continent. Euler set out to find a classification of polyhedra. He did not achieve a complete classification, but he discovered, in fact, a new branch of mathematics, topology, to which his theorem on polyhedra properly belongs.

4. Miscellaneous remarks

We have seen a way which Euler may have taken in discovering his theorem on polyhedra, and this was our main topic. There remain several connected points worth considering, but we must consider them very briefly.

(a) Almost a year after his first Note, Euler presented to the Russian Academy a second Note on polyhedra in which he gave a proof of his theorem.[9] His proof is invalid, and, looking back at it from our present standpoint, we can easily see why it was fated to be invalid.

It hinges on the concept of "polyhedron." What is a polyhedron? A part of three-dimensional space enclosed by plane faces. Take the case of Figure 4, a cube with a cubical cavity. Between the outer, larger cubical surface and the inner, smaller one there is a part of space, enclosed by plane faces. Is it a polyhedron? It has twice as many faces, vertices and edges as an ordinary cube, namely

$$F = 12, \quad V = 16, \quad E = 24,$$

and these numbers do not satisfy Euler's relation,

$$24 \neq 12 + 16 - 2.$$

(b) After Euler's death there arose a debate.[10] Some mathematicians were for Euler's theorem, others against it. The opponents produced counterexamples, such as Figure 4. The partisans answered with invective: Figure 4 is not a polyhedron; it is a monster. It is indecent; it is obscene: The big cube with the small cube inside looks like an expectant mother with her unborn baby. An expectant mother is not a counterexample to the proposition that each person has just one head, and, similarly, Figure 4 is inadmissable as a counterexample.

Which shows that in the heat of debate mathematicians can sound as silly as politicians.

(c) In fact, Euler's theorem cannot be really and truly proved before it is satisfactorily stated. There are two ways to state it more satisfactorily, either by generalizing, by introducing the relevant topological concepts unknown in Euler's time,[11] or by specializing, by restricting it to convex polyhedra.

(d) Analogy suggested the approach to Euler's theorem considered in the foregoing sections. Yet analogy is a many-sided thing, and it can suggest other approaches too.

The sum of the angles in an n-sided polygon is $(n-2)180°$. Looking for an analogous fact about polyhedra we can be led to Euler's theorem, and even to a valid proof of it

---

[9] [1], pp. 94–108.

[10] [4] uses the history of Euler's theorem to discuss topics in epistemology and methodology. Rewriting this discussion so that it becomes simpler and more accessible, even if less witty, would be a rewarding task.

[11] See [5], especially pp. 258–259.

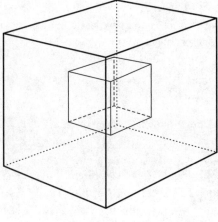

Figure 4.

for convex polyhedra, which is intuitive and can be conveniently presented on the high school level.[12]

(e) I said at the beginning that the "rediscovery" of Euler's theorem can be proposed as a project to an intelligent teenager. A more obvious occasion is to develop the theorem in class discussion; the teacher should lead the discussion so that the students have a fair share in the "rediscovery" of the theorem.

(f) In individual projects, or in class discussion, Euler's theorem could and should be used to introduce young people to the *scientific method*, to *inductive research*, to the fascination of *analogy*. What is "scientific method"? Philosophers and non-philosophers have discussed this question and have not yet finished discussing it. Yet as a first introduction it can be described in three syllables:

Guess and test.

Mathematicians too follow this advice in their research although they sometimes refuse to confess it. They have, however, something which the other scientists cannot really have. For mathematicians the advice is

First guess, then prove.

References

1. Leonhard Euler, *Opera Omnia*, series 1, vol. 26, 1953.

2. G. Pólya, *Mathematics and Plausible Reasoning*, Princeton University Press, 1954.

3. ——, *Mathematical Discovery*, Wiley, 1962/4.

4. I. Lakatos, Proofs and Refutations, *British J. Philos. Sci.*, 14, (1963–64) 53, 54, 55, 56.

5. R. Courant and H. Robbins, *What is Mathematics?* Oxford University Press, 1941.

---

[12]See [3], vol. 2, pp. 149–156. This approach, although not the attached proof, was also known to Euler. See [1], especially p. 90, Propositio 9.

The Truth about Königsberg[1]

Brian Hopkins and Robin J. Wilson

Euler's 1736 paper on the bridges of Königsberg is widely regarded as the earliest contribution to graph theory—yet Euler's solution made no mention of graphs. In this paper we place Euler's views on the Königsberg bridges problem in their historical context, present his method of solution, and trace the development of the present-day solution.

What Euler didn't do

A well-known recreational puzzle concerns the bridges of Königsberg. It is claimed that in the early eighteenth century the citizens of Königsberg used to spend their Sunday afternoons walking around their beautiful city. The city itself consisted of four land areas separated by branches of the river Pregel over which there were seven bridges, as illustrated in Figure 1. The problem that the citizens set themselves was to walk around the city, crossing each of the seven bridges exactly once and, if possible, returning to their starting point.

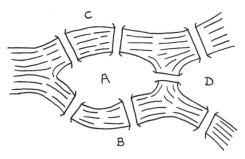

Figure 1. Königsberg

If you look in some books on recreational mathematics, or listen to some graph-theorists who should know better, you will 'learn' that Leonhard Euler investigated the Königsberg bridges problem by drawing a graph of the city, as in Figure 2, with a vertex representing

---

[1]Reprinted from *College Math Journal*, Vol. 35, May, 2004, pp. 198–207. This article received the George Pólya Award in 2005.

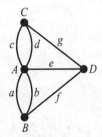

Figure 2. The Königsberg graph

each of the four land areas and an edge representing each of the seven bridges. The problem is then to find a trail in this graph that passes along each edge just once.

But Euler didn't draw the graph in Figure 2—graphs of this kind didn't make their first appearance until the second half of the nineteenth century. So what exactly did Euler do?

The Königsberg bridges problem

In 1254 the Teutonic knights founded the Prussian city of Königsberg (literally, king's mountain). With its strategic position on the river Pregel, it became a trading center and an important medieval city. The river flowed around the island of Kneiphof (literally, pub yard) and divided the city into four regions connected by seven bridges: Blacksmith's bridge, Connecting bridge, High bridge, Green bridge, Honey bridge, Merchant's bridge, and Wooden bridge: Figure 3 shows a seventeenth-century map of the city. Königsberg

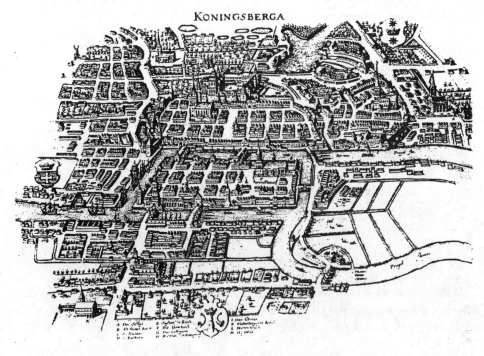

Figure 3. Seventeenth-century Königsberg

later became the capital of East Prussia and more recently became the Russian city of Kaliningrad, while the river Pregel was renamed Pregolya.

In 1727 Leonhard Euler began working at the Academy of Sciences in St. Petersburg. He presented a paper to his colleagues on 26 August 1735 on the solution of 'a problem relating to the geometry of position': this was the Königsberg bridges problem. He also addressed the generalized problem: given any division of a river into branches and any arrangement of bridges, is there a general method for determining whether such a route exists?

In 1736 Euler wrote up his solution in his celebrated paper in the *Commentarii Academiae Scientiarum Imperialis Petropolitanae* under the title 'Solutio problematis ad geometriam situs pertinentis' [2]; Euler's diagram of the Königsberg bridges appears in Figure 4. Although dated 1736, Euler's paper was not actually published until 1741, and was later reprinted in the new edition of the *Commentarii* (*Novi Acta Commentarii...*) which appeared in 1752.

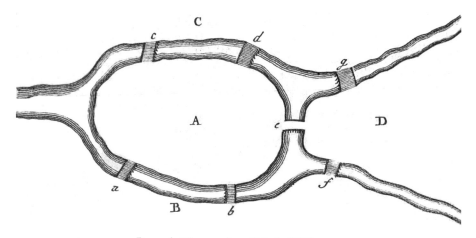

Figure 4. Diagram from Euler's 1736 paper

A full English translation of this paper appears in several places—for example, in [1] and [6]. The paper begins:

1. In addition to that branch of geometry which is concerned with distances, and which has always received the greatest attention, there is another branch, hitherto almost unknown, which Leibniz first mentioned, calling it the geometry of position [*Geometriam situs*]. This branch is concerned only with the determination of position and its properties; it does not involve distances, nor calculations made with them. It has not yet been satisfactorily determined what kinds of problem are relevant to this geometry of position, or what methods should be used in solving them. Hence, when a problem was recently mentioned which seemed geometrical but was so constructed that it did not require the measurement of distances, nor did calculation help at all, I had no doubt that it was concerned with the geometry of position—especially as its solution involved only position, and no calculation was of any use. I have therefore decided to give here the method which I have found for solving this problem, as an example of the geometry of position.

2. The problem, which I am told is widely known, is as follows: in Königsberg
...

This reference to Leibniz and the geometry of position dates back to 8 September 1679, when the mathematician and philosopher Gottfried Wilhelm Leibniz wrote to Christiaan Huygens as follows [5]:

> I am not content with algebra, in that it yields neither the shortest proofs nor the most beautiful constructions of geometry. Consequently, in view of this, I consider that we need yet another kind of analysis, geometric or linear, which deals directly with position, as algebra deals with magnitudes ...

Leibniz introduced the term *analysis situs* (or *geometria situs*), meaning the analysis of situation or position, to introduce this new area of study. Although it is sometimes claimed that Leibniz had vector analysis in mind when he coined this phrase (see, for example, [8] and [11]), it was widely interpreted by his eighteenth-century followers as referring to topics that we now consider 'topological'—that is, geometrical in nature, but with no reference to metrical ideas such as distance, length or angle.

Euler's Königsberg letters

It is not known how Euler became aware of the Königsberg bridges problem. However, as we shall see, three letters from the Archive Collection of the Academy of Sciences in St. Petersburg [3] shed some light on his interest in the problem (see also [10]).

Carl Leonhard Gottlieb Ehler was the mayor of Danzig in Prussia (now Gdansk in Poland), some 80 miles west of Königsberg. He corresponded with Euler from 1735 to 1742, acting as intermediary for Heinrich Kühn, a local mathematics professor. Their initial communication has not been recovered, but a letter of 9 March 1736 indicates they had discussed the problem and its relation to the 'calculus of position':

> You would render to me and our friend Kühn a most valuable service, putting us greatly in your debt, most learned Sir, if you would send us the solution, which you know well, to the problem of the seven Königsberg bridges, together with a proof. It would prove to be an outstanding example of the calculus of position [*Calculi Situs*], worthy of your great genius. I have added a sketch of the said bridges ...

Euler replied to Ehler on 3 April 1736, outlining more clearly his own attitude to the problem and its solution:

> ... Thus you see, most noble Sir, how this type of solution bears little relationship to mathematics, and I do not understand why you expect a mathematician to produce it, rather than anyone else, for the solution is based on reason alone, and its discovery does not depend on any mathematical principle. Because of this, I do not know why even questions which bear so little relationship to mathematics are solved more quickly by mathematicians than by others. In the meantime, most noble Sir, you have assigned this question to the geometry of position, but I am ignorant as to what this new discipline involves, and as to which types of problem Leibniz and Wolff expected to see expressed in this way ...

Around the same time, on 13 March 1736, Euler wrote to Giovanni Marinoni, an Italian mathematician and engineer who lived in Vienna and was Court Astronomer in the court of Kaiser Leopold I. He introduced the problem as follows (see Figure 5):

> A problem was posed to me about an island in the city of Königsberg, surrounded by a river spanned by seven bridges, and I was asked whether someone could traverse the separate bridges in a connected walk in such a way that each bridge is crossed only once. I was informed that hitherto no-one had demonstrated the possibility of doing this, or shown that it is impossible. This question is so banal,

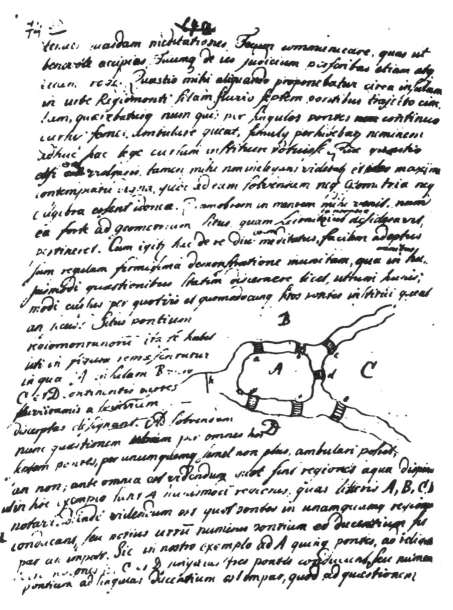

Figure 5. Euler's letter to Marinoni

<"></>

but seemed to me worthy of attention in that geometry, nor algebra, nor even the art of counting was sufficient to solve it. In view of this, it occurred to me to wonder whether it belonged to the geometry of position [*geometriam Situs*], which Leibniz had once so much longed for. And so, after some deliberation, I obtained a simple, yet completely established, rule with whose help one can immediately decide for all examples of this kind, with any number of bridges in any arrangement, whether such a round trip is possible, or not ...

Euler's 1736 paper

Euler's paper is divided into twenty-one numbered paragraphs, of which the first ascribes the problem to the geometry of position as we saw above, the next eight are devoted to the solution of the Königsberg bridges problem itself, and the remainder are concerned with the general problem. More specifically, paragraphs 2–21 deal with the following topics (see also [12]):

Paragraph 2.   Euler described the problem of the Königsberg bridges and its generalization: 'whatever be the arrangement and division of the river into branches, and however many bridges there be, can one find out whether or not it is possible to cross each bridge exactly once?'

Paragraph 3.   In principle, the original problem could be solved exhaustively by checking all possible paths, but Euler dismissed this as 'laborious' and impossible for configurations with more bridges.

Paragraphs 4–7.   The first simplification is to record paths by the land regions rather than bridges. Using the notation in Figure 4, going south from Kneiphof would be notated AB whether one used the Green Bridge or the Blacksmith's Bridge. The final path notation will need to include an adjacent A and B twice; the particular assignment of bridges a and b is irrelevant. A path signified by n letters corresponds to crossing $n - 1$ bridges, so a solution to the Königsberg problem requires an eight-letter path with two adjacent A/B pairs, two adjacent A/C pairs, one adjacent A/D pair, etc.

Paragraph 8.   What is the relation between the number of bridges connecting a land mass and the number of times the corresponding letter occurs in the path? Euler developed the answer from a simpler example (see Figure 6). If there is an odd number k of bridges, then the letter must appear $(k + 1)/2$ times.

Figure 6. A simple case

Paragraph 9. This is enough to establish the impossibility of the desired Königsberg tour. Since Kneiphof is connected by five bridges, the path must contain three As. Similarly, there must be two Bs, two Cs, and two Ds. In *Paragraph 14*, Euler records these data in a table.

region	A	B	C	D
bridges	5	3	3	3
frequency	3	2	2	2

Summing the final row gives nine required letters, but a path using each of the seven bridges exactly once can have only eight letters. Thus there can be no Königsberg tour.

Paragraphs 10–12. Euler continued his analysis from *Paragraph 8:* if there is an even number k of bridges connecting a land mass, then the corresponding letter appears $k/2+1$ times if the path begins in that region, and $k/2$ times otherwise.

Paragraphs 13–15. The general problem can now be addressed. To illustrate the method Euler constructed an example with two islands, four rivers, and fifteen bridges (see Figure 7).

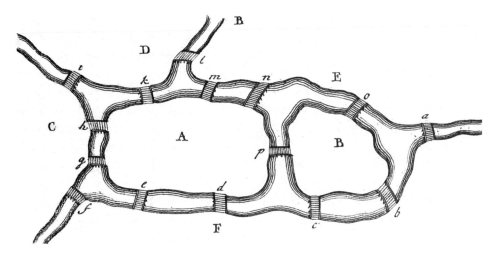

Figure 7. A more complicated example

This system has the following table, where an asterisk indicates a region with an even number of bridges.

region	A*	B*	C*	D	E	F*
bridges	8	4	4	3	5	6
frequency	4	2	2	2	3	3

The frequencies of the letters in a successful path are determined by the rules for even and odd numbers of bridges, developed above. Since there can be only one initial region, he records $k/2$ for the asterisked regions. If the frequency sum is one less than the required number of letters, there is a path using each bridge exactly once that begins in an asterisked region. If the frequency sum equals the required number of letters, there is a path that begins in an unasterisked region. This latter possibility is the case here: the frequency sum is 16, exactly the number of letters required for a path using 15 bridges. Euler exhibited a particular path, including the bridges:

$$E\,a\,F\,b\,B\,c\,F\,d\,A\,e\,F\,f\,C\,g\,A\,h\,C\,i\,D\,k\,A\,m\,E\,n\,A\,p\,B\,o\,E\,l\,D.$$

Paragraph 16–19.   Euler continued with a simpler technique, observing that:

> ... the number of bridges written next to the letters A, B, C, etc. together add up to twice the total number of bridges. The reason for this is that, in the calculation where every bridge leading to a given area is counted, each bridge is counted twice, once for each of the two areas which it joins.

This is the earliest version known of what is now called the *handshaking lemma*. It follows that in the bridge sum, there must be an even number of odd summands.

Paragraph 20.   Euler stated his main conclusions:

> If there are more than two areas to which an odd number of bridges lead, then such a journey is impossible.

> If, however, the number of bridges is odd for exactly two areas, then the journey is possible if it starts in either of these two areas.

> If, finally, there are no areas to which an odd number of bridges lead, then the required journey can be accomplished starting from any area.

Paragraph 21.   Euler concluded by saying:

> When it has been determined that such a journey can be made, one still has to find how it should be arranged. For this I use the following rule: let those pairs of bridges which lead from one area to another be mentally removed, thereby considerably reducing the number of bridges; it is then an easy task to construct the required route across the remaining bridges, and the bridges which have been removed will not significantly alter the route found, as will become clear after a little thought. I do not therefore think it worthwhile to give any further details concerning the finding of the routes.

Note that this final paragraph does not prove the existence of a journey when one is possible, apparently because Euler did not consider it necessary. So Euler provided a rigorous proof only for the first of the three conclusions. The first satisfactory proof of the other two results did not appear until 1871, in a posthumous paper by Carl Hierholzer (see [1] and [4]).

The modern solution

The approach mentioned in the first section developed through diagram-tracing puzzles discussed by Louis Poinsot [7] and others in the early-nineteenth century. The object is to determine whether a figure can be drawn with a single stroke of the pen in such a way that no edge is repeated. Considering the figure to be drawn as a graph, the general conditions in *Paragraph 20* take the following form:

> If there are more than two vertices of odd degree, then such a drawing is impossible.

> If, however, exactly two vertices have odd degree, then the drawing is possible if it starts with either of these two vertices.

> If, finally, there are no vertices of odd degree, then the required drawing can be accomplished starting from any vertex.

So the 4-vertex graph shown in Figure 2, with one vertex of degree 5 and three vertices of degree 3, cannot be drawn with a single stroke of the pen so that no edge is repeated. In contemporary terminology, we say that this graph is not Eulerian. The arrangement of bridges in Figure 7 can be similarly represented by the graph in Figure 8, with six vertices and fifteen edges. Exactly two vertices (E and D) have odd degree, so there is a drawing that starts at E and ends at D, as we saw above. This is sometimes called an Eulerian trail.

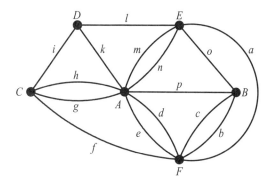

Figure 8. The graph of the bridges in Figure 8

However, it was some time until the connection was made between Euler's work and diagram-tracing puzzles. The 'Königsberg graph' of Figure 2 made its first appearance in W. W. Rouse Ball's *Mathematical Recreations and Problems of Past and Present Times* [9] in 1892.

Background information, including English translations of the papers of Euler [2] and Hierholzer [4], can be found in [1]; an English translation of Euler's paper also appears in [6].

References

1. N. L. Biggs, E. K. Lloyd and R. J. Wilson, *Graph Theory 1736–1936*, reissued paperback edition, Clarendon Press, Oxford, 1998.

2. L. Euler, Solutio problematis ad geometriam situs pertinentis, *Commentarii Academiae Scientiarum Imperialis Petropolitanae* **8** (1736) 128–140 = *Opera Omnia (1) 7* (1911–56), 1–10.

3. ——, *Pis'ma k ucenym*, Izd. Akademii Nauk SSSR, Moscow-Leningrad (1963), F.1, op.3, d.21, L.35–37 ob; F.1, op.3, d.22, L.33–41; F.136, op.2, d.3, L.74–75.

4. C. Hierholzer, Ueber die Möglichkeit, einen Linienzug ohne Wiederholung und ohne Unterbrechnung zu umfahren, *Math. Ann.* **6** (1873) 30–32.

5. G. W. Leibniz, *Mathematische Schriften (1)*, Vol. 2, Berlin (1850), 18–19.

6. J. R. Newman (ed.), *The World of Mathematics*, Vol. 1, Simon and Schuster, New York, 1956.

7. L. Poinsot, Sur les polygones et les polyèdres, *J. École Polytech.* **4** (cah. 10) (1810) 16–48.

8. J.-C. Pont, *La Topologie Algébrique des Origines à Poincaré*, Bibl. De Philos. Contemp., Presses Universitaires de France, Paris, 1974.

9. W. W. Rouse Ball, *Mathematical Recreations and Problems of Past and Present Times*, 1st ed. (later entitled *Mathematical Recreations and Essays*), Macmillan, London, 1892.

10. H. Sachs, M. Stiebitz and R. J. Wilson, An historical note: Euler's Königsberg letters, *J. Graph Theory* **12** (1988) 133–139.

11. R. J. Wilson, Analysis situs, *Graph Theory and its Applications to Algorithms and Computer Science* (ed. Y. Alavi *et al.*), John Wiley & Sons, New York (1985), 789–800.

12. ——, An Eulerian trail through Königsberg, *J. Graph Theory* **10** (1986) 265–275.

Graeco-Latin Squares and a Mistaken Conjecture of Euler[1]

Dominic Klyve and Lee Stemkoski

Introduction

Late in his long and productive career, Leonhard Euler published a hundred-page paper detailing the properties of a new mathematical structure: Graeco-Latin squares. In this paper, Euler claimed that a Graeco-Latin square of size n could never exist for any n of the form $4k + 2$, although he was not able to prove it. In the end, his difficulty was validated. Over a period of 200 years, more than twenty researchers from five countries worked on the problem. Even then, they succeeded only after using techniques from many branches of mathematics, including group theory, finite fields, projective geometry, and statistical and block designs; eventually, modern computers were employed to finish the job.

A *Latin square* (of order n) is an n-by-n array of n distinct symbols (usually the integers $1, 2, \ldots, n$) in which each symbol appears exactly once in each row and column. Some examples appear in Figure 1.

1	2	3
3	1	2
2	3	1

4	3	1	2
3	4	2	1
1	2	4	3
2	1	3	4

1	2	4	3	5
4	5	2	1	3
3	4	1	5	2
2	3	5	4	1
5	1	3	2	4

Figure 1. Latin squares of orders 3, 4, and 5

A *Graeco-Latin square* (of order n) is an n-by-n array of ordered pairs from a set of n symbols such that in each row and each column of the array, each symbol appears

[1]Reprinted from *College Math Journal*, Vol. 37, January, 2006, pp. 2–15.

(1,1)	(2,5)	(3,4)	(4,3)	(5,2)
(2,2)	(3,1)	(4,5)	(5,4)	(1,3)
(3,3)	(4,2)	(5,1)	(1,5)	(2,4)
(4,4)	(5,3)	(1,2)	(2,1)	(3,5)
(5,5)	(1,4)	(2,3)	(3,2)	(4,1)

Figure 2. A Graeco-Latin square of order 5

exactly once in each coordinate, and each of the n^2 possible pairs appears exactly once in the entire square. Figure 2 shows one such example.

Leonhard Euler published two papers concerning Graeco-Latin squares. The first, entitled *De Quadratis Magicis* [9], was written in 1776. In this short paper (seven pages in Euler's *Opera Omnia*), Euler considered magic squares, which are closely related to Graeco-Latin squares. He shows that a Graeco-Latin square of order n can be turned into a magic square by the following simple algorithm: replace the pair (a, b) with the number $(a - 1)n + b$. For example, under this transformation, the Graeco-Latin square in Figure 2 becomes the magic square in Figure 3, all of whose rows and columns sum to 65. (Additional requirements are imposed if we require the diagonals to sum to 65.)

1	10	14	18	22
7	11	20	24	3
13	17	21	5	9
19	23	2	6	15
25	4	8	12	16

Figure 3. A magic square of order 5

Euler used Graeco-Latin squares of orders 3, 4, and 5 to construct magic squares. For an order-6 magic square, however, he used a different method. Perhaps because he was unable to construct an order-6 Graeco-Latin square, he was motivated to investigate their existence in a second paper [10], *Recherches sur une Nouvelle Espèce de Quarrés Magiques*. (Fans of Euler trivia should note that this was the only paper of Euler's originally published in a Dutch journal.) This was the first published mathematical analysis of Graeco-Latin squares.[2]

A lengthy paper (101 pages in the *Opera Omnia*), *Recherches* addressed many questions regarding Latin and Graeco-Latin squares. In this paper, we are primarily concerned with Euler's conclusions about the existence of Graeco-Latin squares of specific orders. In particular, he conjectured that there can be no such square of size $4k + 2$ for any integer k. As we shall see, Euler was unable to prove this, although he did give plausibility arguments for squares of order 6, and he believed that his argument for squares of order 6 would generalize to the order $4k + 2$ case. We begin our survey of the history of Euler's conjecture by carefully considering this paper and examining his results.

---

[2]Graeco-Latin squares had appeared in print earlier. In his *Sources in Recreational Mathematics*, David Singmaster adds the following: "... there are pairs of orthogonal 4 by 4 squares in Ozanam [18] and Alberti [1].... a magic square of al-Buni, c1200, indicates knowledge of two orthogonal 4 by 4 Latin squares."

Euler

By 1782, when *Recherches* was published, Euler had returned to the St. Petersburg Academy, which was enjoying a modest renaissance under the patronage of Catherine the Great. Legend has it that Euler in fact first considered Graeco-Latin squares as a result of a question posed to him by the Empress: given 36 officers, six each of six different ranks and from six different regiments, can they be placed in a square such that exactly one officer of each rank and from each regiment appears in each row and column? Although this is the question with which Euler begins the paper, there is no mention of Catherine the Great and the attribution is probably apocryphal. He immediately claims that there is no solution, and then begins a hundred-page meandering path which eventually leads him to, if not a proof, then at least a plausibility argument for this claim.

As we begin our survey of this paper, we mention Euler's notation for Graeco-Latin squares. In his second paragraph, Euler introduced the Latin and Greek notation (hence the name). Each cell of the square contains one Latin and one Greek letter, forming two Latin squares, such that the orthogonality condition is satisfied: each Latin-Greek letter pair appears just once. He gives the example depicted in Figure 4, meant to demonstrate something very close to a solution of the 36-officer problem. Although the Latin letters and the Greek letters independently form Latin squares, this example is not a solution because the pairs bζ and dε occur twice, while bε and dζ do not occur at all.

aα	bζ	cδ	dε	eγ	fβ
bβ	cα	fε	eδ	aζ	dγ
cγ	dε	aβ	bζ	fδ	eα
dδ	fγ	eζ	cβ	bα	aε
eε	aδ	bγ	fα	dβ	aζ
fζ	eβ	dα	aγ	cε	bδ

Figure 4. Almost a Graeco-Latin square

By paragraph 5, however, Euler abandons this unwieldy notation, and instead opts to use integers for both sets of entries, writing one set as bases, and the other as exponents. An example of this notation appears in Figure 5. Euler uses this notation in defining *formules directrices*, or guiding formulas. A guiding formula for a given n is a list of the columns in which n appears as an exponent, starting from the first row and reading down. For example, to find a guiding formula for the exponent 1 in Figure 5: in the first row, the exponent 1 appears in column 1; in the second row, it appears in column 2; and so forth. Thus, the guiding formula for 1 is $(1, 2, 3, 4, 5)$. Similarly, the guiding formula for the exponent 2 is $(5, 1, 2, 3, 4)$.

1^1	2^5	3^4	4^3	5^2
2^2	3^1	4^5	5^4	1^3
3^3	4^2	5^1	1^5	2^4
4^4	5^5	1^2	2^1	3^5
5^5	1^4	2^3	3^2	4^1

Figure 5. An order-5 Graeco-Latin square, using Euler's preferred notation

In considering the 36-officer problem, Euler organizes Latin squares into categories. He then tries to find a general method for "completing" the squares in each category, that is, by adding exponents to create a Graeco-Latin square.

Single-Step Latin Squares

In a single-step Latin square, the first row is simply $1, 2, \ldots, n$. The remaining rows are formed by cyclically shifting the elements in the previous row one place to the left, as shown in Figure 6.

$$
\begin{array}{cccccc}
1 & 2 & 3 & \cdots & n-1 & n \\
2 & 3 & \cdots & n-1 & n & 1 \\
3 & \cdots & n-1 & n & 1 & 2 \\
\vdots & & & & & \vdots \\
n & 1 & 2 & 3 & \cdots & n-1
\end{array}
$$

Figure 6. A Single-Step Latin Square

In this simple case, Euler was able to complete Latin squares of orders 3, 5, 7, and 9. More importantly, he proved that a single-step Latin square of even order can never be completed. As the proof of this is both easily understood and indicative of the style of reasoning Euler employs throughout his paper, it is worth considering here. His reasoning makes use of the previously defined guiding formulas. If one can show that a guiding formula cannot exist for some exponent of a given square, then one deduces that the square cannot be completed. In particular, Euler often simply proves that no guiding formula can exist for the exponent 1, which is sufficient to show that the given square cannot be completed.

Theorem. *No single-step Latin square of even order can be completed.*

Proof. Suppose there is a such a square of even order n. Without loss of generality, the entry 1^1 is in the first cell. Suppose that there is a guiding formula for the number 1: $(1, a, b, c, d, e, \ldots)$. Denote the consecutive bases of which 1 is an exponent (from top to bottom) by $(1, \alpha, \beta, \gamma, \delta, \varepsilon, \ldots)$. Thus we have a situation similar to that depicted in Figure 7, where blank spaces denote unknown entries.

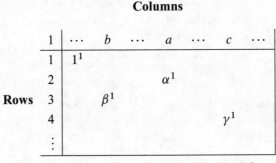

Figure 7. An example of labeling in Euler's proof

Since 1 occurs exactly once in each row and column, the lists contain the same entries, merely in a different order. Thus we have

$$a + b + c + \cdots \equiv \alpha + \beta + \gamma + \cdots \pmod{n}.$$

According to the labeling of the entries in the square, the base located in row 2, column a is α. Moreover, in the second row of a single-step Latin square, the bases have been (cyclically) shifted one position to the left, as compared with the first row. Thus the base in row 2, column a is also congruent to $a + 1 \pmod{n}$ ("modulo n" due to the shift being cyclic). Thus we have that

$$\alpha \equiv a + 1 \pmod{n}.$$

Similarly, since the entries in row r have been shifted $r - 1$ spaces to the left (relative to the first row), we obtain the equations

$$\beta \equiv b + 2 \pmod{n}, \quad \gamma \equiv c + 3 \pmod{n}, \quad \ldots .$$

Adding these $n - 1$ congruences, we see that

$$1 + 2 + \cdots + (n - 1) \equiv 0 \pmod{n}.$$

That is, $n(n - 1)/2$ must be an integer multiple of n, so n must be odd. Since n was assumed to be even, there can be no guiding formula for the exponent 1, so no single-step Latin square of even order can be completed. ■

Multiple-Step Latin squares

For m a divisor of n, an m-step Latin square of order n is defined as follows: partition the n-by-n square into m-by-m blocks. In the first row of blocks, let each block contain an m-by-m single-step Latin square, where the first block uses the numbers 1 through m, the second block uses $m + 1$ through $2m$, and so forth. The remaining rows of blocks are formed by cyclically shifting the blocks in the previous row one place to the left. Figure 8 shows a 2-step (or double-step) Latin square and a 3-step (or triple-step) Latin square.

Recall that Euler was interested in solving the 36-officer problem, which is equivalent to finding an order-6 Graeco-Latin square, or completing an order-6 Latin square. An order-6 Latin square can be an m-step square only for $m = 1, 2, 3$. We have seen Euler's proof that a single-step Latin square of order 6 cannot be completed. Later in his paper, Euler proves that a double-step Latin square of order n can be completed only when n is a multiple of 4. He also gives a proof that a triple-step Latin square of order 6 cannot be completed. Therefore, Euler concludes that an order-6 Graeco-Latin square cannot be constructed by completing an m-step Latin square.

At the beginning of paragraph 140 of *Recherches*, Euler wrote

Ayant vu que toutes [les] méthodes que nous avons exposées jusqu'ici ne sauroient fournir aucun quarré magique pour le cas de $n = 6$ et que la même conclusion semble s'étendre à tous les nombres impairement pairs de n, on pourroit croire que, si de tels quarrés sont possibles, les quarrés latins qui leur servent de base, ne suivant aucun des ordres que nous venons de considérer, seroient tout à fait

1	2	3	4	5	6	$\cdots$	$\cdots$	$n-1$	n
2	1	4	3	6	5	$\cdots$	$\cdots$	n	$n-1$
3	4	5	6	$\cdots$	$\cdots$	$n-1$	n	1	2
4	3	6	5	$\cdots$	$\cdots$	n	$n-1$	2	1
5	6	$\cdots$	$\cdots$	$n-1$	n	1	2	3	4
6	5	$\cdots$	$\cdots$	n	$n-1$	2	1	4	3
$\vdots$	$\vdots$							$\vdots$	$\vdots$
$\vdots$	$\vdots$							$\vdots$	$\vdots$
$n-1$	n	1	2	3	4	$\cdots$	$\cdots$	$n-3$	$n-2$
n	$n-1$	2	1	4	3	$\cdots$	$\cdots$	$n-2$	$n-3$

1	2	3	4	5	6	$\cdots$	$\cdots$	$\cdots$	$n-2$	$n-1$	n
2	3	1	5	6	4	$\cdots$	$\cdots$	$\cdots$	$n-1$	n	$n-2$
3	1	2	6	4	5	$\cdots$	$\cdots$	$\cdots$	n	$n-2$	$n-1$
4	5	6	$\cdots$	$\cdots$	$\cdots$	$n-2$	$n-1$	n	1	2	3
5	6	4	$\cdots$	$\cdots$	$\cdots$	$n-1$	n	$n-2$	2	3	1
6	4	5	$\cdots$	$\cdots$	$\cdots$	n	$n-2$	$n-1$	2	3	1
$\vdots$	$\vdots$	$\vdots$							$\vdots$	$\vdots$	$\vdots$
$\vdots$	$\vdots$	$\vdots$							$\vdots$	$\vdots$	$\vdots$
$\vdots$	$\vdots$	$\vdots$							$\vdots$	$\vdots$	$\vdots$
$n-2$	$n-1$	n	1	2	3	$\cdots$	$\cdots$	$\cdots$	$n-5$	$n-4$	$n-3$
$n-1$	n	$n-2$	2	3	1	$\cdots$	$\cdots$	$\cdots$	$n-4$	$n-3$	$n-5$
n	$n-2$	$n-1$	2	3	1	$\cdots$	$\cdots$	$\cdots$	$n-3$	$n-5$	$n-4$

Figure 8. A Double-Step Latin Square and a Triple-Step Latin Square

irréguliers. Il faudroit donc examiner tous les cas possibles de tels quarrés latins pour le cas de $n = 6$, dont le nombre est sans doute extrèmement grand.[3]

Because the number of cases was too large to check directly, Euler developed a set of transformations between Latin squares that preserved their ability to be completed. Obvious transformations include the swapping of two rows or two columns. Less obvious "completeness-preserving" transformations include finding a subrectangle of numbers with opposite corners matching, and then swapping the two corner numbers, as in Figure 9. Euler proved that if one of these squares can be completed, then both can.

---

[3] Having seen that all methods that we have shown so far do not give any magic squares for the case of $n = 6$ and that the same conclusion seems to apply to any number of the form $4k + 2$, we could believe that if such squares are possible, the Latin squares that serve as their base, not following any order that we have just considered, would be completely irregular. Thus, it would be necessary to examine all the possible cases of such Latin squares for the case of $n = 6$, the number of which is undoubtedly extremely large.

1	2	3	4	5	6
2	[3]	4	5	[6]	1
3	4	5	6	1	2
4	5	6	1	2	3
5	[6]	1	2	[3]	4
6	1	2	3	4	5

becomes

1	2	3	4	5	6
2	[6]	4	5	[3]	1
3	4	5	6	1	2
4	5	6	1	2	3
5	[3]	1	2	[6]	4
6	1	2	3	4	5

Figure 9. Example of a "completeness-preserving" transformation

By using these and other clever transformations, Euler was able to represent by a single Latin square as many as 720 equivalent squares, thus dramatically reducing the amount of searching necessary to determine whether an order 6 square was possible. Here, however, Euler seems to have abandoned rigor in the face of the enormous number of cases he still needed to check. In paragraph 148, he writes

> De là il est clair que, s'il existoit un seul quarré magique complet de 36 cases, on en pourroit déduire plusieurs autres moyennant ces transformations, qui satisferient également aux conditions du problème. Or, ayant examiné un grand nombre de tels quarrés sans avoir rencontré un seul, il est plus que probable qu'il n'y en ait aucun ... l'on voit que le nombre des variations pour le cas de $n = 6$ ne sauroit être si prodigieux, que le nombre de 50 ou 60 que je pourrois avoir examinés n'en fût qu'une petite partie. J'observe encore à cette occasion que le parfait dénombrement de tous les cas possibles de variations semblables seroit un objet digne de l'attention de Géomètres.[4]

Although he had not provided rigorous demonstration of all his claims, Euler still ends his paper with a fascinating and prescient conclusion:

> ... à voir s'il y a des moyens pour achever l'énumerations de tous les cas possibles, ce qui paroît fournir *un vaste champ pour des recherches nouvelles et intéressantes* [emphasis added]. Je mets fin ici aux miennes sur une question qui, quoique en elle-même de peu d'utilité, nous a conduit à des observations assez importantes tant pour la doctrine des combinaisons que pour la théorie générale des quarrés magique.[5]

Early "Proofs" of the 36-Officer Problem

The first proof of the 36-officer problem was apparently by Thomas Clausen [23], an assistant to Heinrich Schumacher, a nineteenth-century astronomer in Altona, Germany.

---

[4]From here it is clear that if there existed a single complete magic square with 36 entries, we could derive several others using these transformations that would also satisfy the conditions of the problem. But, having examined a large number of such squares without having encountered a single one, it is most likely that there are none at all.... we see that the number of variations for the case of $n = 6$ cannot be so prodigious that the 50 or 60 that I have examined were but a small part. I observe further here that the exact count of all the possible cases of similar variations would be an object worthy of the attention of Geometers [mathematicians].

[5]... seeing if there are methods of achieving the enumeration of all the possible cases would seem to provide a vast field for new and interesting research. Here, I bring to an end my [work] on a question that, although is of little use itself, has led us to some observations just as important for the doctrine of combinations as for the general theory of magic squares.

Schumacher and Carl Gauss, then Astronomer in Göttingen, enjoyed a brief correspondence; in a letter dated August 10, 1842, Schumacher wrote that Clausen had proved the nonexistence of orthogonal Latin squares of order 6. Apparently Clausen proved this by dividing all Latin squares of order 6 into 17 families and then proving that each in turn could not be completed. Clausen also believed, as Euler did, that a similar result was possible for order-10 squares, but he reported that:

> Der Beweis der vermutheten Unmöglichkeit für 10, so geführt wie er ihn für 6 geführt hat, würde wie er sagt, vielleicht für menschliche Kräfte unausführbar seyn.[6]

Sadly, although Clausen published over 150 papers during his scientific career, few of them survive, and no record of his alleged proof can be found. Thus, in order to establish precedence in the proof of the 36-officer problem, which is tantamount to determining whether Clausen gave a correct proof, we can only consider his record as a scientist and a mathematician in order to assess his claim.

The definitive published study on Clausen is by Biermann [2], who describes him as "a remarkable man." By the age of 23, Clausen had mastered Latin, Greek, French, English, and Italian, and had gained sufficient notoriety in mathematics and astronomy to earn him an appointment at the Altona Observatory in 1824. He was well known to the leading scientists of his day, including Gauss. Not known for generously praising others, Gauss nonetheless described Clausen as a man of "outstanding talents." Clausen won the prize of the Copenhagen Academy for his paper on the comet of 1770 [7]. Perhaps more impressive was his factorization of the 6th Fermat number, $2^{64} + 1$, showing that it was not prime. It is still not known how, without the aid of modern computational devices, Clausen was able to do this (the smallest factor is 274,177). In his article, Biermann writes that "He possessed an enormous facility for calculation, a critical eye, and perseverance and inventiveness in his methodology."

Certainly these facts give Clausen a strong degree of authenticity in his claims. Further evidence for his claim is the fact that his method of breaking Latin squares into 17 families directly foreshadows the earliest surviving proof, published in 1900. All of this leads the authors to believe that the claims of priority for the first correct proof of the 36-officer problem rightly belong to Thomas Clausen.

The first surviving proof of this problem is that of Gaston Tarry, a French school teacher, who published his work [25] in 1900. Tarry's paper was necessarily quite lengthy; he proved the non-existence of an order-6 Graeco-Latin square by individually considering not only 17 families, but 9408 separate cases. Thus did Tarry fulfill Euler's 118-year-old request for a "complete enumeration" of all possible cases.

After the appearance of Tarry's paper, mathematicians began to search for a more clever proof. In 1902, Petersen published *Les 36 officiers* [21], in which he attempted to provide a proof using a geometrical argument. He constructed simplicial complexes from Latin squares, and used a generalization of (fittingly enough) Euler's polyhedron formula[7] to construct impossibility relations between the numbers of 0-, 1-, and 2-cells in his complexes to prove that the order-6 Latin squares could not be completed.

---

[6]The proof that [order] 10 is impossible, based on the proof of the [order] 6 [square], is perhaps impractical for human forces.

[7]On the relation between the number of faces f, edges e, and vertices v of a polyhedron, namely $f - e + v = 2$.

Then, in 1910, Wernicke published *Das Problem der 36 Offiziere* [26], in which he shows that Petersen's proof is incomplete. He goes on to use a group-theoretic technique to put limits on the maximum possible number of mutually orthogonal Latin squares of order n. He purports to show that there do not exist two orthogonal Latin squares of order 6; in other words, there is no Graeco-Latin square of order 6.

A Resolution of Euler's Conjecture

Recall that Euler believed not only that Graeco-Latin squares of order 6 could not exist, but in general could not exist for any order of the form $4k + 2$. Euler was unable to resolve this conjecture with techniques then available. However, as time passed, a variety of new tools became available that could be used to investigate Graeco-Latin squares.

The first modern reformulation involved endowing Latin squares with an algebraic structure, as follows: a *quasigroup* is a set Q with a binary relation $\circ$ such that for all elements a and b, the equations $a \circ x = b$ and $y \circ a = b$ have unique solutions. For example, let $Q = \{0, 1, 2\}$ and $a \circ b = 2a + b + 2 \pmod 3$. The multiplication table for this operation is given in Figure 10. Note that this is in fact a Latin square. It turns out that this is true in general: the multiplication table of a quasigroup is a Latin square (and vice versa). In particular, the multiplication table of a group is a Latin square, since a group is an associative quasigroup with an identity element.

$\circ$	0	1	2
0	2	0	1
1	1	2	0
2	0	1	2

Figure 10. Multiplication table for a quasigroup

The first application of group-theoretic techniques to Latin squares was implemented by MacNeish [15] in 1922. He also disproved Wernicke's earlier results, just as Wernicke had disproven Petersen's. MacNeish's greatest contribution was the introduction of the concept of the direct product of Latin squares, a method for combining two Latin squares to make a third, whose order is the product of the orders of the original two. To get an idea of how this construction works, consider the two Latin squares and their direct product in Figure 11. In a sense, it is as though we have superimposed the pattern of the second Latin square on each entry of the first.

A useful property of this construction is that if we have a pair of orthogonal Latin squares A and B (necessarily of the same size), and another orthogonal pair C and D, then $A \times C$ and $B \times D$ are orthogonal! This allows us to build large Graeco-Latin squares from smaller ones. Unfortunately, this could not be used to construct a Graeco-Latin square or order 6 or 10, since there is no such square of order 2.

From this method, MacNeish proved the following result: Let $N(n)$ be the number of mutually orthogonal Latin squares of order n. Then,

$$N(ab) \geq \min\{N(a), N(b)\}.$$

A	B	C	D
B	C	D	A
C	D	A	B
D	A	B	C

1	2	3
2	3	1
3	1	2

A1	A2	A3	B1	B2	B3	C1	C2	C3	D1	D2	D3
A2	A3	A1	B2	B3	B1	C2	C3	C1	D2	D3	D1
A3	A1	A2	B3	B1	B2	C3	C1	C2	D3	D1	D2
B1	B2	B3	C1	C2	C3	D1	D2	D3	A1	A2	A3
B2	B3	B1	C2	C3	C1	D2	D3	D1	A2	A3	A1
B3	B1	B2	C3	C1	C2	D3	D1	D2	A3	A1	A2
C1	C2	C3	D1	D2	D3	A1	A2	A3	B1	B2	B3
C2	C3	C1	D2	D3	D1	A2	A3	A1	B2	B3	B1
C3	C1	C2	D3	D1	D2	A3	A1	A2	B3	B1	B2
D1	D2	D3	A1	A2	A3	B1	B2	B3	C1	C2	C3
D2	D3	D1	A2	A3	A1	B2	B3	B1	C2	C3	C1
D3	D1	D2	A3	A1	A2	B3	B1	B2	C3	C1	C2

Figure 11. Two Latin squares and their direct product

As a plausibility argument, we present the following example: Let L_1, L_2, and L_3 be mutually orthogonal Latin squares of order a, and let M_1, M_2, M_3, and M_4 be mutually orthogonal Latin squares of order b. Then, $L_1 \times M_1$, $L_2 \times M_2$, and $L_3 \times M_3$ are all mutually orthogonal Latin squares of order ab. MacNeish then proved a stronger result: If the prime factorization of n is $p_1^{e_1} p_2^{e_2} \cdots p_k^{e_k}$, then $N(n) \geq \min\{p_i^{e_i} - 1\}$. (To prove this, he used group-theoretic techniques to construct large numbers of mutually orthogonal Latin squares of prime power order.) Finally, MacNeish conjectured that equality holds; that is, the number of mutually orthogonal Latin squares is actually equal to $\min\{p_i^{e_i} - 1\}$. If true, this would imply Euler's conjecture, since 2^1 is the smallest prime power in the factorization of $4k + 2$.

The next surge of research on Latin squares was motivated by practical applications. In the late 1930s, Fisher and Yates began to advocate the use of Latin squares and sets of mutually orthogonal Latin squares in the statistical design of experiments [11]. For example, suppose that we wish to test five different fertilizers but only have a single plot of land on which to do so. There may be unknown characteristics of the land, such as soil variation or a moisture gradient, that may bias the results of the experiment. To minimize the effects of such position-dependent factors, we divide the plot of land into a five-by-five grid, number the subplot as a Latin square, and place each type of fertilizer in those subplots with a particular number.

Sets of mutually orthogonal Latin squares have their uses as well. A set of k orthogonal Latin squares of size n gives a schedule for an experiment with k groups of n subjects each such that

1. Each subject meets every subject in each of the other groups exactly once;

2. Each subject is tested once at each location (to remove location-dependent bias).

For example, say that we want to test two groups of laboratory mice (an experimental group and a control group) in a series of n mazes so that each mouse races against each one in the other group, and no mouse runs in the same maze twice. A schedule for the tests can be developed using a Graeco-Latin square of order n.

After Yates constructed sets of mutually orthogonal Latin squares of orders 4, 8, and 9, Fisher conjectured during a seminar at the Indian Statistical Institute that a maximum set of orthogonal Latin squares of order n (i.e. a set of $n-1$) exists for each prime power order. This was proved soon after by Bose [3] in 1938, using finite fields (sometimes called Galois fields). Until this point, mathematicians had used groups—algebraic structures with a single binary operation—to construct Latin squares. One of Bose's great contributions was that he developed a method that used fields—algebraic structures with *two* binary operations. In essence, one operation allows the construction of a Latin square, and the second enables the permutation of the entries to create other squares orthogonal to it. More precisely, given a field $\mathbb{F}$ of n elements, $\mathbb{F} = \{g_1, g_2, \ldots, g_n\}$, choose some nonzero element g in $\mathbb{F}$. Define an order-n Latin square L_g by assigning to the position in the ith row and the jth column the element $(g \cdot g_i) + g_j$. Furthermore, it is also true that if g and h are different nonzero elements in $\mathbb{F}$, then L_g and L_h are orthogonal! For example, consider the field $\mathbb{F} = \{0, 1, 2, 3, 4\}$ with the operations of addition and multiplication modulo 5. Then the Latin squares L_2 and L_3, shown in Figure 12, are orthogonal.

$$
L_2 = \begin{matrix}
0 & 1 & 2 & 3 & 4 \\
2 & 3 & 4 & 0 & 1 \\
4 & 0 & 1 & 2 & 3 \\
1 & 2 & 3 & 4 & 0 \\
3 & 4 & 0 & 1 & 2
\end{matrix}
\qquad
L_3 = \begin{matrix}
0 & 1 & 2 & 3 & 4 \\
3 & 4 & 0 & 1 & 2 \\
1 & 2 & 3 & 4 & 0 \\
4 & 0 & 1 & 2 & 3 \\
2 & 3 & 4 & 0 & 1
\end{matrix}
$$

Figure 12. Orthogonal Latin squares constructed from a field

Bose was also able to construct orthogonal Latin squares using projective geometry. A *projective plane of order n* is a set of $n^2 + n + 1$ elements (where $n \geq 2$) called *points* and a collection of subsets called *lines* that satisfy two conditions: each pair of points lies on exactly one line, and each pair of lines meets in exactly one point. (From these conditions it follows that there are $n + 1$ points on each line, and each point is on $n + 1$ lines.) Bose developed a method to turn a finite projective plane of order n into a set of $n - 1$ mutually orthogonal Latin squares of order n, and conversely. For a fully worked-out example when $n = 3$ (which is rather lengthy), we refer the reader to [17].

Projective planes are only known to exist when n is the power of a prime, so they cannot be used to yield any Graeco-Latin squares of orders not constructible by the field method. For example, although we could use a projective plane of order 125 to build a Graeco-Latin square of order 125, we could have just as easily used a pair of orthogonal Latin squares of order 5 (such as those in Figure 12) and the direct product construction three times (since $5^3 = 125$). Nevertheless, the equivalence of the two problems is in itself interesting.

At this point, using the methods we have discussed so far, we can now construct Graeco-Latin squares of every order n except those values for which the prime factorization of n contains only a single factor of 2; equivalently, we can construct exactly those Graeco-Latin squares that Euler stated were constructible. The next step in settling Euler's conjecture was to look at the methods of construction, as done by Mann [16] in 1942. He introduced a general framework in which to view all the work that preceded him.

Assume that a given Latin square is in standard form, that is, the first row contains the numbers 1 through n in order from left to right. Let σ_i be the permutation of $1, 2, \ldots, n$ that sends j to the element of the Latin square in row i, column j. In this manner we associate a permutation with each row. Since we require that the entries be in standard form, the first row is associated with the identity permutation. For an example, see Figure 13.

$$
\begin{array}{|ccccc|}
\hline
1 & 2 & 3 & 4 & 5 \\
2 & 3 & 4 & 5 & 1 \\
3 & 4 & 5 & 1 & 2 \\
4 & 5 & 1 & 2 & 3 \\
5 & 1 & 2 & 3 & 4 \\
\hline
\end{array}
\begin{array}{l}
\Rightarrow (1)(2)(3)(4)(5) \\
\Rightarrow (12345) \\
\Rightarrow (13524) \\
\Rightarrow (14253) \\
\Rightarrow (15432)
\end{array}
$$

Figure 13. A Latin square and the permutations associated with its rows

If these permutations form a group G, the Latin square is said to be based on G. For example, the Latin square in Figure 13 is based on the subgroup of S_5 consisting of the permutations

$$(1)(2)(3)(4)(5), (12345), (13524), (14253), (15432).$$

Mann noted that all constructions up to this point (those of Euler, Yates, Bose, etc.) had been based on groups. He went on to prove that for all group-based Latin squares, MacNeish's conjecture is true, and thus Euler's conjecture is true. However, Mann demonstrates that not all sets of orthogonal Latin squares are based on groups, and he gives an example of two such squares of order 12 in [17]. Therefore, any counterexample to Euler's conjecture must involve constructing Latin squares in a way entirely different from those that had been considered up to this point.

Not for another 17 years did someone succeed in methodically constructing Latin squares using methods not based on groups. In 1959, E. T. Parker [19] began to use orthogonal arrays to represent sets of mutually orthogonal Latin squares. An orthogonal array of order n is a k-by-n^2 matrix filled with the symbols $1, 2, \ldots, n$ so that in any 2-by-n^2 submatrix, each of the possible n^2 pairs of symbols from $\{1, 2, \ldots, n\}$ occurs exactly once. Orthogonal arrays can encode the information present in a set of mutually orthogonal Latin squares: the first row of the array represents row indices (of the Latin square), the second row represents column indices, and the remaining rows represent the entries in a given cell. Figure 14 shows a Graeco-Latin square and its corresponding orthogonal array. To see how this correspondence works, consider the seventh column of the orthogonal array, $(3, 1, 3, 2)$. This means that in row 3 column 1 of its corresponding Graeco-Latin square, we will find the symbol 3, then 2. One of the advantages to working with orthogonal arrays is that permuting the data is easier: we could in fact take any two rows to represent row and column indices and still obtain a valid set of orthogonal Latin squares.

To determine which k-by-n^2 matrices correspond to orthogonal Latin squares, Parker

$$\begin{array}{|ccc|}
\hline
(1,1) & (2,2) & (3,3) \\
(2,3) & (3,1) & (1,2) \\
(3,2) & (1,3) & (2,1) \\
\hline
\end{array}$$

$$\begin{pmatrix}
1 & 1 & 1 & 2 & 2 & 2 & 3 & 3 & 3 \\
1 & 2 & 3 & 1 & 2 & 3 & 1 & 2 & 3 \\
1 & 2 & 3 & 2 & 3 & 1 & 3 & 1 & 2 \\
1 & 2 & 3 & 3 & 1 & 2 & 2 & 3 & 1
\end{pmatrix}$$

Figure 14. A Graeco-Latin square and its corresponding orthogonal array

used the incidence properties of block designs—combinatorial designs similar to projective planes, but with fewer structural restrictions. Recall that Bose had used projective planes to produce Latin squares earlier, but since block designs have greater flexibility, Parker was able to use them to produce Latin squares that were not based on groups, and he was thus able to circumvent the limitations discovered earlier by Mann. In particular, Parker constructed four orthogonal Latin squares of order 21 using this method, thus disproving MacNeish's conjecture (since $N(21) \geq 4$ but $\min\{3 - 1, 7 - 1\} = 2$). Though this cast some doubt on Euler's conjecture (by disproving the major conjecture that supported it), the conjecture was still at least plausible. No Graeco-Latin square of order $4k + 2$ had been found in 180 years of searching.

After the appearance of Parker's paper, a flurry of correspondence ensued between Parker, Bose, and Shrikhande; this eventually resulted in the publication of a series of papers that completely refuted Euler's conjecture. Bose and Shrikhande expanded on Parker's results and used block designs to produce a Graeco-Latin square of order 22, the first counterexample to Euler's conjecture ([4], [5]). Parker then constructed one of order 10 (the minimum remaining possible order of a counterexample) using orthogonal arrays [20]. The components of the columns were elements of a field, permuted via an algorithm similar to that in Bose's 1938 paper, with the exception of nine columns that corresponded to a 3-by-3 Graeco-Latin subsquare. Parker attributed the inspiration to Bose and Shrikhande. All three authors collaborated on a final paper in which counterexamples are given for all orders $n = 4k + 2 \geq 10$ [6]. Their proof involves the use of block designs (in a lengthy case-by-case analysis) and techniques from their earlier papers. A modern description of these techniques can be found in [14].

Thus, by 1960, Euler's conjecture had been settled, and it was shown to be almost entirely incorrect. However, a good problem is never truly finished, and work continued on the conjecture for years afterward. The most significant contribution to the refinement of the disproof of Euler's conjecture was by Sade [22]. He developed a singular direct product construction for quasigroups (recall that the multiplication tables of quasigroups are equivalent to Latin squares), and this provided counterexamples to Euler's conjecture via purely algebraic methods. However, this result was mostly overlooked at the time since Bose, Shrikhande, and Parker had just completed their seminal paper.

The singular direct product (SDP) of Latin squares requires three Latin squares, one each of orders k, n, and $n + m$, the last containing a Latin square of order m, and produces a Latin square of order $m + nk$. An example of this construction with squares of size

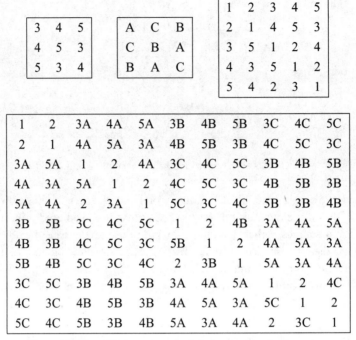

Figure 15. Three Latin squares and their singular direct product

$k = 3$, $n = 3$, and $n + m = 5$ appears in Figure 15.

As with the direct product, if the process is performed on sets of squares that are orthogonal, the resulting squares will also be orthogonal. Details aside, the important point is that the result is a Latin square whose order is not necessarily a multiple of any of the orders of the input squares. Using previous methods (such as MacNeish's direct product), one could not construct a Graeco-Latin square of order $4k + 2$ from smaller squares because a Graeco-Latin square of order 2 does not exist. Sade's SDP, however, allowed him to construct many such squares, and in fact an infinite number of counterexamples to Euler's conjecture via purely algebraic methods.

A few subsequent contributions to the Graeco-Latin square problem are worth noting. In 1975, Crampin and Hilton [8] showed that if one starts with Latin squares of orders 10, 14, 18, 26, and 62, Sade's construction yields a complete set of counterexamples to Euler's conjecture. Using a computer, they also showed that the SDP can be used to construct self-orthogonal Latin squares (Latin squares orthogonal to their transpose) of all but 217 sizes. In 1984, Stinson [24] gave a modern mathematical *tour de force* by proving the 36-officer problem in only three pages by using a transversal design, finite vector spaces, and graph theory. Finally, in 1982, Zhu Lie [13] published what is considered by many to be the most elegant disproof of Euler's conjecture, using the SDP, a related construction of his own, and nothing else.

A good measure of the value of a mathematical problem is the number of interesting results generated by attempts to solve it. By this measure, Euler's conjecture of 1782 surely must rank among the most fertile problems in the history of mathematics. Although he was mistaken in his conjecture, it is a testimony to Euler's mathematical insight that he understood the importance of investigating such a simple problem.

References

[1] G.A. Alberti, *Giuochi Numerici Fatti Arcani Palesati*, Bologna, Bartolomea Borghi (1747).

[2] K.R. Biermann, Thomas Clausen, Mathematiker und Astronom, *J. für die reine und angewandte Mathematik*, 216 (1964) 159–198.

[3] R.C. Bose, On the Application of the Properties of Galois Fields to the Problem of Constructions of Hyper-Graeco-Latin Squares, *Sankhya: The Indian Journal of Statistics,* 3 (1938) 323–338.

[4] R.C. Bose and S.S. Shrikhande, On the falsity of Euler's conjecture about the non-existence of two orthogonal Latin squares of order $4t + 2$, *Proc. Nat. Acad. Sci. U.S.A.,* 45 (1959) 734–737.

[5] R. C. Bose and S.S. Shrikhande, On the construction of sets of mutually orthogonal Latin squares and the falsity of a conjecture of Euler, *Trans. Amer. Math. Soc.,* 95 (1960) 191–209.

[6] R.C. Bose, S.S. Shrikhande, and E.T. Parker, Further results on the construction of mutually orthogonal Latin squares and the falsity of Euler's conjecture, *Canad. J. Math.,* 12 (1960) 189–203.

[7] T. Clausen, Determination of the Path of the 1770 Comet. *Astronomische Nachrichten,* 19 (1842), 121–168.

[8] D.J. Crampin and A.J. Hilton, Remarks on Sade's disproof of the Euler conjecture with an application to Latin squares orthogonal to their transpose, *J. Comb. Theory Ser. A,* 18 (1975) 47–59.

[9] L. Euler, De Quadratis Magicis. *Opera Omnia,* Ser. I, Vol 7: 441–457, *Commentationes arithmeticae* 2 (1849), 593–602. Also available online at http://www.eulerarchive.org.

[10] L. Euler, Recherches sur une nouvelle espèce de quarrés magiques. *Opera Omnia,* Ser. I, Vol 7, 291–392, *Verhandelingen uitgegeven door het zeeuwsch Genootschap der Wetenschappen te Vlissingen* 9 (1782), 85–239. Also available online at http://www.eulerarchive.org.

[11] Sir R. A. Fisher, *The Design of Experiments*, 9th ed. Hafner Press (Macmillan), 1971.

[12] M. Gardner, *New Mathematical Diversions*, Simon and Schuster, 1966.

[13] Z. Lie, A short disproof of Euler's conjecture concerning orthogonal Latin squares, *Ars Combinatoria,* 14 (1982) 47–55.

[14] C.C. Lindner and C.A. Rodger, *Design theory,* CRC Press, 1997.

[15] H.F. MacNeish, Euler Squares, *Annals of Mathematics,* 23 (1922) 221–227.

[16] H.B. Mann, The construction of orthogonal Latin squares, *Ann. Math. Stat.,* 13 (1942) 418–423.

[17] H.B. Mann, On orthogonal Latin squares, *Bull. Amer. Math. Soc.,* 50 (1950) 249–257.

[18] J. Ozanam. *Récréations Mathématiques,* Paris, Jombert, 1697.

[19] E.T. Parker, Construction of some sets of mutually orthogonal latin squares. *Proc. Amer. Math. Soc.,* 10 (1959) 946–994.

[20] E. T. Parker, Orthogonal Latin Squares. *Proc. Nat. Acad. Sci. U.S.A.,* 45 (1959) 859–862.

[21] J. Petersen, Les 36 officiers, *Annuaire des mathématiciens,* (1902) 413–426.

[22] A. Sade, Produit direct-singulier de quasigroupes orthogonaux et anti-abéliens, *Ann. Soc. Sci. Bruxelles Sér. I,* 74 (1960) 91–99.

[23] Letter from Shumacher to Gauss, regarding Thomas Clausen, August 10, 1842. Gauss, *Werke* Bd. 12, p.16. Göttingen, Dieterich, (1863).

[24] D.R. Stinson, A short proof of the nonexistence of a pair of orthogonal Latin squares of order six, *J. Combin. Theory Ser. A,* 36 (1984) 373–376.

[25] G. Tarry, Le problème des 36 officiers, *Comptes Rendus Assoc. France Av. Sci.* 29, part 2 (1900) 170–203.

[26] P. Wernicke, Das Problem der 36 Offiziere, *Jahresbericht der deutschen Mathematiker-Vereinigung,* 19 (1910) 264.

Glossary[1]

Leonhard Euler's vast contribution to mathematics can be glimpsed in the many terms, formulas, equations, and theorems which today bear his name. An exhaustive list of such terms would be difficult to compile; even harder would be a list all of whose entries could be carefully verified as having originated in Euler's work. This compilation contains those items which can be readily found in mathematics texts and reference works; although impressive, certainly the list is incomplete.

The glossary was produced through the efforts of Karl Anderson and Jeff Ondich, students at St. Olaf College, with the assistance of Lynn Steen, Gerald Alexanderson, the editor (Doris Schattschneider) and members of the editorial board of *Mathematics Magazine*. Definitions of several of the entries vary according to the source consulted; we have chosen descriptions that seem most common. Each entry in the glossary can be found in one or more of the reference works or in one of the texts listed in the References.

Although the symbols do not bear his name, Euler introduced many of the modern conventions of mathematical notation—most notably, the symbol $f(x)$ for a function, the notations $\sin x$, $\cos x$ for sine and cosine functions, the symbols $\sum$ for summation, Δy, $\Delta^2 y$, etc., for finite differences, e for the base of the natural logarithm, and i for $\sqrt{-1}$. In addition, it is easy to point out numerous mathematical terms and theorems missing from our list (because they bear some other mathematician's name), but which are rightfully attributed to Euler. R. A. Raimi has noted, "There is ample precedent for naming laws and theorems for persons other than their discoverers, else half of analysis would be named for Euler" (*Amer. Math. Monthly*, 83 (1976) 522).

Terms

Euler Angles.   These three angles are commonly used to fix the directions of a new set of rectangular space coordinates x^*, y^*, z^* with reference to an old set x, y, z. They are usually taken as the angle between the z^* and z-axes, the angle between the x-axis and the line l of intersection of the x^*y^* and xy-planes (l is called the nodal line), and the angle between the x^*-axis and l. (See Figure 1; see Euler's theorem for rotation of coordinate system.)

---

[1]Reprinted from *Mathematics Magazine*, Vol. 56, November, 1983, pp. 315–325.

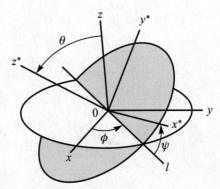

Figure 1. The xyz coordinate axes rotate into the $x^*y^*z^*$ coordinate axes by successive rotations through the Euler angles ϕ, θ, ψ.

The Euler Characteristic. For a polyhedral surface, this is the number $\chi = V - E + F$, where V is the number of vertices, E the number of edges, and F the number of faces. More generally, for an n-dimensional simplicial complex K, the Euler (or Euler-Poincaré) characteristic is defined by

$$\chi = \sum_{i=0}^{n}(-1)^i s(i)$$

where $s(i)$ is the number of i-dimensional simplexes in K.

Euler's (Nine-point) Circle. This circle passes through the midpoints of the sides of a triangle, the feet of its altitudes, and the midpoints of the line segments between its vertices and its orthocenter; these last 3 points are called the **Euler points** of the triangle. See Figure 2. Euler proved that *for any triangle, the feet of its altitudes and the midpoints of its sides all lie on one circle*, but it was Poncelet (1788–1867) who showed that the Euler points also lie on this same circle. Poncelet named this circle the nine-point circle; it is also called Feuerbach's circle.

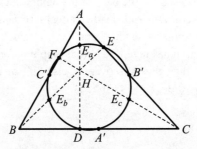

Figure 2. The midpoints of the sides of $\triangle ABC$ are A', B', C', the feet of its altitudes are D, E, F, and the point of their intersection, the orthocenter, is H. The Euler points, E_a, E_b, E_c, are the midpoints of segments AH, BH, and CH, respectively.

Euler Circuit. This is a closed path of edges in a graph in which each edge appears exactly once. Euler's theorem for graphs states that *an undirected graph has an Euler circuit if and only if it is connected and all of its vertices have even degree* (the degree of a vertex is the number of edges that meet at that vertex). An 18th century Sunday pastime—to stroll a continuous path, attempting to cross each of the seven bridges of Konigsberg exactly once—was the source of the theorem.

Euler's Constant γ.   The constant $\gamma = .577215665\ldots$, which was calculated by Euler to 16 decimal places, is defined by the limit

$$\gamma = \lim_{n \to \infty} \left(1 + \frac{1}{2} + \frac{1}{3} + \cdots + \frac{1}{n} - \ln n \right).$$

Sometimes called the Euler-Mascheroni constant, it can also be defined by the integral

$$\gamma = \int_{\infty}^{0} e^{-t} \ln t \, dt.$$

It is not known if γ is an irrational number.

Euler (-Venn) Diagram.   Such a diagram consists of closed curves, used to represent relations between logical propositions or sets. See Figure 3.

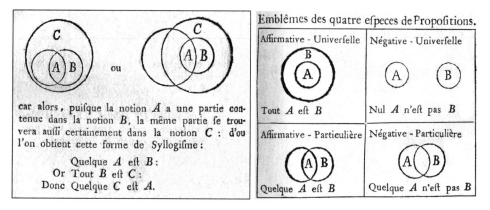

Figure 3. Euler diagrams from letters 103, 104 of *Lettres à une Princesse d'Allemagne*....

Euler Line.   The line defined by the centroid (the intersection of the medians), the orthocenter (the intersection of the altitudes), and the circumcenter (the intersection of the perpendicular bisectors) of a triangle. In addition to the remarkable collinearity of these three points, the distance between the centroid and the circumcenter is equal to half the distance between the centroid and the orthocenter. See Figure 4.

Euler Numbers.   The Euler numbers E_n can be defined by the infinite series

$$\frac{1}{\cos z} = \sum_{n=0}^{\infty} (-1)^n \frac{E_{2n}}{(2n)!} z^{2n};$$

alternately, they can be defined using the Euler polynomial $E_n(x)$ (see **Functions** below):

$$E_n = 2^n E_n(1/2).$$

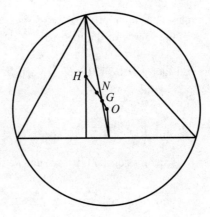

Figure 4. The orthocenter H, centroid G and circumcenter O determine the Euler line of a triangle. The center N of the nine-point circle is also on this line, and $3(\overline{OG}) = \overline{OH}$, $2(\overline{ON}) = \overline{OH}$.

Some properties are: $E_0 = 1$, $E_{2k+1} = 0$ for all k, $E_2 = -1$, $E_4 = 5$, $E_6 = 61$;

$$\sum_{j=0}^{n} \binom{2n}{2j} E_{2j} = 0, \quad n \geq 1.$$

Eulerian Numbers.   The numbers $A_{n,k}$ were defined by Euler as follows: If $H_n(\lambda)$ is the rational function of λ defined by the generating function

$$\frac{1-\lambda}{e^t - \lambda} = \sum_{n=0}^{\infty} H_n(\lambda)\frac{t^n}{n!}, \quad \lambda \neq 1,$$

then the Eulerian numbers $A_{n,k}$ are defined by the polynomial

$$(\lambda - 1)^n H_n(\lambda) = \sum_{k=1}^{n} A_{n,k}\lambda^{k-1}.$$

(The polynomials $A_n(\lambda) = \sum_{k=1}^{n} A_{n,k}\lambda^k$ are called **Eulerian polynomials**.) The numbers $A_{n,k}$ have important combinatorial properties; a combinatorial formula due to Euler is

$$A_{n,k} = \sum_{j=0}^{k}(-1)^j \binom{n+1}{j}(k-j)^n, \quad k = 0, 1, \ldots, n,$$

and another (Worpetzky, 1883) is

$$x^n = \sum_{k=1}^{n} A_{n,k}\binom{x+k-1}{n}.$$

Today, the number $A_{n,k}$ is commonly defined as the number of permutations of the set $\{1, 2, \ldots, n\}$ having $k - 1$ descents (a descent of a permutation

$$\begin{pmatrix} 1 & 2 & \ldots & n \\ a_1 & a_2 & \ldots & a_n \end{pmatrix}$$

is a pair a_i, a_{i+1} with $a_i > a_{i+1}$). From this, the properties $A_{n,k} = A_{n,n-k+1}$ $(1 \leq k \leq n)$ and $\sum_{k=1}^{n} A_{n,k} = n!$ are evident.

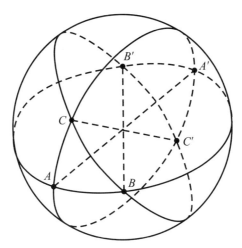

Figure 5. Euler spherical triangles.

Euler Spherical Triangles. If three points A, B, C are on a sphere such that no pair of these are diametrically opposite, then they determine three great circles, each of which joins a pair of the points. These three great circles divide the sphere into eight spherical triangles, whose sides are arcs of great circles, and have length less than π; these are called Euler spherical triangles. See Figure 5.

Euler Triangle. Given any triangle, its Euler triangle has as vertices the three Euler points of the triangle (see Figure 2).

Functions

Euler's First Integral (Beta function). The beta function is defined by the integral

$$B(z, w) = \int_0^1 t^{z-1}(1-t)^{w-1} \, dt, \quad R(z) > 0, \ R(w) > 0.$$

It is related to Euler's second integral, the gamma function, by the equation

$$B(z, w) = \frac{\Gamma(z)\Gamma(w)}{\Gamma(z+w)} = B(w, z).$$

Euler's Second Integral (Gamma function). Euler's integral defines the gamma function as

$$\Gamma(z) = \int_0^\infty e^{-t} t^{z-1} \, dt, \quad R(z) > 0.$$

This function satisfies the equation $\Gamma(z+1) = z\Gamma(z)$ for all $R(z) > 0$, hence $\Gamma(n+1) = n!$ for all positive integers n. Other formulas for $\Gamma(z)$ also due to Euler are

$$\Gamma(z) = \lim_{n \to \infty} \frac{n! n^z}{z(z+1)(z+2)\cdots(z+n)}, \quad z \neq 0, -1, -2, \ldots$$

and

$$\Gamma(z) = \frac{1}{z} \prod_{n=1}^{\infty} \left(1 + \frac{1}{n}\right)^z \left(1 + \frac{z}{n}\right)^{-1}, \quad z \neq 0, -1, -2, \dots.$$

Euler Polynomials. The polynomials $E_n(x)$ are defined by the generating function

$$\frac{2e^{xt}}{e^t + 1} = \sum_{n=0}^{\infty} E_n(x) \frac{t^n}{n!}, \quad |t| < \pi.$$

(The Bernoulli polynomials $B_n(x)$ are defined by the generating function

$$\frac{te^{xt}}{e^t - 1} = \sum_{n=0}^{\infty} B_n(x) \frac{t^n}{n!};$$

there are many equations relating the $E_n(x)$ and $B_n(x)$.) Some properties of the Euler polynomials are:

$$E_n'(x) = n E_{n-1}(x)$$

$$E_n(x+1) + E_n(x) = 2x^n$$

$$\sum_{k=1}^{m} (-1)^{m-k} k^n = \frac{1}{2} \left[E_n(m+1) + (-1)^m E_n(0) \right].$$

Euler Product. Under certain conditions, the Dirichlet series

$$F(s) = \sum_{n=1}^{\infty} \frac{f(n)}{n^s} \quad (s \text{ real or complex})$$

has a representation as a formal product

$$F(s) = \prod_{p \text{ prime}} F_p(s),$$

called an Euler product; the functions

$$F_p(s) = 1 + f(p)p^{-s} + f(p^2)p^{-2s} + f(p^3)p^{-3s} + \cdots$$

are called **Euler factors**. The premier example is the zeta function, for which $f(n)$ is the constant function 1 (see Euler's identity, below). Another example, derived from this, is

$$\frac{\zeta(2s)}{\zeta(s)} = \prod_p \left(\frac{1 - p^{-s}}{1 - p^{-2s}} \right) = \prod_p \left(1 + \frac{1}{p^s} \right)^{-1}$$

$$= \prod_p (1 - p^{-s} + p^{-2s} - p^{-3s} + \cdots)$$

$$= \sum_{n=1}^{\infty} \frac{\lambda(n)}{n^s},$$

where $\lambda(n) = (-1)^\rho$, ρ defined as $\sum_{i=1}^{k} \alpha_i$ if $n = p_1^{\alpha_1} \cdots p_k^{\alpha_k}$ is the prime factorization of n.

Euler's ϕ function (Euler's totient function). For each positive integer n, $\phi(n)$ is defined as the number of positive integers less than n and relatively prime to n. If $p_1, p_2, \ldots, p_k$ are the distinct prime factors of n, then

$$\phi(n) = n \prod_{i=1}^{k} \left(1 - \frac{1}{p_i}\right).$$

Euler's Identity (Zeta function). For $R(s) > 1$,

$$\zeta(s) = \sum_{n=1}^{\infty} \frac{1}{n^s} = \prod_p \left(1 - \frac{1}{p^s}\right)^{-1},$$

the product taken over all primes. The sum on the left is known today as the Riemann zeta function.

Formulas

Euler-Binet formula. The rational solutions to the equation

$$w^3 + 3w(x^2 + y^2 + z^2) + 6xyz = 0$$

are given by

$$w = -6pabc, \qquad x = pa(a^2 + 3b^2 + 3c^2),$$
$$y = pb(a^2 + 3b^2 + 9c^2), \qquad z = 3pc(a^2 + b^2 + 3c^2).$$

Here, $(a, b, c) = 1$ and p is rational. The Euler-Binet formula is used to find solutions to the equation

$$x^3 + y^3 + z^3 + w^3 = 0.$$

Euler Force (critical load) for a beam or column. The Euler force is the maximum axial load that a long, slender beam or column can carry without buckling. This critical force is given by the formula

$$K\pi^2 \frac{YI}{L^2},$$

where I is the beam's moment of inertia of cross-sectional area, L is its unsupported length, Y is its stiffness, and K is a constant that depends on the conditions of end support of the beam ($K = 1$ if both ends are free to turn).

Euler's formula for $e^{i\theta}$. This fundamental rule links trigonometric to exponential functions:

$$e^{i\theta} = \cos\theta + i\sin\theta.$$

When $\theta = \pi$ and 2π, Euler's formula yields the famous, surprising results

$$e^{i\pi} = -1 \quad \text{and} \quad e^{i2\pi} = 1.$$

Euler-Fourier formulas.   In the Fourier series expansion of the function $F(x)$,

$$F(x) = \frac{1}{2}a_0 + \sum_{k=1}^{\infty}(a_k \cos kx + b_k \sin kx), \quad -\pi < x < \pi,$$

where the Euler-Fourier coefficients are

$$a_k = \frac{1}{\pi}\int_{-\pi}^{\pi} F(x) \cos kx \, dx, \quad b_k = \frac{1}{\pi}\int_{-\pi}^{\pi} F(x) \sin kx \, dx.$$

Euler-Maclaurin sum formula.   This summation formula is one of the most important in the calculus of finite differences; it was discovered by Euler and independently by Maclaurin in the decade 1732–42. One version of the formula is: If $f(x)$ has its first $2n$ derivatives continuous on an interval $[0, m]$, m an integer, then

$$\sum_{k=0}^{m} f(k) = \int_0^m f(x) \, dx + \frac{1}{2}\big(f(0) + f(m)\big) + \sum_{k=1}^{n} \frac{B_{2k}}{(2k)!}\big(f^{(2k-1)}(m) - f^{(2k-1)}(0)\big) + R_n,$$

where the B_{2n} are the Bernoulli numbers (see: Euler numbers), and the remainder term R_n may be given in several forms. One form is

$$R_n = \frac{B_{2n}}{(2n)!}\sum_{k=0}^{m-1} f^{(2n)}(k + \theta), \quad \text{for some } 0 < \theta < 1;$$

another is given by M. Kline, see p. 106.

   In its simplest form, the summation formula for a function $f(x)$ with a continuous derivative on $[0, m]$ is

$$\sum_{k=0}^{m} f(k) = \int_0^m f(x) \, dx + \frac{1}{2}\big(f(0) + f(m)\big) + \int_0^m \left(x - [x] - \frac{1}{2}\right) f'(x) \, dx,$$

where $[x]$ is the greatest integer function.

Equations

Euler's equation in the calculus of variations.   Problem: Find a curve $y(x)$ joining the points (x_1, y_1) and (x_2, y_2) such that the integral

$$\int_{x_1}^{x_2} I(x, y, y') \, dx$$

is a maximum or minimum (the integral has a stationary value). A necessary condition for $y(x)$ to be a solution to the problem is that y satisfy the Euler (-Lagrange) equation:

$$\frac{\partial I}{\partial y} - \frac{d}{dx}\left[\frac{\partial I}{\partial y'}\right] = 0, \quad \text{where } y' = \frac{dy}{dx}.$$

Euler's (equidimensional) equation. Also called the Euler-Cauchy equation, this is an nth order differential equation of the form

$$z^n w^{(n)} + b_1 z^{n-1} w^{(n-1)} + \cdots + b_n w = 0,$$

where $w(z)$ is a function of z, and the b_i are constants. The substitution $z = e^s$ transforms this into an equation with constant coefficients, i.e., if $\tilde{w}(s) = w(e^s)$, then the equation becomes

$$\tilde{w}^{(n)} + c_1 \tilde{w}^{(n-1)} + \cdots + c_n \tilde{w} = 0,$$

the c_i constants. The second-order equation is most often cited, i.e.,

$$x^2 y'' + p x y' + q y = 0.$$

Euler's equation of motion for an ideal compressible or incompressible fluid. The Eulerian method of analyzing fluid flow focuses on each position in space and observes how the fluid motion varies over time at that position. Euler's equation is

$$\frac{\partial \mathbf{v}}{\partial t} + (\mathbf{v} \cdot \nabla) \mathbf{v} = \mathbf{F} - \frac{1}{\rho} \nabla p.$$

where $\mathbf{v}$ is the velocity field, p is the pressure, ρ is the density, and $\mathbf{F}$ is the external force per unit of mass of the fluid.

Euler's equations (of motion) for the rotation of a rigid body. The basic equations for the rotation of a rigid body are

$$N_1 = I_1 \frac{d\omega_1}{dt} + (I_3 - I_2) \omega_3 \omega_2$$

$$N_2 = I_2 \frac{d\omega_2}{dt} + (I_1 - I_3) \omega_1 \omega_3$$

$$N_3 = I_3 \frac{d\omega_3}{dt} + (I_2 - I_1) \omega_2 \omega_1$$

where the N_i are torques, the I_i are the principal moments of inertia, and the ω_i are angular velocities about the three coordinate axes.

Euler's equation on normal curvature. If κ_1 and κ_2 are the principal normal curvatures at point P on a surface S, and κ is the normal curvature in the direction making an angle θ with the direction having normal curvature κ_1, then

$$\kappa = \kappa_1 \cos^2 \theta + \kappa_2 \sin^2 \theta.$$

Techniques

Euler multiplier method of solving a differential equation. If a differential equation $y' g(x, y) + h(x, y) = 0$ is not exact, the equation is multiplied by a function $\mu(x, y)$, called an Euler multiplier, or integrating factor, so that the product $y' g\mu + h\mu$ is a perfect derivative. A partial differential equation for the determination of $\mu(x, y)$ is

$$\frac{\partial (g\mu)}{\partial x} = \frac{\partial (h\mu)}{\partial y}.$$

Euler's numerical method for the solution of differential equations. This iterative scheme is used to find a numerical solution of an ordinary differential equation, $y' = f(x, y)$. The recursion formula is

$$y_{n+1} = y_n + hf(x_n, y_n)$$

where $h = x_{n+1} - x_n$ is the step size between successive points. The error in this method is often substantial, and the *improved Euler method* is used, which is based on approximation by the trapezoidal rule. The recursion formula is

$$y_{n+1} = y_n + \frac{h}{2}[f(x_n, y_n) + f(x_{n+1}, z_{n+1})],$$

where $z_{n+1} = y_n + hf(x_n, y_n)$.

Euler's transformation of a series. The series $\sum_{k=0}^{\infty}(-1)^k a_k$ is transformed into the series

$$\sum_{k=0}^{\infty} \frac{(-1)^k}{2^{k+1}} \Delta^k a_0, \text{ where } \Delta^0 a_0 = a_0, \text{ and } \Delta^k a_0 = \sum_{m=0}^{k} (-1)^m \binom{k}{m} a_{k-m}, \ k \geq 1.$$

If the original series converges, the transformed series converges (often more quickly) to the same sum.

Theorems

Euler's Addition Theorem for elliptic integrals. *If $g(x) = (1 - x^2)(1 - k^2 x^2)$, then*

$$\int_0^a \frac{dx}{\sqrt{g(x)}} + \int_0^b \frac{dx}{\sqrt{g(x)}} = \int_0^c \frac{dx}{\sqrt{g(x)}}$$

where

$$c = \frac{b\sqrt{g(a)} + a\sqrt{g(b)}}{\sqrt{1 - k^2 a^2 b^2}}.$$

Euler's Criterion for quadratic residues. An integer a is a quadratic residue of an integer b if there is a solution to the congruence $x^2 \equiv a \bmod b$. Euler's criterion says that *a is a quadratic residue of the odd prime p if and only if $a^{(p-1)/2} \equiv 1 \pmod{p}$*.

Euler's generalization of Fermat's Theorem. *If n and a are positive integers which are relatively prime, then*

$$a^{\phi(n)} \equiv 1 \pmod{n}.$$

Euler's theorem for homogeneous functions. *If $f : R^n \to R$ is a function that satisfies*

$$f(\lambda \mathbf{x}) = \lambda^k f(\mathbf{x})$$

for all $\mathbf{x} \in R^n$ and $\lambda \in R$, then

$$\mathbf{x} \cdot \nabla f(\mathbf{x}) = k f(\mathbf{x}).$$

Euler-Lagrange Theorem.   *Any positive integer can be expressed as the sum of at most four squares.*

Euler's Officer Problem.   Euler's analysis of the 36 officer problem—the assignment of six different officers from each of six regiments to six squads, each including an officer of every rank and a member of every regiment—led him to conjecture that there exist no such pairs of orthogonal Latin squares of order $n \equiv 2$ (mod 4). Although Euler was right for $n = 6$, his conjecture has been proved false for all $n > 6$.

Euler's theorems on partitions.   Many theorems in the theory of partitions were first proved by Euler and bear his name; we give two of these. *The number of partitions of a positive integer n into odd parts is equal to the number of partitions of n into distinct parts.* For example, $n = 6$ is partitioned into odd parts in four ways: $5 + 1$, $3 + 3$, $3 + 1 + 1 + 1$, $1 + 1 + 1 + 1 + 1 + 1$, and into distinct parts in four ways: 6, $5 + 1$, $4 + 2$, $3 + 2 + 1$.

Let $p(k)$ denote the number of partitions of k for $k \geq 1$, and define $p(k) = 0$ for k negative, and $p(0) = 1$. If $n > 0$, then

$$p(n) - p(n - 1) - p(n - 2) + p(n - 5) + p(n - 7) + \cdots$$
$$+ (-1)^m p\left(n - \frac{m}{2}(3m - 1)\right) + (-1)^m p\left(n - \frac{m}{2}(3m + 1)\right) + \cdots = 0.$$

Euler's theorem for pentagonal numbers.   *If $|x| < 1$, then*

$$\prod_{m=1}^{\infty} (1 - x^m) = 1 - x - x^2 + x^5 + x^7 - x^{12} - x^{15} + \cdots$$

$$= 1 + \sum_{n=-\infty}^{\infty} (-1)^n x^{\omega(n)}$$

where $\omega(n) = \dfrac{3n^2 - n}{2}$ is the nth pentagonal number.

Euler's theorem for points on a line.   *If the points P, Q, R and S are collinear, then*
$$PQ \cdot RS + PR \cdot SQ + PS \cdot QR = 0.$$

Euler's theorem for polyhedra.   *For any convex polyhedron, the Euler characteristic is equal to 2; that is, $V - E + F = 2$.*

Euler's theorem for primes.   *The sum $\sum(1/p)$, and the product $\prod\left[1 - (1/p)\right]^{-1}$ are both divergent, as p runs through all the primes. It follows that there are an infinite number of primes.*

Euler's theorem for rotation of coordinate system (or rigid body).   *If two rectangular coordinate systems with the same origin and arbitrary directions of axes are given in 3-space, there exists a line through the origin such that one coordinate system is transformed into the other by a rotation about this line.* (The transformation can also be achieved by three successive rotations through the Euler angles ϕ, θ and ψ; see Figure 1.)

Euler's recursion theorem for the sums of divisors. *Let $\sigma(k)$ be the sum of the divisors of the positive integer k, and define $\sigma(k) = 0$ for $k \leq 0$. Define the sum $S(n)$:*

$$S(n) = \sigma(n) - \sigma(n-1) - \sigma(n-2) + \sigma(n-5) + \sigma(n-7)$$
$$- \sigma(n-12) - \sigma(n-15) + \cdots + (-1)^m \sigma\left(n - \frac{m}{2}(3m-1)\right)$$
$$+ (-1)^m \sigma\left(n - \frac{m}{2}(3m+1)\right) + \cdots .$$

Then $S(n) = (-1)^{m-1} n$ if $n = \frac{m}{2}(3m \pm 1)$ for some m, and $S(n) = 0$ otherwise.

Euler's theorem for a triangle. *The distance d between the circumcenter and incenter of a triangle is given by the equation $d^2 = R(R - 2r)$, where R, r are the circumradius and inradius, respectively.* See Figure 6.

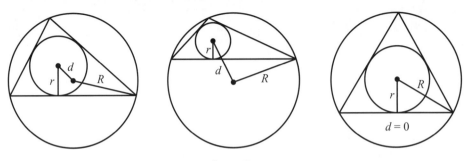

Figure 6.

References

Encyclopedias and Dictionaries

1. *Dictionary of Mathematics*, C. C. T. Baker, ed., Hart Publishing Co., New York, 1966.

2. *Encyclopedic Dictionary of Mathematics for Engineers and Applied Scientists*, I. N. Sneddon, ed., Pergamon, 1976.

3. *Handbook of Mathematical Functions*, M. Abramowitz and I. A. Stegun, eds., National Bureau of Standards, Appl. Math. Series, 55, 1964.

4. *International Dictionary of Applied Mathematics*, Van Nostrand, 1960.

5. *Mathematics Dictionary*, G. James and R. James, eds., D. Van Nostrand, New York, 1949.

6. *Mathematical Handbook for Scientists and Engineers*, G. Korn and T. Korn, eds., McGraw-Hill, 1961.

7. *McGraw-Hill Dictionary of Scientific and Technical Terms*, D. N. Lapedes, ed., McGraw-Hill, 1974.

8. *McGraw-Hill Encyclopedia of Science and Technology*, 5th ed., McGraw-Hill, 1982.

9. *VNR Concise Encyclopedia of Mathematics*, W. Gellert, H. Kiinster, M. Hellwich, H. Kastner, eds., Van Nostrand Reinhold, New York, 1977.

10. *Van Nostrand's Scientific Encyclopedia*, 6th ed., D. M. Considine and G. D. Considine, eds., Van Nostrand Reinhold Co., 1983.

Books and Articles

11. N. Altshiller-Court, *College Geometry*, Johnson Pub. Co., 1925.

12. G. E. Andrews, *The Theory of Partitions*, Encyclopedia of Mathematics and its Applications, vol. 2, Addison-Wesley, 1976.

13. G. Birkhoff, *A Source Book in Classical Analysis*, Harvard U. Press, 1973.

14. C. B. Boyer, *A History of Mathematics*, Wiley, 1968.

15. L. Carlitz, Eulerian numbers and polynomials, *Math. Mag.*, 32 (1959) 247–260.

16. H. S. M. Coxeter, *Introduction to Geometry*, 2nd ed., Wiley, 1969.

17. H. Eves, *An Introduction to the History of Mathematics*, 5th ed., Saunders, 1983.

18. ——, *A Survey of Geometry*, vols. I, II, Allyn and Bacon, Boston, 1963.

19. G. M. Ewing, *Calculus of Variations with Applications*, Norton, 1969.

20. H. H. Goldstine, *A History of Numerical Analysis from the 16th through the 19th Century*, Springer-Verlag, New York, 1977.

21. G. H. Hardy and E. M. Wright, *An Introduction to the Theory of Numbers*, 4th ed., Oxford, 1960.

22. M. Kline, *Mathematical Thought from Ancient to Modern Times*, Oxford, 1972.

23. K. Knopp, *Theory and Application of Infinite Series*, Hafner, New York, 1947.

24. G. E. Martin, *Transformation Geometry, An Introduction to Symmetry*, Springer-Verlag, 1982.

25. J. Riordan, *Introduction to Combinatorial Analysis*, Wiley, New York, 1958.

26. G. F. Simmons, *Differential Equations with Applications and Historical Notes*, McGraw-Hill, 1972.

List of Photos

Index

About the Editor

William Dunham, who received his BS (1969) from the University of Pittsburgh and his MS (1970) and PhD (1974) from The Ohio State University, is the Truman Koehler Professor of Mathematics at Muhlenberg College. He is the author of four books on mathematics and its history: *Journey Through Genius: The Great Theorems of Mathematics* (Wiley, 1990), *The Mathematical Universe* (Wiley, 1994), *Euler: The Master of Us All* (MAA, 1999), and *The Calculus Gallery: Masterpieces from Newton to Lebesgue* (Princeton University Press, 2005). In addition, he has contributed articles to various publications, and his writing has been recognized by the MAA with its George Pólya Award (1992), Trevor Evans Award (1997), and Lester R. Ford Award (2006).